Means Residential Square Foot Costs

Contractor's Pricing Guide 2007

Senior Editor
Robert W. Mewis, CCC

Contributing Editors
Christopher Babbit
Ted Baker
Barbara Balboni
Robert A. Bastoni
John H. Chiang, PE
Gary W. Christensen
Cheryl Elsmore
Robert J. Kuchta
Robert C. McNichols
Melville J. Mossman, PE
Jeannene D. Murphy
Stephen C. Plotner
Eugene R. Spencer
Marshall J. Stetson
Phillip R. Waier, PE

Senior Engineering Operations Manager
John H. Ferguson, PE

Senior Vice President & General Manager
John Ware

Vice President of Direct Response
John M. Shea

Director of Product Development
Thomas J. Dion

Production Manager
Michael Kokernak

Production Coordinator
Wayne D. Anderson

Technical Support
Jonathan Forgit
Mary Lou Geary
Jill Goodman
Gary L. Hoitt
Genevieve Medeiros
Paula Reale-Camelio
Kathryn S. Rodriguez
Sheryl A. Rose

Book & Cover Design
Norman R. Forgit

RSMeans

D1452308

Means Residential Square Foot Costs

Contractor's Pricing Guide 2007

- Residential Cost Models for All Standard Building Classes
- Costs for Modifications & Additions
- Costs for Hundreds of Residential Building Systems & Components
- Cost Adjustment Factors for Your Location
- Illustrations

$39.95 per copy (in United States).
Price subject to change without prior notice.

Copyright © 2006

RSMeans
Construction Publishers & Consultants
63 Smiths Lane
Kingston, MA 02364-0800
(781) 422-5000

The authors, editors and engineers of Reed Construction Data's RSMeans product line apply diligence and judgment in locating and using reliable sources for the information published. **However, RSMeans makes no express or implied warranty or guarantee in connection with the content of the information contained herein, including the accuracy, correctness, value, sufficiency, or completeness of the data, methods and other information contained herein. RSMeans makes no express or implied warranty of merchantability or fitness for a particular purpose.** RSMeans shall have no liability to any customer or third party for any loss, expense, or damage, including consequential, incidental, special or punitive damage, including lost profits or lost revenue, caused directly or indirectly by any error or omission, or arising out of, or in connection with, the information contained herein.

No part of this publication may be reproduced, stored in a retrieval system, or transmitted in any form or by any means without prior written permission of Reed Construction Data.

Printed in the United States of America.

10 9 8 7 6 5 4 3 2 1

ISSN 1074-049X

ISBN 0-87629-874-9

Foreword

Reed Construction Data's portfolio of project information products and services includes national, regional, and local construction data, project leads, and project plans, specifications, and addenda available online or in print. Reed Bulletin (www.reedbulletin.com) and Reed CONNECT™ (www.reedconnect.com) deliver the most comprehensive, timely, and reliable project information to support contractors, distributers, and building product manufacturers in identifying, bidding, and tracking projects. The Reed First Source (www.reedfirstsource.com) suite of products used by design professionals for the search, selection, and specification of nationally available building projects consists of the First Source™ annual, SPEC-DATA™, MANU-SPEC™, First Source CAD, and manufacturer catalogs. Reed Design Registry, a database of more than 30,000 U.S. architectural firms, is also published by Reed Construction Data. RSMeans (www.rsmeans.com) provides construction cost data, training, and consulting services in print, CD-ROM, and online.

Associated Construction Publications (www.reedpubs.com) reports on heavy, highway, and non-residential construction through a network of 14 regional construction magazines. Reed Construction Data (www.reedconstructiondata.com), headquartered in Atlanta, is a subsidiary of Reed Business Information (www.reedbusinessinformation.com), North America's largest business-to-business information provider. With more than 80 market-leading publications and 55 websites, Reed Business Information's wide range of services also includes research, business development, direct marketing lists, training and development programs, and technology solutions. Reed Business Information is a member of the Reed Elsevier plc group (NYSE: RUK and ENL)—a leading provider of global information-driven services and solutions in the science and medical, legal, education, and business-to-business industry sectors.

Our Mission

Since 1942, RSMeans has been actively engaged in construction cost publishing and consulting throughout North America. Today, over 60 years after RSMeans began, our primary objective remains the same: to provide you, the construction and facilities professional, with the most current and comprehensive construction cost data possible.

Whether you are a contractor, an owner, an architect, an engineer, a facilities manager, or anyone else who needs a reliable construction cost estimate, you'll find this publication to be a highly useful and necessary tool.

Today, with the constant flow of new construction methods and materials, it's difficult to find the time to look at and evaluate all the different construction cost possibilities. In addition, because labor and material costs keep changing, last year's cost information is not a reliable basis for today's estimate or budget.

That's why so many construction professionals turn to RSMeans. We keep track of the costs for you, along with a wide range of other key information, from city cost indexes . . . to productivity rates . . . to crew composition . . . to contractor's overhead and profit rates.

RSMeans performs these functions by collecting data from all facets of the industry and organizing it in a format that is instantly accessible to you. From the preliminary budget to the detailed unit price estimate, you'll find the data in this book useful for all phases of construction cost determination.

The Staff, the Organization, and Our Services

When you purchase one of RSMeans' publications, you are, in effect, hiring the services of a full-time staff of construction and engineering professionals.

Our thoroughly experienced and highly qualified staff works daily at collecting, analyzing, and disseminating comprehensive cost information for your needs. These staff members have years of practical construction experience and engineering training prior to joining the firm. As a result, you can count on them not only for the cost figures, but also for additional background reference information that will help you create a realistic estimate.

The RSMeans organization is always prepared to help you solve construction problems through its five major divisions: Construction and Cost Data Publishing, Electronic Products and Services, Consulting and Research Services, Insurance Services, and Professional Development Services.

Besides a full array of construction cost estimating books, RSMeans also publishes a number of other reference works for the construction industry. Subjects include construction estimating and project and business management; special topics such as HVAC, roofing, plumbing, and hazardous waste remediation; and a library of facility management references.

In addition, you can access all of our construction cost data electronically using *Means CostWorks®* CD or on the Web.

What's more, you can increase your knowledge and improve your construction estimating and management performance with an RSMeans Construction Seminar or In-House Training Program. These two-day seminar programs offer unparalleled opportunities for everyone in your organization to get updated on a wide variety of construction-related issues.

RSMeans is also a worldwide provider of construction cost management and analysis services for commercial and government owners, and of claims and valuation services for insurers.

In short, RSMeans can provide you with the tools and expertise for constructing accurate and dependable construction estimates and budgets in a variety of ways.

Robert Snow Means Established a Tradition of Quality That Continues Today

Robert Snow Means spent years building RSMeans, making certain he always delivered a quality product.

Today, at RSMeans, we do more than talk about the quality of our data and the usefulness of our books. We stand behind all of our data, from historical cost indexes to construction materials and techniques to current costs.

If you have any questions about our products or services, please call us toll-free at 1-800-334-3509. Our customer service representatives will be happy to assist you. You can also visit our Web site at www.rsmeans.com.

Table of Contents

Foreword	v
How the Book Is Built: An Overview	ix
How To Use the Book: The Details	x
Square Foot Cost Section	1
Assemblies Cost Section	81
Location Factors	264
Abbreviations	269
Index	272
Other RSMeans Products and Services	275
Labor Trade Rates including Overhead & Profit	Inside Back Cover

How the Book is Built: An Overview

A Powerful Construction Tool

You have in your hands one of the most powerful construction tools available today. A successful project is built on the foundation of an accurate and dependable estimate. This book will enable you to construct just such an estimate.

For the casual user the book is designed to be:

- quickly and easily understood so you can get right to your estimate.
- filled with valuable information so you can understand the necessary factors that go into the cost estimate.

For the regular user, the book is designed to be:

- a handy desk reference that can be quickly referred to for key costs.
- a comprehensive, fully reliable source of current construction costs so you'll be prepared to estimate any project.
- a source book for preliminary project cost, product selections, and alternate materials and methods.

To meet all of these requirements we have organized the book into the following clearly defined sections.

Square Foot Cost Section

This section lists Square Foot costs for typical residential construction projects. The organizational format used divides the projects into basic building classes. These classes are defined at the beginning of the section. The individual projects are further divided into ten common components of construction. A Table of Contents, an explanation of square foot prices, and an outline of a typical page layout are located at the beginning of the section.

Assemblies Cost Section

This section uses an Assemblies (sometimes referred to as systems) format grouping all the functional elements of a building into nine construction divisions.

At the top of each Assemblies cost table is an illustration, a brief description, and the design criteria used to develop the cost. Each of the components and its contributing cost to the system is shown.

Material: These cost figures include a standard 10% markup for profit. They are national average material costs as of January of the current year and include delivery to the job site.

Installation: The installation costs include labor and equipment, plus a markup for the installing contractor's overhead and profit.

For a complete breakdown and explanation of a typical Assemblies page, see "How to Use the Assemblies Section" at the beginning of the Assemblies Section.

Location Factors: You can adjust total project costs to over 900 locations throughout the U.S. and Canada by using the data in this section.

Abbreviations: A listing of the abbreviations used throughout this book, along with the terms they represent, is included in this section.

Index

A comprehensive listing of all terms and subjects in this book to help you find what you need quickly.

The Scope of This Book

This book is designed to be as comprehensive and as easy to use as possible. To that end we have made certain assumptions and limited its scope in three key ways:

1. We have established material prices based on a national average.
2. We have computed labor costs based on a seven major-region average of residential wage rates.

Project Size

This book is intended for use by those involved primarily in Residential construction costing less than $850,000. This includes the construction of homes, row houses, townhouses, condominiums, and apartments.

With reasonable exercise of judgment the figures can be used for any building work. For other types of projects, such as repair and remodeling or commercial buildings, consult the appropriate RSMeans publication for more information.

How to Use the Book: The Details

What's Behind the Numbers? The Development of Cost Data

The staff at RSMeans continuously monitors developments in the construction industry in order to ensure reliable, thorough, and up-to-date cost information.

While **overall** construction costs may vary relative to general economic conditions, price fluctuations within the industry are dependent upon many factors. Individual price variations may, in fact, be opposite to overall economic trends. Therefore, costs are continually monitored and complete updates are published yearly. Also, new items are frequently added in response to changes in materials and methods.

Costs – $ (U.S.)

All costs represent U.S. national averages and are given in U.S. dollars. The RSMeans Location Factors can be used to adjust costs to a particular location. The Location Factors for Canada can be used to adjust U.S. national averages to local costs in Canadian dollars. No exchange rate conversion is necessary.

Material Costs

The RSMeans staff contacts manufacturers, dealers, distributors, and contractors all across the U.S. and Canada to determine national average material costs. If you have access to current material costs for your specific location, you may wish to make adjustments to reflect differences from the national average. Included within material costs are fasteners for a normal installation. RSMeans engineers use manufacturers' recommendations, written specifications, and/or standard construction practice for size and spacing of fasteners. Adjustments to material costs may be required for your specific application or location. Material costs do not include sales tax.

Labor Costs

Labor costs are based on the average of residential wages from across the U.S. for the current year. Rates, along with overhead and profit markups, are listed on the inside back cover of this book.

- If wage rates in your area vary from those used in this book, or if rate increases are expected within a given year, labor costs should be adjusted accordingly.

Labor costs reflect productivity based on actual working conditions. These figures include time spent during a normal workday on tasks other than actual installation, such as material receiving and handling, mobilization at site, site movement, breaks, and cleanup.

Productivity data is developed over an extended period so as not to be influenced by abnormal variations and reflects a typical average.

Equipment Costs

Equipment costs include not only rental, but also operating costs for equipment under normal use. Equipment and rental rates are obtained from industry sources throughout North America—contractors, suppliers, dealers, manufacturers, and distributers.

Factors Affecting Costs

Costs can vary depending upon a number of variables. Here's how we have handled the main factors affecting costs.

Quality—The prices for materials and the workmanship upon which productivity is based represent sound construction work. They are also in line with U.S. government specifications.

Overtime—We have made no allowance for overtime. If you anticipate premium time or work beyond normal working hours, be sure to make an appropriate adjustment to your labor costs.

Productivity—The productivity, daily output, and labor-hour figures for each line item are based on working an eight-hour day in daylight hours in moderate temperatures. For work that extends beyond normal work hours or is performed under adverse conditions, productivity may decrease.

Size of Project—The size, scope of work, and type of construction project will have a significant impact on cost. Economies of scale can reduce costs for large projects. Unit costs can often run higher for small projects. Costs in this book are intended for the size and type of project as previously described in "How the Book Is Built: An Overview." Costs for projects of a significantly different size or type should be adjusted accordingly.

Location—Material prices in this book are for metropolitan areas. However, in dense urban areas, traffic and site storage limitations may increase costs. Beyond a 20-mile radius of large cities, extra trucking or transportation charges may also increase the material costs slightly. On the other hand, lower wage rates may be in effect. Be sure to consider both of these factors when preparing an estimate, particularly if the job site is located in a central city or remote rural location.

In addition, highly specialized subcontract items may require travel and per-diem expenses for mechanics.

Other Factors–

- season of year
- contractor management
- weather conditions
- local union restrictions
- building code requirements
- availability of:
    - adequate energy
    - skilled labor
    - building materials
- owner's special requirements/ restrictions
- safety requirements
- environmental considerations

General Conditions—The "Square Foot Cost" and "Assemblies" sections of this book use costs that include the installing contractor's overhead and profit (O&P). An allowance covering the general contractor's markup must be added to these figures. The general contractor can include this price in the bid with a normal markup ranging from 5% to 15%. The markup depends on economic conditions plus the supervision and troubleshooting expected by the general contractor. For purposes of this book, it is best for a general contractor to add an allowance of 10% to the figures in the Square Foot Costs and Assemblies sections.

Overhead & Profit—For square foot costs and systems costs, simply add 10% to the estimate for general contractor's profit.

Unpredictable Factors—General business conditions influence "in-place" costs of all items. Substitute materials and construction methods may have to be employed. These may affect the installed cost and/or life cycle costs. Such factors may be difficult to evaluate and cannot necessarily be predicted on the basis of the job's location in a particular section of the country. Thus, where these factors apply, you may find significant but unavoidable cost variations for which you will have to apply a measure of judgment to your estimate.

Rounding of Costs

In general, all unit prices in excess of $5.00 have been rounded to make them easier to use and still maintain adequate precision of the results. The rounding rules we have chosen are in the following table.

Prices from . . .	Rounded to the nearest . . .
$.01 to $5.00	$.01
$5.01 to $20.00	$.05
$20.01 to $100.00	$.50
$100.01 to $300.00	$1.00
$300.01 to $1,000.00	$5.00
$1,000.01 to $10,000.00	$25.00
$10,000.01 to $50,000.00	$100.00
$50,000.01 and above	$500.00

Final Checklist

Estimating can be a straightforward process provided you remember the basics. Here's a checklist of some of the steps you should remember to complete before finalizing your estimate.

Did you remember to . . .

- factor in the Location Factor for your locale?
- take into consideration which items have been marked up and by how much?
- mark up the entire estimate sufficiently for your purposes?
- include all components of your project in the final estimate?
- double check your figures for accuracy?
- call RSMeans if you have any questions about your estimate or the data you've found in our publications?

Remember, RSMeans stands behind its publications. If you have any questions about your estimate . . . about the costs you've used from our books . . . or even about the technical aspects of the job that may affect your estimate, feel free to call the RSMeans editors at 1-800-334-3509.

Square Foot Cost Section

Table of Contents

	Page
How to Use S.F. Cost Pages	3,4,5
Building Classes	6
Building Types	6,7
Building Materials	8
Building Configurations	9
Garage Types	10
Estimate Form	11,12

Building Types

Economy

	Page
Illustrations	13
1 Story	14
1-1/2 Story	16
2 Story	18
Bi-Level	20
Tri-Level	22
Wings and Ells	24

Average

	Page
Illustrations	25
1 Story	26
1-1/2 Story	28
2 Story	30
2-1/2 Story	32
3 Story	34
Bi-Level	36
Tri-Level	38
Solid Wall (log home) 1 Story	40
Solid Wall (log home) 2 Story	42
Wings and Ells	44

Custom

	Page
Illustrations	45
1 Story	46
1-1/2 Story	48
2 Story	50
2-1/2 Story	52
3 Story	54
Bi-Level	56
Tri-Level	58
Wings and Ells	60

Building Types

Luxury

	Page
Illustrations	61
1 Story	62
1-1/2 Story	64
2 Story	66
2-1/2 Story	68
3 Story	70
Bi-Level	72
Tri-Level	74
Wings and Ells	76

Residential Modifications

	Page
Kitchen Cabinets	77
Kitchen Counter Tops	77
Vanity Bases	78
Solid Surface Vanity Tops	78
Fireplaces & Chimneys	78
Windows & Skylights	78
Dormers	78
Appliances	79
Breezeway	79
Porches	79
Finished Attic	79
Alarm System	79
Sauna, Prefabricated	79
Garages	80
Swimming Pools	80
Wood & Coal Stoves	80
Sidewalks	80
Fencing	80
Carport	80

How to Use the Square Foot Cost Pages

Introduction: This section contains costs per square foot for four classes of construction in seven building types. Costs are listed for various exterior wall materials which are typical of the class and building type. There are cost tables for wings and ells with modification tables to adjust the base cost of each class of building. Non-standard items can easily be added to the standard structures.

Accompanying each building type in each class is a list of components used in a typical residence. The components are divided into ten primary estimating divisions. The divisions correspond with the "Assemblies" section of this manual.

Cost estimating for a residence is a three-step process:
(1) Identification
(2) Listing dimensions
(3) Calculations

Guidelines and a sample cost estimating form are shown on the following pages.

Identification: To properly identify a residential building, the class of construction, type, and exterior wall material must be determined. Located at the beginning of this section are drawings and guidelines for determining the class of construction. There are also detailed specifications accompanying each type of building, along with additional drawings at the beginning of each set of tables, to further aid in proper building class and type identification.

Sketches for seven types of residential buildings and their configurations are shown along with definitions of living area next to each sketch. Sketches and definitions of garage types follow the residential buildings.

Living Area: Base cost tables are prepared as costs per square foot of living area. The living area of a residence is that area which is suitable and normally designed for full-time living. It does not include basement recreation rooms or finished attics, although these areas are often considered full-time living areas by owners.

Living area is calculated from the exterior dimensions without the need to adjust for exterior wall thickness. When calculating the living area of a 1-1/2-story, two-story, three-story or tri-level residence, overhangs and other differences in size and shape between floors must be considered.

Only the floor area with a ceiling height of six feet or more in a 1-1/2 story-residence is considered living area. In bi-levels and tri-levels, the areas that are below grade are considered living area, even when these areas may not be completely finished.

Base Tables and Modifications: Base cost tables show the base cost per square foot without a basement, with one full bath and one full kitchen. Adjustments for finished and unfinished basements are part of the base cost tables. Adjustments for multi-family residences, additional bathrooms, townhouses, alternative roofs, and air conditioning and heating systems are listed in Modifications, Adjustments, and Alternatives tables below the base cost tables.

The component list for each residence type should also be consulted when preparing an estimate. If the components listed are not appropriate, modifications can be made by consulting the "Assemblies" section of this manual.

Costs for other modifications, adjustments, and alternatives, including garages, breezeways, and site improvements, are in the modification tables at the end of this section.

Listing of Dimensions: To use this section of the manual, only the dimensions used to calculate the horizontal area of the building and additions and modifications are needed. The dimensions, normally the length and width, can come from drawings or field measurements. For ease in calculation, consider measuring in tenths of feet, i.e., 9 ft. 6 in. = 9.5 ft., 9 ft. 4 in. = 9.3 ft. In all cases, make a sketch of the building. Any protrusions or other variations in shape should be noted on the sketch with dimensions.

Calculations: The calculations portion of the estimate is a two-step activity:
(1) The selection of appropriate costs from the tables
(2) Computations

Selection of Appropriate Costs: To select the appropriate cost from the base tables the following information is needed:
(1) Class of construction
(2) Type of residence
(3) Occupancy
(4) Building configuration
(5) Exterior wall construction
(6) Living area

Consult the tables and accompanying information to make the appropriate selections. Modifications, adjustments, and alternatives are classified by class, type and size. Further modifications can be made using the "Assemblies" Section.

Computations: The computation process should take the following sequence:
(1) Multiply the base cost by the area
(2) Add or subtract the modifications
(3) Apply the location modifier

When selecting costs, interpolate or use the cost that most nearly matches the structure under study. This applies to size, exterior wall construction, and class.

How to Use the Square Foot Cost Pages

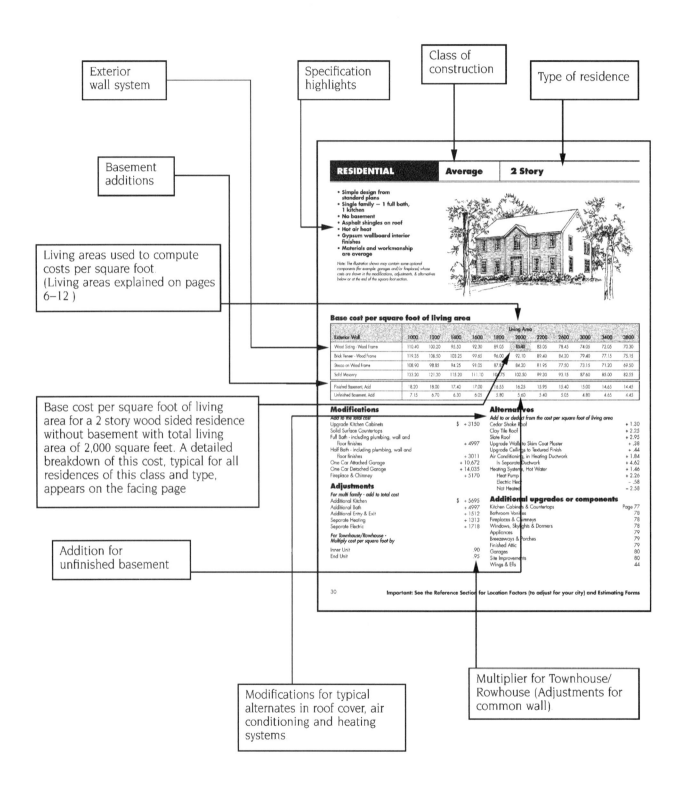

Components
This page contains the ten components needed to develop the complete square foot cost of the typical dwelling specified. All components are defined with a description of the materials and/or task involved. Use cost figures from each component to estimate the cost per square foot of that section of the project.

Specifications
The parameters for an example dwelling from the facing page are listed here. Included are the square foot dimensions of the proposed building. LIVING AREA takes into account the number of floors and other factors needed to define a building's TOTAL SQUARE FOOTAGE. Perimeter and partition dimensions are defined in terms of linear feet.

Line Totals
The extreme right-hand column lists the sum of two figures. Use this total to determine the sum of MATERIAL COST plus INSTALLATION COST. The result is a convenient total cost for each of the ten components.

Average 2 Story
Living Area - 2000 S.F.
Perimeter - 135 L.F.

#	Component	Description	Labor-Hours	Mat.	Labor	Total
1	Site Work	Site preparation for slab; 4' deep trench excavation for foundation wall.	.034		.66	.66
2	Foundation	Continuous reinforced concrete footing 8" deep x 18" wide; dampproofed and insulated reinforced concrete foundation wall, 8" thick, 4' deep, 4" concrete slab on 4" crushed stone base and polyethylene vapor barrier, trowel finish.	.066	2.91	3.35	6.26
3	Framing	Exterior walls - 2" x 4" wood studs, 16" O.C.; 1/2" plywood sheathing; 2" x 6" rafters 16" O.C. with 1/2" plywood sheathing, 4 in 12 pitch; 2" x 6" ceiling joists 16" O.C.; 2" x 8" floor joists 16" O.C. with 5/8" plywood subfloor; 1/2" plywood subfloor on 1" x 2" wood sleepers 16" O.C.	.131	6.33	7.37	13.70
4	Exterior Walls	Beveled wood siding and building paper on insulated wood frame walls; 6" attic insulation; double hung windows; 3 flush solid core wood exterior doors with storms.	.111	10.02	4.97	14.99
5	Roofing	25 year asphalt shingles; #15 felt building paper; aluminum gutters, downspouts, drip edge and flashings.	.024	.67	1.01	1.68
6	Interiors	Walls and ceilings, 1/2" taped and finished gypsum wallboard, primed and painted with 2 coats; painted baseboard and trim, finished hardwood floor 40%, carpet with 1/2" underlayment 40%, vinyl tile with 1/2" underlayment 15%, ceramic tile with 1/2" underlayment 5%; hollow core and louvered	.232	13.21	12.21	25.42
7	Specialties	Average grade kitchen cabinets - 14 L.F. wall and base with solid surface counter top and kitchen sink; 40 gallon electric water heater.	.021	1.73	.74	2.47
8	Mechanical	1 lavatory, white, wall hung; 1 water closet, white; 1 bathtub with shower; enameled steel, white; gas fired warm air heat.	.060	2.84	2.41	5.25
9	Electrical	200 Amp. service; romex wiring; incandescent lighting fixtures, switches, receptacles.	.039	1.20	1.38	2.58
10	Overhead	Contractor's overhead and profit and plans.		6.59	5.80	12.39
		Total		45.50	39.90	85.40

Labor-hours
Use this column to determine the unit of measure in LABOR-HOURS needed to perform a task. This figure will give the builder LABOR-HOURS PER SQUARE FOOT of the building. The TOTAL LABOR-HOURS PER COMPONENT are determined by multiplying the LIVING AREA times the LABOR-HOURS listed on that line. (TOTAL LABOR-HOURS PER COMPONENT = LIVING AREA x LABOR-HOURS).

Installation
The labor rates included here incorporate the overhead and profit costs for the installing contractor. The average mark-up used to create these figures is 69.5% over and above BARE LABOR COST including fringe benefits.

Bottom Line Total
This figure is the complete square foot cost for the construction project. To determine TOTAL PROJECT COST, multiply the BOTTOM LINE TOTAL times the LIVING AREA. (TOTAL PROJECT COST = BOTTOM LINE TOTAL x LIVING AREA).

Materials
This column gives the unit needed to develop the COST OF MATERIALS. Note: The figures given here are not BARE COSTS. Ten percent has been added to BARE MATERIAL COST to cover handling.

*NOTE
The components listed on this page are typical of all the sizes of residences from the facing page. Specific quantities of components required would vary with the size of the dwelling and the exterior wall system.

Building Classes

Given below are the four general definitions of building classes. Each building — Economy, Average, Custom and Luxury — is common in residential construction. All four are used in this book to determine costs per square foot.

Economy Class

An economy class residence is usually mass-produced from stock plans. The materials and workmanship are sufficient only to satisfy minimum building codes. Low construction cost is more important than distinctive features. Design is seldom other than square or rectangular.

Average Class

An average class residence is simple in design and is built from standard designer plans. Material and workmanship are average, but often exceed the minimum building codes. There are frequently special features that give the residence some distinctive characteristics.

Custom Class

A custom class residence is usually built from a designer's plans which have been modified to give the building a distinction of design. Material and workmanship are generally above average with obvious attention given to construction details. Construction normally exceeds building code requirements.

Luxury Class

A luxury class residence is built from an architect's plan for a specific owner. It is unique in design and workmanship. There are many special features, and construction usually exceeds all building codes. It is obvious that primary attention is placed on the owner's comfort and pleasure. Construction is supervised by an architect.

Residential Building Types

One Story

This is an example of a one-story dwelling. The living area of this type of residence is confined to the ground floor. The headroom in the attic is usually too low for use as a living area.

One-and-a-half Story

The living area in the upper level of this type of residence is 50% to 90% of the ground floor. This is made possible by a combination of this design's high-peaked roof and/or dormers. Only the upper level area with a ceiling height of 6' or more is considered living area. The living area of this residence is the sum of the ground floor area plus the area on the second level with a ceiling height of 6' or more.

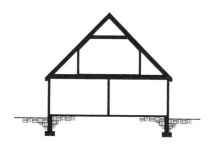

Residential Building Types

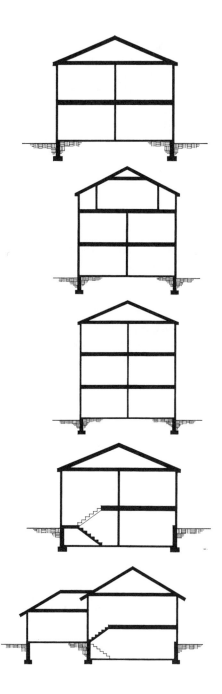

Two Story

This type of residence has a second floor or upper level area which is equal or nearly equal to the ground floor area. The upper level of this type of residence can range from 90% to 110% of the ground floor area, depending on setbacks or overhangs. The living area is the sum of the ground floor area and the upper level floor area.

Two-and-one-half Story

This type of residence has two levels of equal or nearly equal area and a third level which has a living area that is 50% to 90% of the ground floor. This is made possible by a high peaked roof, extended wall heights and/or dormers. Only the upper level area with a ceiling height of 6 feet or more is considered living area. The living area of this residence is the sum of the ground floor area, the second floor area and the area on the third level with a ceiling height of 6 feet or more.

Three Story

This type of residence has three levels which are equal or nearly equal. As in the 2 story residence, the second and third floor areas may vary slightly depending on the setbacks or overhangs. The living area is the sum of the ground floor area and the two upper level floor areas.

Bi-Level

This type of residence has two living areas, one above the other. One area is about 4 feet below grade and the second is about 4 feet above grade. Both are equal in size. The lower level in this type of residence is originally designed and built to serve as a living area and not as a basement. Both levels have full ceiling heights. The living area is the sum of the lower level area and the upper level area.

Tri-Level

This type of residence has three levels of living area. One is at grade level, the second is about 4 feet below grade, and the third is about 4 feet above grade. All levels are originally designed to serve as living areas. All levels have full ceiling heights. The living area is the sum of the areas of each of the three levels.

Exterior Wall Construction

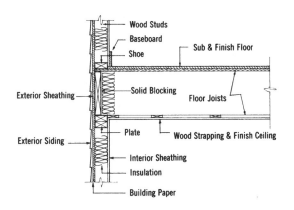

Typical Frame Construction

Typical wood frame construction consists of wood studs with insulation between them. A typical exterior surface is made up of sheathing, building paper and exterior siding consisting of wood, vinyl, aluminum or stucco over the wood sheathing.

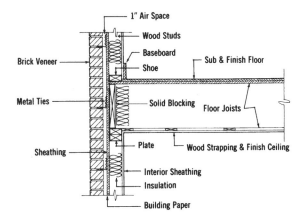

Brick Veneer

Typical brick veneer construction consists of wood studs with insulation between them. A typical exterior surface is sheathing, building paper and an exterior of brick tied to the sheathing with metal strips.

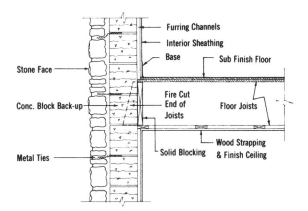

Stone

Typical solid masonry construction consists of a stone or block wall covered on the exterior with brick, stone or other masonry.

Residential Configurations

Detached House
This category of residence is a freestanding separate building with or without an attached garage. It has four complete walls.

Town/Row House
This category of residence has a number of attached units made up of inner units and end units. The units are joined by common walls. The inner units have only two exterior walls. The common walls are fireproof. The end units have three walls and a common wall. Town houses/row houses can be any of the five types.

Semi-Detached House
This category of residence has two living units side-by-side. The common wall is a fireproof wall. Semi-detached residences can be treated as a row house with two end units. Semi-detached residences can be any of the five types.

Residential Garage Types

Attached Garage
Shares a common wall with the dwelling. Access is typically through a door between dwelling and garage.

Built-In Garage
Constructed under the second floor living space and above basement level of dwelling. Reduces gross square feet of living area.

Basement Garage
Constructed under the roof of the dwelling but below the living area.

Detached Garage
Constructed apart from the main dwelling. Shares no common area or wall with the dwelling.

RESIDENTIAL COST ESTIMATE

OWNER'S NAME: _____ APPRAISER: _____

RESIDENCE ADDRESS: _____ PROJECT: _____

CITY, STATE, ZIP CODE: _____ DATE: _____

CLASS OF CONSTRUCTION	RESIDENCE TYPE	CONFIGURATION	EXTERIOR WALL SYSTEM
☐ ECONOMY	☐ 1 STORY	☐ DETACHED	☐ WOOD SIDING - WOOD FRAME
☐ AVERAGE	☐ 1-1/2 STORY	☐ TOWN/ROW HOUSE	☐ BRICK VENEER - WOOD FRAME
☐ CUSTOM	☐ 2 STORY	☐ SEMI-DETACHED	☐ STUCCO ON WOOD FRAME
☐ LUXURY	☐ 2-1/2 STORY		☐ PAINTED CONCRETE BLOCK
	☐ 3 STORY	OCCUPANCY	☐ SOLID MASONRY (AVERAGE & CUSTOM)
	☐ BI-LEVEL	☐ ONE FAMILY	☐ STONE VENEER - WOOD FRAME
	☐ TRI-LEVEL	☐ TWO FAMILY	☐ SOLID BRICK (LUXURY)
		☐ THREE FAMILY	☐ SOLID STONE (LUXURY)
		☐ OTHER _____	

* LIVING AREA (Main Building)		* LIVING AREA (Wing or Ell) ( )		* LIVING AREA (WING or ELL) ( )	
First Level	_____ S.F.	First Level	_____ S.F.	First Level	_____ S.F.
Second level	_____ S.F.	Second level	_____ S.F.	Second level	_____ S.F.
Third Level	_____ S.F.	Third Level	_____ S.F.	Third Level	_____ S.F.
Total	_____ S.F.	Total	_____ S.F.	Total	_____ S.F.

* Basement Area is not part of living area.

MAIN BUILDING	COSTS PER S.F. LIVING AREA
Cost per Square Foot of Living Area, from Page _____	$
Basement Addition: _____ % Finished, _____ % Unfinished	+
Roof Cover Adjustment: _____ Type, Page _____ (Add or Deduct)	( )
Central Air Conditioning: ☐ Separate Ducts ☐ Heating Ducts, Page _____	+
Heating System Adjustment: _____ Type, Page _____ (Add or Deduct)	( )
Main Building: Adjusted Cost per S.F. of Living Area	$

MAIN BUILDING TOTAL COST $ _____ /S.F. x _____ S.F. x _____ = $ _____
Cost per S.F. Living Area   Living Area   Town/Row House Multiplier   TOTAL COST
(Use 1 for Detached)

WING OR ELL ( ) _____ STORY	COSTS PER S.F. LIVING AREA
Cost per Square Foot of Living Area, from Page _____	$
Basement Addition: _____ % Finished, _____ % Unfinished	+
Roof Cover Adjustment: _____ Type, Page _____ (Add or Deduct)	( )
Central Air Conditioning: ☐ Separate Ducts ☐ Heating Ducts, Page _____	+
Heating System Adjustment: _____ Type, Page _____ (Add or Deduct)	( )
Wing or Ell ( ): Adjusted Cost per S.F. of Living Area	$

WING OR ELL ( ) TOTAL COST $ _____ /S.F. x _____ S.F. x = $ _____
Cost per S.F. Living Area   Living Area   TOTAL COST

WING OR ELL ( ) _____ STORY	COSTS PER S.F. LIVING AREA
Cost per Square Foot of Living Area, from Page _____	$
Basement Addition: _____ % Finished, _____ % Unfinished	+
Roof Cover Adjustment: _____ Type, Page _____ (Add or Deduct)	( )
Central Air Conditioning: ☐ Separate Ducts ☐ Heating Ducts, Page _____	+
Heating System Adjustment: _____ Type, Page _____ (Add or Deduct)	( )
Wing or Ell ( ): Adjusted Cost per S.F. of Living Area	$

WING OR ELL ( ) TOTAL COST $ _____ /S.F. x _____ S.F. x = $ _____
Cost per S.F. Living Area   Living Area   TOTAL COST

TOTAL THIS PAGE _____

RESIDENTIAL COST ESTIMATE

Total Page 1					$
			QUANTITY	UNIT COST	
Additional Bathrooms: _____ Full _____ Half					
Finished Attic: _____ Ft. x _____ Ft.			S.F.		+
Breezeway: ☐ Open ☐ Enclosed _____ Ft. x _____ Ft.			S.F.		+
Covered Porch: ☐ Open ☐ Enclosed _____ Ft. x _____ Ft.			S.F.		+
Fireplace: ☐ Interior Chimney ☐ Exterior Chimney					
☐ No. of Flues ☐ Additional Fireplaces					+
Appliances:					+
Kitchen Cabinets Adjustments: (±)					
☐ Garage ☐ Carport: _____ Car(s) Description _____ (±)					
Miscellaneous:					+
			ADJUSTED TOTAL BUILDING COST		$

REPLACEMENT COST

ADJUSTED TOTAL BUILDING COST	$ _____
Site Improvements	
(A) Paving & Sidewalks	$ _____
(B) Landscaping	$ _____
(C) Fences	$ _____
(D) Swimming Pools	$ _____
(E) Miscellaneous	$ _____
TOTAL	$ _____
Location Factor	x _____
Location Replacement Cost	$ _____
Depreciation	- $ _____
LOCAL DEPRECIATED COST	$ _____

INSURANCE COST

ADJUSTED TOTAL BUILDING COST	$ _____
Insurance Exclusions	
(A) Footings, Site work, Underground Piping	- $ _____
(B) Architects Fees	- $ _____
Total Building Cost Less Exclusion	$ _____
Location Factor	x _____
LOCAL INSURABLE REPLACEMENT COST	$ _____

SKETCH AND ADDITIONAL CALCULATIONS

RESIDENTIAL | Economy | Illustrations

1 Story

1-1/2 Story

2 Story

Bi-Level

Tri-Level

13

RESIDENTIAL | Economy | 1 Story

- Mass produced from stock plans
- Single family — 1 full bath, 1 kitchen
- No basement
- Asphalt shingles on roof
- Hot air heat
- Gypsum wallboard interior finishes
- Materials and workmanship are sufficient to meet codes

Note: The illustration shown may contain some optional components (for example: garages and/or fireplaces) whose costs are shown in the modifications, adjustments, & alternatives below or at the end of the square foot section.

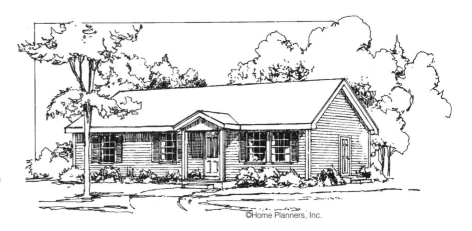
©Home Planners, Inc.

Base cost per square foot of living area

Exterior Wall	Living Area										
	600	800	1000	1200	1400	1600	1800	2000	2400	2800	3200
Wood Siding - Wood Frame	101.85	92.40	85.15	79.30	74.10	70.90	69.30	67.15	62.55	59.30	57.10
Brick Veneer - Wood Frame	109.40	99.10	91.25	84.80	79.10	75.55	73.80	71.40	66.45	62.90	60.40
Stucco on Wood Frame	97.95	88.90	81.95	76.45	71.55	68.45	67.00	64.95	60.60	57.45	55.35
Painted Concrete Block	102.05	92.55	85.35	79.45	74.25	70.95	69.45	67.25	62.65	59.40	57.20
Finished Basement, Add	25.60	24.15	23.00	22.05	21.30	20.75	20.45	20.05	19.50	19.05	18.60
Unfinished Basement, Add	11.50	10.30	9.50	8.70	8.10	7.70	7.45	7.15	6.65	6.30	6.00

Modifications

Add to the total cost

Upgrade Kitchen Cabinets	$ + 724
Solid Surface Countertops	+ 688
Full Bath - including plumbing, wall and floor finishes	+ 3982
Half Bath - including plumbing, wall and floor finishes	+ 2399
One Car Attached Garage	+ 9770
One Car Detached Garage	+ 12,594
Fireplace & Chimney	+ 4650

Adjustments

For multi family - add to total cost

Additional Kitchen	$ + 2930
Additional Bath	+ 3982
Additional Entry & Exit	+ 1512
Separate Heating	+ 1313
Separate Electric	+ 978

For Townhouse/Rowhouse - Multiply cost per square foot by

Inner Unit	.95
End Unit	.97

Alternatives

Add to or deduct from the cost per square foot of living area

Composition Roll Roofing	– .75
Cedar Shake Roof	+ 2.95
Upgrade Walls and Ceilings to Skim Coat Plaster	+ .52
Upgrade Ceilings to Textured Finish	+ .44
Air Conditioning, in Heating Ductwork	+ 2.94
In Separate Ductwork	+ 5.61
Heating Systems, Hot Water	+ 1.53
Heat Pump	+ 1.86
Electric Heat	– 1.31
Not Heated	– 3.31

Additional upgrades or components

Kitchen Cabinets & Countertops	Page 77
Bathroom Vanities	78
Fireplaces & Chimneys	78
Windows, Skylights & Dormers	78
Appliances	79
Breezeways & Porches	79
Finished Attic	79
Garages	80
Site Improvements	80
Wings & Ells	24

Important: See the Reference Section for Location Factors (to adjust for your city) and Estimating Forms

Economy 1 Story

Living Area - 1200 S.F.
Perimeter - 146 L.F.

			Labor-Hours	Cost Per Square Foot Of Living Area		
				Mat.	Labor	Total
1	Site Work	Site preparation for slab; 4' deep trench excavation for foundation wall.	.060		1.06	1.06
2	Foundation	Continuous reinforced concrete footing, 8" deep x 18" wide; dampproofed and insulated 8" thick reinforced concrete block foundation wall, 4' deep; 4" concrete slab on 4" crushed stone base and polyethylene vapor barrier, trowel finish.	.131	5.29	6.00	11.29
3	Framing	Exterior walls - 2" x 4" wood studs, 16" O.C.; 1/2" insulation board sheathing; wood truss roof frame, 24" O.C. with 1/2" plywood sheathing, 4 in 12 pitch.	.098	4.32	5.16	9.48
4	Exterior Walls	Metal lath reinforced stucco exterior on insulated wood frame walls; 6" attic insulation; sliding sash wood windows; 2 flush solid core wood exterior doors with storms.	.110	6.23	5.38	11.61
5	Roofing	20 year asphalt shingles; #15 felt building paper; aluminum gutters, downspouts, drip edge and flashings.	.047	1.30	1.96	3.26
6	Interiors	Walls and ceilings, 1/2" taped and finished gypsum wallboard, primed and painted with 2 coats; painted baseboard and trim; rubber backed carpeting 80%, asphalt tile 20%; hollow core wood interior doors.	.243	8.43	10.14	18.57
7	Specialties	Economy grade kitchen cabinets - 6 L.F. wall and base with plastic laminate counter top and kitchen sink; 30 gallon electric water heater.	.004	1.70	.78	2.48
8	Mechanical	1 lavatory, white, wall hung; 1 water closet, white; 1 bathtub, enameled steel, white; gas fired warm air heat.	.086	3.70	2.83	6.53
9	Electrical	100 Amp. service; romex wiring; incandescent lighting fixtures, switches, receptacles.	.036	.97	1.23	2.20
10	Overhead	Contractor's overhead and profit		4.81	5.16	9.97
		Total		36.75	39.70	**76.45**

RESIDENTIAL | Economy | 1-1/2 Story

- Mass produced from stock plans
- Single family — 1 full bath, 1 kitchen
- No basement
- Asphalt shingles on roof
- Hot air heat
- Gypsum wallboard interior finishes
- Materials and workmanship are sufficient to meet codes

Note: The illustration shown may contain some optional components (for example: garages and/or fireplaces) whose costs are shown in the modifications, adjustments, & alternatives below or at the end of the square foot section.

Base cost per square foot of living area

Exterior Wall	Living Area										
	600	800	1000	1200	1400	1600	1800	2000	2400	2800	3200
Wood Siding - Wood Frame	116.55	97.20	86.75	82.05	78.65	73.40	70.90	68.20	62.60	60.50	58.25
Brick Veneer - Wood Frame	127.00	104.80	93.80	88.65	84.95	79.10	76.30	73.35	67.10	64.80	62.20
Stucco on Wood Frame	111.15	93.25	83.10	78.65	75.40	70.45	68.05	65.55	60.20	58.25	56.15
Painted Concrete Block	116.85	97.40	87.00	82.25	78.85	73.55	71.00	68.35	62.70	60.60	58.35
Finished Basement, Add	19.70	16.70	15.95	15.35	14.95	14.40	14.05	13.75	13.15	12.90	12.55
Unfinished Basement, Add	10.10	7.80	7.15	6.70	6.35	5.90	5.65	5.40	4.95	4.75	4.45

Modifications

Add to the total cost

Upgrade Kitchen Cabinets	$ + 724
Solid Surface Countertops	+ 688
Full Bath - including plumbing, wall and floor finishes	+ 3982
Half Bath - including plumbing, wall and floor finishes	+ 2399
One Car Attached Garage	+ 9770
One Car Detached Garage	+ 12,594
Fireplace & Chimney	+ 4650

Adjustments

For multi family - add to total cost

Additional Kitchen	$ + 2930
Additional Bath	+ 3982
Additional Entry & Exit	+ 1512
Separate Heating	+ 1313
Separate Electric	+ 978

For Townhouse/Rowhouse - Multiply cost per square foot by

Inner Unit	.95
End Unit	.97

Alternatives

Add to or deduct from the cost per square foot of living area

Composition Roll Roofing	– .55
Cedar Shake Roof	+ 2.10
Upgrade Walls and Ceilings to Skim Coat Plaster	+ .52
Upgrade Ceilings to Textured Finish	+ .44
Air Conditioning, in Heating Ductwork	+ 2.20
In Separate Ductwork	+ 4.93
Heating Systems, Hot Water	+ 1.47
Heat Pump	+ 2.05
Electric Heat	– 1.05
Not Heated	– 3.05

Additional upgrades or components

Kitchen Cabinets & Countertops	Page 77
Bathroom Vanities	78
Fireplaces & Chimneys	78
Windows, Skylights & Dormers	78
Appliances	79
Breezeways & Porches	79
Finished Attic	79
Garages	80
Site Improvements	80
Wings & Ells	24

Economy 1-1/2 Story

Living Area - 1600 S.F.
Perimeter - 135 L.F.

			Labor-Hours	Cost Per Square Foot Of Living Area		
				Mat.	Labor	Total
1	Site Work	Site preparation for slab; 4' deep trench excavation for foundation wall.	.041		.80	.80
2	Foundation	Continuous reinforced concrete footing, 8" deep x 18" wide; dampproofed and insulated 8" thick reinforced concrete block foundation wall, 4' deep; 4" concrete slab on 4" crushed stone base and polyethylene vapor barrier, trowel finish.	.073	3.54	4.06	7.60
3	Framing	Exterior walls - 2" x 4" wood studs, 16" O.C.; 1/2" insulation board sheathing; 2" x 6" rafters, 16" O.C. with 1/2" plywood sheathing, 8 in 12 pitch; 2" x 8" floor joists 16" O.C. with bridging and 5/8" plywood subfloor.	.090	4.38	5.78	10.16
4	Exterior Walls	Beveled wood siding and building paper on insulated wood frame walls; 6" attic insulation; double hung windows; 2 flush solid core wood exterior doors with storms.	.077	9.13	4.47	13.60
5	Roofing	20 year asphalt shingles; #15 felt building paper; aluminum gutters, downspouts, drip edge and flashings.	.029	.81	1.23	2.04
6	Interiors	Walls and ceilings, 1/2" taped and finished gypsum wallboard, primed and painted with 2 coats; painted baseboard and trim; rubber backed carpeting 80%, asphalt tile 20%; hollow core wood interior doors.	.204	9.27	10.88	20.15
7	Specialties	Economy grade kitchen cabinets - 6 L.F. wall and base with plastic laminate counter top and kitchen sink; 30 gallon electric water heater.	.020	1.28	.58	1.86
8	Mechanical	1 lavatory, white, wall hung; 1 water closet, white; 1 bathtub, enameled steel, white; gas fired warm air heat.	.079	3.08	2.54	5.62
9	Electrical	100 Amp. service; romex wiring; incandescent lighting fixtures, switches, receptacles.	.033	.88	1.11	1.99
10	Overhead	Contractor's overhead and profit.		4.88	4.70	9.58
		Total		37.25	36.15	**73.40**

RESIDENTIAL | Economy | 2 Story

- Mass produced from stock plans
- Single family — 1 full bath, 1 kitchen
- No basement
- Asphalt shingles on roof
- Hot air heat
- Gypsum wallboard interior finishes
- Materials and workmanship are sufficient to meet codes

Note: The illustration shown may contain some optional components (for example: garages and/or fireplaces) whose costs are shown in the modifications, adjustments, & alternatives below or at the end of the square foot section.

Base cost per square foot of living area

Exterior Wall	\multicolumn{11}{c}{Living Area}
	1000	1200	1400	1600	1800	2000	2200	2600	3000	3400	3800
Wood Siding - Wood Frame	92.30	83.35	79.30	76.55	73.80	70.50	68.40	64.45	60.45	58.70	57.20
Brick Veneer - Wood Frame	100.40	90.85	86.30	83.20	80.10	76.60	74.20	69.65	65.30	63.25	61.55
Stucco on Wood Frame	88.10	79.40	75.65	73.05	70.55	67.35	65.40	61.70	57.90	56.30	54.95
Painted Concrete Block	92.50	83.50	79.50	76.70	74.00	70.65	68.55	64.60	60.60	58.80	57.35
Finished Basement, Add	13.40	12.85	12.40	12.05	11.75	11.55	11.30	10.85	10.60	10.35	10.15
Unfinished Basement, Add	6.20	5.75	5.45	5.15	4.90	4.75	4.55	4.20	3.95	3.80	3.60

Modifications
Add to the total cost

Upgrade Kitchen Cabinets	$ + 724
Solid Surface Countertops	+ 688
Full Bath - including plumbing, wall and floor finishes	+ 3982
Half Bath - including plumbing, wall and floor finishes	+ 2399
One Car Attached Garage	+ 9770
One Car Detached Garage	+ 12,594
Fireplace & Chimney	+ 5140

Adjustments
For multi family - add to total cost

Additional Kitchen	$ + 2930
Additional Bath	+ 3982
Additional Entry & Exit	+ 1512
Separate Heating	+ 1313
Separate Electric	+ 978

For Townhouse/Rowhouse - Multiply cost per square foot by

Inner Unit	.93
End Unit	.96

Alternatives
Add to or deduct from the cost per square foot of living area

Composition Roll Roofing	– .40
Cedar Shake Roof	+ 1.45
Upgrade Walls and Ceilings to Skim Coat Plaster	+ .53
Upgrade Ceilings to Textured Finish	+ .44
Air Conditioning, in Heating Ductwork	+ 1.78
In Separate Ductwork	+ 4.51
Heating Systems, Hot Water	+ 1.43
Heat Pump	+ 2.17
Electric Heat	– .91
Not Heated	– 2.88

Additional upgrades or components

Kitchen Cabinets & Countertops	Page 77
Bathroom Vanities	78
Fireplaces & Chimneys	78
Windows, Skylights & Dormers	78
Appliances	79
Breezeways & Porches	79
Finished Attic	79
Garages	80
Site Improvements	80
Wings & Ells	24

Economy 2 Story

Living Area - 2000 S.F.
Perimeter - 135 L.F.

			Labor-Hours	Cost Per Square Foot Of Living Area		
				Mat.	Labor	Total
1	Site Work	Site preparation for slab; 4' deep trench excavation for foundation wall.	.034		.64	.64
2	Foundation	Continuous reinforced concrete footing, 8" deep x 18" wide; dampproofed and insulated 8" thick reinforced concrete block foundation wall, 4' deep; 4" concrete slab on 4" crushed stone base and polyethylene vapor barrier, trowel finish.	.069	2.83	3.26	6.09
3	Framing	Exterior walls - 2" x 4" wood studs, 16" O.C.; 1/2" insulation board sheathing; wood truss roof frame, 24" O.C. with 1/2" plywood sheathing, 4 in 12 pitch; 2" x 8" floor joists 16" O.C. with bridging and 5/8" plywood subfloor.	.112	4.46	6.05	10.51
4	Exterior Walls	Beveled wood siding and building paper on insulated wood frame walls; 6" attic insulation; double hung windows; 2 flush solid core wood exterior doors with storms.	.107	9.33	4.58	13.91
5	Roofing	20 year asphalt shingles; #15 felt building paper; aluminum gutters, downspouts, drip edge and flashings.	.024	.65	.98	1.63
6	Interiors	Walls and ceilings, 1/2" taped and finished gypsum wallboard, primed and painted with 2 coats; painted baseboard and trim; rubber backed carpeting 80%, asphalt tile 20%; hollow core wood interior doors.	.219	9.21	10.88	20.09
7	Specialties	Economy grade kitchen cabinets - 6 L.F. wall and base with plastic laminate counter top and kitchen sink; 30 gallon electric water heater.	.017	1.02	.47	1.49
8	Mechanical	1 lavatory, white, wall hung; 1 water closet, white; 1 bathtub, enameled steel, white; gas fired warm air heat.	.061	2.71	2.35	5.06
9	Electrical	100 Amp. service; romex wiring; incandescent lighting fixtures; switches, receptacles.	.030	.84	1.04	1.88
10	Overhead	Contractor's overhead and profit		4.65	4.55	9.20
		Total		35.70	34.80	**70.50**

RESIDENTIAL — Economy | Bi-Level

- Mass produced from stock plans
- Single family — 1 full bath, 1 kitchen
- No basement
- Asphalt shingles on roof
- Hot air heat
- Gypsum wallboard interior finishes
- Materials and workmanship are sufficient to meet codes

Note: The illustration shown may contain some optional components (for example: garages and/or fireplaces) whose costs are shown in the modifications, adjustments, & alternatives below or at the end of the square foot section.

Base cost per square foot of living area

Exterior Wall	1000	1200	1400	1600	1800	2000	2200	2600	3000	3400	3800
Wood Siding - Wood Frame	85.85	77.40	73.70	71.20	68.75	65.70	63.85	60.30	56.65	55.05	53.75
Brick Veneer - Wood Frame	91.90	83.00	79.00	76.25	73.50	70.25	68.15	64.20	60.25	58.50	57.05
Stucco on Wood Frame	82.65	74.40	71.00	68.60	66.35	63.35	61.60	58.25	54.70	53.25	52.10
Painted Concrete Block	86.00	77.50	73.85	71.35	68.95	65.80	63.90	60.40	56.75	55.10	53.85
Finished Basement, Add	13.40	12.85	12.40	12.05	11.75	11.55	11.30	10.85	10.60	10.35	10.15
Unfinished Basement, Add	6.20	5.75	5.45	5.15	4.90	4.75	4.55	4.20	3.95	3.80	3.60

Modifications
Add to the total cost

Upgrade Kitchen Cabinets	$ + 724
Solid Surface Countertops	+ 688
Full Bath - including plumbing, wall and floor finishes	+ 3982
Half Bath - including plumbing, wall and floor finishes	+ 2399
One Car Attached Garage	+ 9770
One Car Detached Garage	+ 12,594
Fireplace & Chimney	+ 4650

Adjustments
For multi family - add to total cost

Additional Kitchen	$ + 2930
Additional Bath	+ 3982
Additional Entry & Exit	+ 1512
Separate Heating	+ 1313
Separate Electric	+ 978

For Townhouse/Rowhouse - Multiply cost per square foot by

Inner Unit	.94
End Unit	.97

Alternatives
Add to or deduct from the cost per square foot of living area

Composition Roll Roofing	– .40
Cedar Shake Roof	+ 1.45
Upgrade Walls and Ceilings to Skim Coat Plaster	+ .50
Upgrade Ceilings to Textured Finish	+ .44
Air Conditioning, in Heating Ductwork	+ 1.78
In Separate Ductwork	+ 4.51
Heating Systems, Hot Water	+ 1.43
Heat Pump	+ 2.17
Electric Heat	– .91
Not Heated	– 2.88

Additional upgrades or components

Kitchen Cabinets & Countertops	Page 77
Bathroom Vanities	78
Fireplaces & Chimneys	78
Windows, Skylights & Dormers	78
Appliances	79
Breezeways & Porches	79
Finished Attic	79
Garages	80
Site Improvements	80
Wings & Ells	24

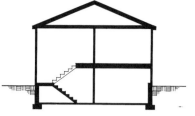

Economy Bi-Level

Living Area - 2000 S.F.
Perimeter - 135 L.F.

		Labor-Hours	Cost Per Square Foot Of Living Area		
			Mat.	Labor	Total
1 Site Work	Excavation for lower level, 4' deep. Site preparation for slab.	.029		.64	.64
2 Foundation	Continuous reinforced concrete footing, 8" deep x 18" wide; dampproofed and insulated 8" thick reinforced concrete block foundation wall, 4' deep; 4" concrete slab on 4" crushed stone base and polyethylene vapor barrier, trowel finish.	.069	2.83	3.26	6.09
3 Framing	Exterior walls - 2" x 4" wood studs, 16" O.C.; 1/2" insulation board sheathing; wood truss roof frame, 24" O.C. with 1/2" plywood sheathing, 4 in 12 pitch; 2" x 8" floor joists 16" O.C. with bridging and 5/8" plywood subfloor.	.107	4.20	5.69	9.89
4 Exterior Walls	Beveled wood siding and building paper on insulated wood frame walls; 6" attic insulation; double hung windows; 2 flush solid core wood exterior doors with storms.	.089	7.32	3.59	10.91
5 Roofing	20 year asphalt shingles; #15 felt building paper; aluminum gutters, downspouts, drip edge and flashings.	.024	.65	.98	1.63
6 Interiors	Walls and ceilings, 1/2" taped and finished gypsum wallboard, primed and painted with 2 coats; painted baseboard and trim, rubber backed carpeting 80%, asphalt tile 20%; hollow core wood interior doors.	.213	8.99	10.54	19.53
7 Specialties	Economy grade kitchen cabinets - 6 L.F. wall and base with plastic laminate counter top and kitchen sink; 30 gallon electric water heater.	.018	1.02	.47	1.49
8 Mechanical	1 lavatory, white, wall hung; 1 water closet, white; 1 bathtub, enameled steel, white; gas fired warm air heat.	.061	2.71	2.35	5.06
9 Electrical	100 Amp. service; romex wiring; incandescent lighting fixtures; switches, receptacles.	.030	.84	1.04	1.88
10 Overhead	Contractor's overhead and profit.		4.29	4.29	8.58
	Total		32.85	32.85	65.70

RESIDENTIAL — Economy — Tri-Level

- Mass produced from stock plans
- Single family — 1 full bath, 1 kitchen
- No basement
- Asphalt shingles on roof
- Hot air heat
- Gypsum wallboard interior finishes
- Materials and workmanship are sufficient to meet codes

Note: The illustration shown may contain some optional components (for example: garages and/or fireplaces) whose costs are shown in the modifications, adjustments, & alternatives below or at the end of the square foot section.

©Design Basics, Inc.

Base cost per square foot of living area

Exterior Wall	1200	1500	1800	2000	2200	2400	2800	3200	3600	4000	4400
Wood Siding - Wood Frame	79.20	72.75	67.95	66.10	63.20	60.95	59.20	56.70	53.85	52.85	50.70
Brick Veneer - Wood Frame	84.80	77.80	72.55	70.45	67.30	64.80	62.95	60.20	57.10	56.00	53.65
Stucco on Wood Frame	76.30	70.15	65.65	63.85	61.00	58.90	57.25	54.95	52.20	51.25	49.15
Solid Masonry	79.35	72.90	68.10	66.20	63.30	61.05	59.30	56.80	53.95	53.00	50.80
Finished Basement, Add*	16.05	15.35	14.70	14.40	14.10	13.85	13.60	13.25	12.95	12.85	12.60
Unfinished Basement, Add*	6.90	6.30	5.80	5.55	5.35	5.10	4.90	4.65	4.45	4.35	4.15

*Basement under middle level only.

Modifications

Add to the total cost

Upgrade Kitchen Cabinets	$ + 724
Solid Surface Countertops	+ 688
Full Bath - including plumbing, wall and floor finishes	+ 3982
Half Bath - including plumbing, wall and floor finishes	+ 2399
One Car Attached Garage	+ 9770
One Car Detached Garage	+ 12,594
Fireplace & Chimney	+ 4650

Adjustments

For multi family - add to total cost

Additional Kitchen	$ + 2930
Additional Bath	+ 3982
Additional Entry & Exit	+ 1512
Separate Heating	+ 1313
Separate Electric	+ 978

For Townhouse/Rowhouse - Multiply cost per square foot by

Inner Unit	.93
End Unit	.96

Alternatives

Add to or deduct from the cost per square foot of living area

Composition Roll Roofing	– .55
Cedar Shake Roof	+ 2.10
Upgrade Walls and Ceilings to Skim Coat Plaster	+ .46
Upgrade Ceilings to Textured Finish	+ .44
Air Conditioning, in Heating Ductwork	+ 1.52
In Separate Ductwork	+ 4.20
Heating Systems, Hot Water	+ 1.36
Heat Pump	+ 2.26
Electric Heat	– .79
Not Heated	– 2.79

Additional upgrades or components

Kitchen Cabinets & Countertops	Page 77
Bathroom Vanities	78
Fireplaces & Chimneys	78
Windows, Skylights & Dormers	78
Appliances	79
Breezeways & Porches	79
Finished Attic	79
Garages	80
Site Improvements	80
Wings & Ells	24

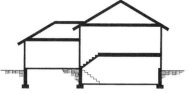

Economy Tri-Level

Living Area - 2400 S.F.
Perimeter - 163 L.F.

			Labor-Hours	Cost Per Square Foot Of Living Area		
				Mat.	Labor	Total
1	Site Work	Site preparation for slab; 4' deep trench excavation for foundation wall, excavation for lower level, 4' deep.	.027		.53	.53
2	Foundation	Continuous reinforced concrete footing, 8" deep x 18" wide; dampproofed and insulated 8" thick reinforced concrete block foundation wall, 4' deep; 4" concrete slab on 4" crushed stone base and polyethylene vapor barrier, trowel finish.	.071	3.21	3.54	6.75
3	Framing	Exterior walls - 2" x 4" wood studs, 16" O.C.; 1/2" insulation board sheathing; wood truss roof frame, 24" O.C. with 1/2" plywood sheathing, 4 in 12 pitch; 2" x 8" floor joists 16" O.C. with bridging and 5/8" plywood subfloor.	.094	3.97	5.14	9.11
4	Exterior Walls	Beveled wood siding and building paper on insulated wood frame walls; 6" attic insulation; double hung windows; 2 flush solid core wood exterior doors with storms.	.081	6.32	3.10	9.42
5	Roofing	20 year asphalt shingles; #15 felt building paper; aluminum gutters, downspouts, drip edge and flashings.	.032	.87	1.31	2.18
6	Interiors	Walls and ceilings, 1/2" taped and finished gypsum wallboard, primed and painted with 2 coats; painted baseboard and trim, rubber backed carpeting 80%, asphalt tile 20%; hollow core wood interior doors.	.177	7.90	9.35	17.25
7	Specialties	Economy grade kitchen cabinets - 6 L.F. wall and base with plastic laminate counter top and kitchen sink; 30 gallon electric water heater.	.014	.86	.39	1.25
8	Mechanical	1 lavatory, white, wall hung; 1 water closet, white; 1 bathtub, enameled steel, white; gas fired warm air heat.	.057	2.45	2.24	4.69
9	Electrical	100 Amp. service; romex wiring; incandescent lighting fixtures, switches, receptacles.	.029	.80	.99	1.79
10	Overhead	Contractor's overhead and profit		3.97	4.01	7.98
		Total		30.35	30.60	**60.95**

RESIDENTIAL | Economy | Wings & Ells

1 Story — Base cost per square foot of living area

Exterior Wall	Living Area							
	50	100	200	300	400	500	600	700
Wood Siding - Wood Frame	140.90	107.25	92.60	77.25	72.50	69.65	67.70	68.30
Brick Veneer - Wood Frame	159.80	120.80	103.90	84.75	79.25	75.95	73.65	74.05
Stucco on Wood Frame	131.05	100.25	86.70	73.40	68.95	66.35	64.55	65.25
Painted Concrete Block	141.45	107.60	92.90	77.45	72.65	69.85	67.85	68.45
Finished Basement, Add	39.15	31.95	28.90	23.90	22.90	22.35	21.95	21.60
Unfinished Basement, Add	21.85	16.30	13.90	10.05	9.30	8.80	8.50	8.30

1-1/2 Story — Base cost per square foot of living area

Exterior Wall	Living Area							
	100	200	300	400	500	600	700	800
Wood Siding - Wood Frame	111.10	89.15	75.85	68.00	64.00	62.00	59.55	58.85
Brick Veneer - Wood Frame	128.00	102.70	87.05	76.75	72.05	69.70	66.80	66.05
Stucco on Wood Frame	102.35	82.15	69.95	63.45	59.80	58.10	55.80	55.15
Painted Concrete Block	111.60	89.55	76.15	68.25	64.20	62.25	59.75	59.00
Finished Basement, Add	26.15	23.15	21.10	18.95	18.35	17.95	17.55	17.50
Unfinished Basement, Add	13.45	11.15	9.60	7.90	7.40	7.10	6.80	6.75

2 Story — Base cost per square foot of living area

Exterior Wall	Living Area							
	100	200	400	600	800	1000	1200	1400
Wood Siding - Wood Frame	113.05	83.95	71.25	58.95	54.85	52.40	50.70	51.45
Brick Veneer - Wood Frame	131.95	97.45	82.45	66.45	61.55	58.65	56.65	57.25
Stucco on Wood Frame	103.25	76.90	65.35	55.10	51.30	49.05	47.55	48.45
Painted Concrete Block	113.55	84.35	71.50	59.15	55.00	52.50	50.85	51.60
Finished Basement, Add	19.60	15.95	14.50	12.05	11.50	11.15	11.00	10.85
Unfinished Basement, Add	10.95	8.15	7.00	5.05	4.70	4.45	4.25	4.15

Base costs do not include bathroom or kitchen facilities. Use Modifications/Adjustments/Alternatives on pages 77-80 where appropriate.

RESIDENTIAL | Average | Illustrations

1 Story

1-1/2 Story

2 Story

2-1/2 Story

Bi-Level

Tri-Level

RESIDENTIAL — Average — 1 Story

- Simple design from standard plans
- Single family — 1 full bath, 1 kitchen
- No basement
- Asphalt shingles on roof
- Hot air heat
- Gypsum wallboard interior finishes
- Materials and workmanship are average

Note: The illustration shown may contain some optional components (for example: garages and/or fireplaces) whose costs are shown in the modifications, adjustments, & alternatives below or at the end of the square foot section.

©Home Planners, Inc.

Base cost per square foot of living area

Exterior Wall	600	800	1000	1200	1400	1600	1800	2000	2400	2800	3200
Wood Siding - Wood Frame	123.65	112.35	103.80	97.00	91.05	87.35	85.35	82.95	77.70	74.00	71.50
Brick Veneer - Wood Frame	142.20	129.95	120.75	113.30	106.85	102.70	100.55	97.85	92.20	88.20	85.40
Stucco on Wood Frame	132.55	121.35	112.95	106.25	100.40	96.75	94.75	92.45	87.25	83.60	81.15
Solid Masonry	155.05	141.40	131.20	122.65	115.40	110.70	108.30	105.10	98.85	94.25	91.05
Finished Basement, Add	31.35	30.30	28.90	27.60	26.55	25.85	25.45	24.90	24.15	23.50	23.00
Unfinished Basement, Add	13.05	11.80	10.90	10.10	9.45	9.05	8.85	8.50	8.00	7.65	7.35

Modifications

Add to the total cost

Upgrade Kitchen Cabinets	$ + 3150
Solid Surface Countertops	
Full Bath - including plumbing, wall and floor finishes	+ 4997
Half Bath - including plumbing, wall and floor finishes	+ 3011
One Car Attached Garage	+ 10,672
One Car Detached Garage	+ 14,035
Fireplace & Chimney	+ 4635

Adjustments

For multi family - add to total cost

Additional Kitchen	$ + 5695
Additional Bath	+ 4997
Additional Entry & Exit	+ 1512
Separate Heating	+ 1313
Separate Electric	+ 1718

For Townhouse/Rowhouse - Multiply cost per square foot by

Inner Unit	.92
End Unit	.96

Alternatives

Add to or deduct from the cost per square foot of living area

Cedar Shake Roof	+ 2.65
Clay Tile Roof	+ 4.50
Slate Roof	+ 5.85
Upgrade Walls to Skim Coat Plaster	+ .32
Upgrade Ceilings to Textured Finish	+ .44
Air Conditioning, in Heating Ductwork	+ 3.04
In Separate Ductwork	+ 5.82
Heating Systems, Hot Water	+ 1.57
Heat Pump	+ 1.93
Electric Heat	- .73
Not Heated	- 2.76

Additional upgrades or components

Kitchen Cabinets & Countertops	Page 77
Bathroom Vanities	78
Fireplaces & Chimneys	78
Windows, Skylights & Dormers	78
Appliances	79
Breezeways & Porches	79
Finished Attic	79
Garages	80
Site Improvements	80
Wings & Ells	44

Important: See the Reference Section for Location Factors (to adjust for your city) and Estimating Forms

Average 1 Story

Living Area - 1600 S.F.
Perimeter - 163 L.F.

		Labor-Hours	Cost Per Square Foot Of Living Area		
			Mat.	Labor	Total
1 Site Work	Site preparation for slab; 4' deep trench excavation for foundation wall.	.048		.82	.82
2 Foundation	Continuous reinforced concrete footing 8" deep x 18" wide; dampproofed and insulated reinforced concrete foundation wall, 8" thick, 4' deep; 4" concrete slab on 4" crushed stone base and polyethylene vapor barrier, trowel finish.	.113	4.95	5.47	10.42
3 Framing	Exterior walls - 2" x 4" wood studs, 16" O.C.; 1/2" plywood sheathing; 2" x 6" rafters 16" O.C. with 1/2" plywood sheathing, 4 in 12 pitch; 2" x 6" ceiling joists 16" O.C.; 1/2" plywood subfloor on 1" x 2" wood sleepers 16" O.C.	.136	5.59	7.64	13.23
4 Exterior Walls	Beveled wood siding and building paper on insulated wood frame walls; 6" attic insulation; double hung windows; 3 flush solid core wood exterior doors with storms.	.098	8.50	4.22	12.72
5 Roofing	25 year asphalt shingles; #15 felt building paper; aluminum gutters, downspouts, drip edge and flashings.	.047	1.34	2.01	3.35
6 Interiors	Walls and ceilings, 1/2" taped and finished gypsum wallboard, primed and painted with 2 coats; painted baseboard and trim, finished hardwood floor 40%, carpet with 1/2" underlayment 40%, vinyl tile with 1/2" underlayment 15%, ceramic tile with 1/2" underlayment 5%; hollow core and louvered	.251	11.64	10.76	22.40
7 Specialties	Average grade kitchen cabinets - 14 L.F. wall and base with solid surface counter top and kitchen sink; 40 gallon electric water heater.	.009	2.15	.91	3.06
8 Mechanical	1 lavatory, white, wall hung; 1 water closet, white; 1 bathtub with shower, enameled steel, white; gas fired warm air heat.	.098	3.24	2.60	5.84
9 Electrical	200 Amp. service; romex wiring; incandescent lighting fixtures, switches, receptacles.	.041	1.30	1.51	2.81
10 Overhead	Contractor's overhead and profit and plans.		6.59	6.11	12.70
	Total		45.30	42.05	**87.35**

RESIDENTIAL — Average — 1-1/2 Story

- Simple design from standard plans
- Single family — 1 full bath, 1 kitchen
- No basement
- Asphalt shingles on roof
- Hot air heat
- Gypsum wallboard interior finishes
- Materials and workmanship are average

Note: The illustration shown may contain some optional components (for example: garages and/or fireplaces) whose costs are shown in the modifications, adjustments, & alternatives below or at the end of the square foot section.

©By Designer

Base cost per square foot of living area

Exterior Wall	\	\	\	\	Living Area	\	\	\	\	\	\
	600	800	1000	1200	1400	1600	1800	2000	2400	2800	3200
Wood Siding - Wood Frame	138.65	116.75	104.60	99.15	95.15	89.20	86.20	83.20	76.85	74.50	71.90
Brick Veneer - Wood Frame	150.25	125.10	112.35	106.40	102.05	95.45	92.20	88.85	81.90	79.25	76.30
Stucco on Wood Frame	136.70	115.35	103.35	97.95	94.00	88.15	85.25	82.25	76.05	73.70	71.20
Solid Masonry	168.10	138.10	124.40	117.70	112.75	105.20	101.45	97.65	89.65	86.60	83.10
Finished Basement, Add	25.70	22.40	21.45	20.70	20.20	19.40	18.95	18.55	17.75	17.40	16.95
Unfinished Basement, Add	11.25	8.85	8.20	7.70	7.40	6.90	6.65	6.40	5.90	5.70	5.40

Modifications

Add to the total cost

Upgrade Kitchen Cabinets	$ + 3150
Solid Surface Countertops	
Full Bath - including plumbing, wall and floor finishes	+ 4997
Half Bath - including plumbing, wall and floor finishes	+ 3011
One Car Attached Garage	+ 10,672
One Car Detached Garage	+ 14,035
Fireplace & Chimney	+ 4635

Adjustments

For multi family - add to total cost

Additional Kitchen	$ + 5695
Additional Bath	+ 4997
Additional Entry & Exit	+ 1512
Separate Heating	+ 1313
Separate Electric	+ 1718

For Townhouse/Rowhouse - Multiply cost per square foot by

Inner Unit	.92
End Unit	.96

Alternatives

Add to or deduct from the cost per square foot of living area

Cedar Shake Roof	+ 1.90
Clay Tile Roof	+ 3.25
Slate Roof	+ 4.25
Upgrade Walls to Skim Coat Plaster	+ .37
Upgrade Ceilings to Textured Finish	+ .44
Air Conditioning, in Heating Ductwork	+ 2.31
In Separate Ductwork	+ 5.09
Heating Systems, Hot Water	+ 1.49
Heat Pump	+ 2.14
Electric Heat	− .66
Not Heated	− 2.66

Additional upgrades or components

Kitchen Cabinets & Countertops	Page 77
Bathroom Vanities	78
Fireplaces & Chimneys	78
Windows, Skylights & Dormers	78
Appliances	79
Breezeways & Porches	79
Finished Attic	79
Garages	80
Site Improvements	80
Wings & Ells	44

Average 1-1/2 Story

Living Area - 1800 S.F.
Perimeter - 144 L.F.

		Labor-Hours	Cost Per Square Foot Of Living Area		
			Mat.	Labor	Total
1 Site Work	Site preparation for slab; 4' deep trench excavation for foundation wall.	.037		.73	.73
2 Foundation	Continuous reinforced concrete footing 8" deep x 18" wide; dampproofed and insulated reinforced concrete foundation wall, 8" thick, 4' deep; 4" concrete slab on 4" crushed stone base and polyethylene vapor barrier, trowel finish.	.073	3.51	4.00	7.51
3 Framing	Exterior walls - 2" x 4" wood studs, 16" O.C.; 1/2" plywood sheathing; 2" x 6" rafters 16" O.C. with 1/2" plywood sheathing, 8 in 12 pitch; 2" x 8" floor joists 16" O.C. with 5/8" plywood subfloor; 1/2" plywood subfloor on 1" x 2" wood sleepers 16" O.C.	.098	6.04	7.13	13.17
4 Exterior Walls	Beveled wood siding and building paper on insulated wood frame walls; 6" attic insulation; double hung windows; 3 flush solid core wood exterior doors with storms.	.078	9.20	4.56	13.76
5 Roofing	25 year asphalt shingles; #15 felt building paper; aluminum gutters, downspouts, drip edge and flashings.	.029	.83	1.26	2.09
6 Interiors	Walls and ceilings, 1/2" taped and finished gypsum wallboard, primed and painted with 2 coats; painted baseboard and trim, finished hardwood floor 40%, carpet with 1/2" underlayment 40%, vinyl tile with 1/2" underlayment 15%, ceramic tile with 1/2" underlayment 5%; hollow core and louvered	.225	13.34	12.21	25.55
7 Specialties	Average grade kitchen cabinets - 14 L.F. wall and base with solid surface counter top and kitchen sink; 40 gallon electric water heater.	.022	1.91	.82	2.73
8 Mechanical	1 lavatory, white, wall hung; 1 water closet, white; 1 bathtub with shower, enameled steel, white; gas fired warm air heat.	.049	3.01	2.49	5.50
9 Electrical	200 Amp. service; romex wiring; incandescent lighting fixtures, switches, receptacles.	.039	1.24	1.43	2.67
10 Overhead	Contractor's overhead and profit and plans.		6.62	5.87	12.49
	Total		45.70	40.50	86.20

RESIDENTIAL | Average | 2 Story

- Simple design from standard plans
- Single family — 1 full bath, 1 kitchen
- No basement
- Asphalt shingles on roof
- Hot air heat
- Gypsum wallboard interior finishes
- Materials and workmanship are average

Note: The illustration shown may contain some optional components (for example: garages and/or fireplaces) whose costs are shown in the modifications, adjustments, & alternatives below or at the end of the square foot section.

Base cost per square foot of living area

Exterior Wall	Living Area										
	1000	1200	1400	1600	1800	2000	2200	2600	3000	3400	3800
Wood Siding - Wood Frame	110.40	100.20	95.50	92.30	89.05	85.40	83.05	78.45	74.05	72.05	70.30
Brick Veneer - Wood Frame	119.35	108.50	103.25	99.65	96.00	92.10	89.40	84.20	79.40	77.15	75.15
Stucco on Wood Frame	108.90	98.85	94.25	91.05	87.85	84.30	81.95	77.50	73.15	71.20	69.50
Solid Masonry	133.20	121.30	115.20	111.10	106.75	102.50	99.30	93.15	87.60	85.00	82.55
Finished Basement, Add	18.20	18.00	17.40	17.00	16.55	16.25	15.95	15.40	15.00	14.65	14.45
Unfinished Basement, Add	7.15	6.70	6.30	6.05	5.80	5.60	5.40	5.05	4.80	4.65	4.45

Modifications

Add to the total cost

Upgrade Kitchen Cabinets	$ + 3150
Solid Surface Countertops	
Full Bath - including plumbing, wall and floor finishes	+ 4997
Half Bath - including plumbing, wall and floor finishes	+ 3011
One Car Attached Garage	+ 10,672
One Car Detached Garage	+ 14,035
Fireplace & Chimney	+ 5170

Adjustments

For multi family - add to total cost

Additional Kitchen	$ + 5695
Additional Bath	+ 4997
Additional Entry & Exit	+ 1512
Separate Heating	+ 1313
Separate Electric	+ 1718

For Townhouse/Rowhouse - Multiply cost per square foot by

Inner Unit	.90
End Unit	.95

Alternatives

Add to or deduct from the cost per square foot of living area

Cedar Shake Roof	+ 1.30
Clay Tile Roof	+ 2.25
Slate Roof	+ 2.95
Upgrade Walls to Skim Coat Plaster	+ .38
Upgrade Ceilings to Textured Finish	+ .44
Air Conditioning, in Heating Ductwork	+ 1.84
In Separate Ductwork	+ 4.62
Heating Systems, Hot Water	+ 1.46
Heat Pump	+ 2.26
Electric Heat	− .58
Not Heated	− 2.58

Additional upgrades or components

Kitchen Cabinets & Countertops	Page 77
Bathroom Vanities	78
Fireplaces & Chimneys	78
Windows, Skylights & Dormers	78
Appliances	79
Breezeways & Porches	79
Finished Attic	79
Garages	80
Site Improvements	80
Wings & Ells	44

Average 2 Story

Living Area - 2000 S.F.
Perimeter - 135 L.F.

		Labor-Hours	Cost Per Square Foot Of Living Area		
			Mat.	Labor	Total
1 Site Work	Site preparation for slab; 4' deep trench excavation for foundation wall.	.034		.66	.66
2 Foundation	Continuous reinforced concrete footing 8" deep x 18" wide; dampproofed and insulated reinforced concrete foundation wall, 8" thick, 4' deep, 4" concrete slab on 4" crushed stone base and polyethylene vapor barrier, trowel finish.	.066	2.91	3.35	6.26
3 Framing	Exterior walls - 2" x 4" wood studs, 16" O.C.; 1/2" plywood sheathing; 2" x 6" rafters 16" O.C. with 1/2" plywood sheathing, 4 in 12 pitch; 2" x 6" ceiling joists 16" O.C.; 2" x 8" floor joists 16" O.C. with 5/8" plywood subfloor; 1/2" plywood subfloor on 1" x 2" wood sleepers 16" O.C.	.131	6.33	7.37	13.70
4 Exterior Walls	Beveled wood siding and building paper on insulated wood frame walls; 6" attic insulation; double hung windows; 3 flush solid core wood exterior doors with storms.	.111	10.02	4.97	14.99
5 Roofing	25 year asphalt shingles; #15 felt building paper; aluminum gutters, downspouts, drip edge and flashings.	.024	.67	1.01	1.68
6 Interiors	Walls and ceilings, 1/2" taped and finished gypsum wallboard, primed and painted with 2 coats; painted baseboard and trim, finished hardwood floor 40%, carpet with 1/2" underlayment 40%, vinyl tile with 1/2" underlayment 15%, ceramic tile with 1/2" underlayment 5%; hollow core and louvered	.232	13.21	12.21	25.42
7 Specialties	Average grade kitchen cabinets - 14 L.F. wall and base with solid surface counter top and kitchen sink; 40 gallon electric water heater.	.021	1.73	.74	2.47
8 Mechanical	1 lavatory, white, wall hung; 1 water closet, white; 1 bathtub with shower; enameled steel, white; gas fired warm air heat.	.060	2.84	2.41	5.25
9 Electrical	200 Amp. service; romex wiring; incandescent lighting fixtures, switches, receptacles.	.039	1.20	1.38	2.58
10 Overhead	Contractor's overhead and profit and plans.		6.59	5.80	12.39
	Total		45.50	39.90	**85.40**

RESIDENTIAL | Average | 2-1/2 Story

- Simple design from standard plans
- Single family — 1 full bath, 1 kitchen
- No basement
- Asphalt shingles on roof
- Hot air heat
- Gypsum wallboard interior finishes
- Materials and workmanship are average

Note: The illustration shown may contain some optional components (for example: garages and/or fireplaces) whose costs are shown in the modifications, adjustments, & alternatives below or at the end of the square foot section.

Base cost per square foot of living area

Exterior Wall	Living Area										
	1200	1400	1600	1800	2000	2400	2800	3200	3600	4000	4400
Wood Siding - Wood Frame	109.70	103.30	94.40	92.65	89.55	84.50	80.35	76.35	74.25	70.45	69.20
Brick Veneer - Wood Frame	119.25	111.90	102.30	100.65	96.95	91.30	86.80	82.15	79.80	75.55	74.15
Stucco on Wood Frame	108.10	101.85	93.10	91.40	88.25	83.45	79.25	75.30	73.35	69.60	68.35
Solid Masonry	133.95	125.25	114.55	112.95	108.50	101.65	96.75	91.10	88.35	83.50	81.85
Finished Basement, Add	15.45	15.10	14.60	14.55	14.15	13.55	13.30	12.85	12.65	12.30	12.15
Unfinished Basement, Add	5.95	5.45	5.15	5.10	4.85	4.50	4.30	4.05	3.90	3.75	3.65

Modifications

Add to the total cost

Upgrade Kitchen Cabinets Solid Surface Countertops	$ + 3150
Full Bath - including plumbing, wall and floor finishes	+ 4997
Half Bath - including plumbing, wall and floor finishes	+ 3011
One Car Attached Garage	+ 10,672
One Car Detached Garage	+ 14,035
Fireplace & Chimney	+ 5875

Adjustments

For multi family - add to total cost

Additional Kitchen	$ + 5695
Additional Bath	+ 4997
Additional Entry & Exit	+ 1512
Separate Heating	+ 1313
Separate Electric	+ 1718

For Townhouse/Rowhouse - Multiply cost per square foot by

Inner Unit	.90
End Unit	.95

Alternatives

Add to or deduct from the cost per square foot of living area

Cedar Shake Roof	+ 1.15
Clay Tile Roof	+ 1.95
Slate Roof	+ 2.55
Upgrade Walls to Skim Coat Plaster	+ .37
Upgrade Ceilings to Textured Finish	+ .44
Air Conditioning, in Heating Ductwork	+ 1.68
In Separate Ductwork	+ 4.46
Heating Systems, Hot Water	+ 1.31
Heat Pump	+ 2.31
Electric Heat	– 1.02
Not Heated	– 3.01

Additional upgrades or components

Kitchen Cabinets & Countertops	Page 77
Bathroom Vanities	78
Fireplaces & Chimneys	78
Windows, Skylights & Dormers	78
Appliances	79
Breezeways & Porches	79
Finished Attic	79
Garages	80
Site Improvements	80
Wings & Ells	44

Average 2-1/2 Story

Living Area - 3200 S.F.
Perimeter - 150 L.F.

		Labor-Hours	Cost Per Square Foot Of Living Area		
			Mat.	Labor	Total
1 Site Work	Site preparation for slab; 4' deep trench excavation for foundation wall.	.046		.41	.41
2 Foundation	Continuous reinforced concrete footing 8" deep x 18" wide, dampproofed and insulated reinforced concrete foundation wall, 8" thick, 4' deep; 4" concrete slab on 4" crushed stone base and polyethylene vapor barrier, trowel finish.	.061	2.10	2.37	4.47
3 Framing	Exterior walls - 2" x 4" wood studs, 16" O.C.; 1/2" plywood sheathing; 2" x 6" rafters 16" O.C. with 1/2" plywood sheathing, 4 in 12 pitch; 2" x 6" ceiling joists 16" O.C.; 2" x 8" floor joists 16" O.C. with 5/8" plywood subfloor; 1/2" plywood subfloor on 1" x 2" wood sleepers 16" O.C.	.127	6.31	7.24	13.55
4 Exterior Walls	Beveled wood siding and building paper on insulated wood frame walls; 6" attic insulation; double hung windows; 3 flush solid core wood exterior doors with storms.	.136	8.39	4.19	12.58
5 Roofing	25 year asphalt shingles; #15 felt building paper; aluminum gutters, downspouts, drip edge and flashings.	.018	.51	.77	1.28
6 Interiors	Walls and ceilings, 1/2" taped and finished gypsum wallboard, primed and painted with 2 coats; painted baseboard and trim, finished hardwood floor 40%, carpet with 1/2" underlayment 40%, vinyl tile with 1/2" underlayment 15%, ceramic tile with 1/2" underlayment 5%; hollow core and louvered in	.286	12.95	11.79	24.74
7 Specialties	Average grade kitchen cabinets - 14 L.F. wall and base with solid surface counter top and kitchen sink; 40 gallon electric water heater.	.030	1.08	.46	1.54
8 Mechanical	1 lavatory, white, wall hung; 1 water closet, white; 1 bathtub with shower, enameled steel, white; gas fired warm air heat.	.072	2.25	2.14	4.39
9 Electrical	200 Amp. service; romex wiring; incandescent lighting fixtures, switches, receptacles.	.046	1.05	1.21	2.26
10 Overhead	Contractor's overhead and profit and plans.		5.91	5.22	11.13
	Total		40.55	35.80	**76.35**

RESIDENTIAL | Average | 3 Story

- Simple design from standard plans
- Single family — 1 full bath, 1 kitchen
- No basement
- Asphalt shingles on roof
- Hot air heat
- Gypsum wallboard interior finishes
- Materials and workmanship are average

Note: The illustration shown may contain some optional components (for example: garages and/or fireplaces) whose costs are shown in the modifications, adjustments, & alternatives below or at the end of the square foot section.

Base cost per square foot of living area

Exterior Wall	\multicolumn{11}{c}{Living Area}
	1500	1800	2100	2500	3000	3500	4000	4500	5000	5500	6000
Wood Siding - Wood Frame	101.95	92.70	88.75	85.60	79.65	77.10	73.45	69.45	68.25	66.80	65.40
Brick Veneer - Wood Frame	110.90	101.00	96.45	93.00	86.35	83.45	79.20	74.75	73.40	71.75	70.05
Stucco on Wood Frame	100.50	91.35	87.45	84.40	78.55	76.05	72.55	68.55	67.40	65.95	64.60
Solid Masonry	124.80	113.80	108.45	104.40	96.80	93.25	88.20	83.00	81.40	79.45	77.30
Finished Basement, Add	13.30	13.15	12.80	12.50	12.10	11.75	11.40	11.15	11.05	10.95	10.75
Unfinished Basement, Add	4.85	4.55	4.25	4.15	3.85	3.65	3.50	3.30	3.20	3.10	3.00

Modifications
Add to the total cost

Upgrade Kitchen Cabinets	$ + 3150
Solid Surface Countertops	
Full Bath - including plumbing, wall and floor finishes	+ 4997
Half Bath - including plumbing, wall and floor finishes	+ 3011
One Car Attached Garage	+ 10,672
One Car Detached Garage	+ 14,035
Fireplace & Chimney	+ 5875

Adjustments
For multi family - add to total cost

Additional Kitchen	$ + 5695
Additional Bath	+ 4997
Additional Entry & Exit	+ 1512
Separate Heating	+ 1313
Separate Electric	+ 1718

For Townhouse/Rowhouse - Multiply cost per square foot by

Inner Unit	.88
End Unit	.94

Alternatives
Add to or deduct from the cost per square foot of living area

Cedar Shake Roof	+ .90
Clay Tile Roof	+ 1.50
Slate Roof	+ 1.95
Upgrade Walls to Skim Coat Plaster	+ .38
Upgrade Ceilings to Textured Finish	+ .44
Air Conditioning, in Heating Ductwork	+ 1.68
In Separate Ductwork	+ 4.46
Heating Systems, Hot Water	+ 1.31
Heat Pump	+ 2.31
Electric Heat	– .79
Not Heated	– 2.80

Additional upgrades or components

Kitchen Cabinets & Countertops	Page 77
Bathroom Vanities	78
Fireplaces & Chimneys	78
Windows, Skylights & Dormers	78
Appliances	79
Breezeways & Porches	79
Finished Attic	79
Garages	80
Site Improvements	80
Wings & Ells	44

Average 3 Story

Living Area - 3000 S.F.
Perimeter - 135 L.F.

			Labor-Hours	Cost Per Square Foot Of Living Area		
				Mat.	Labor	Total
1	Site Work	Site preparation for slab; 4' deep trench excavation for foundation wall.	.038		.44	.44
2	Foundation	Continuous reinforced concrete footing 8" deep x 18" wide, dampproofed and insulated reinforced concrete foundation wall, 8" thick, 4' deep; 4" concrete slab on 4" crushed stone base and polyethylene vapor barrier, trowel finish.	.053	1.95	2.23	4.18
3	Framing	Exterior walls - 2" x 4" wood studs, 16" O.C.; 1/2" plywood sheathing; 2" x 6" rafters 16" O.C. with 1/2" plywood sheathing, 4 in 12 pitch; 2" x 6" ceiling joists 16" O.C.; 2" x 8" floor joists 16" O.C. with 5/8" plywood subfloor; 1/2" plywood subfloor on 1" x 2" wood sleepers 16" O.C.	.128	6.55	7.45	14.00
4	Exterior Walls	Horizontal beveled wood siding; building paper; 3-1/2" batt insulation; wood double hung windows; 3 flush solid core wood exterior doors; storms and screens.	.139	9.56	4.76	14.32
5	Roofing	25 year asphalt shingles; #15 felt building paper; aluminum gutters, downspouts, drip edge and flashings.	.014	.44	.67	1.11
6	Interiors	Walls and ceilings, 1/2" taped and finished gypsum wallboard, primed and painted with 2 coats; painted baseboard and trim, finished hardwood floor 40%, carpet with 1/2" underlayment 40%, vinyl tile with 1/2" underlayment 15%, ceramic tile with 1/2" underlayment 5%; hollow core and louvered	.280	13.38	12.25	25.63
7	Specialties	Average grade kitchen cabinets - 14 L.F. wall and base with solid surface counter top and kitchen sink; 40 gallon electric water heater.	.025	1.15	.48	1.63
8	Mechanical	1 lavatory, white, wall hung; 1 water closet, white; 1 bathtub with shower, enameled steel, white; gas fired warm air heat.	.065	2.32	2.17	4.49
9	Electrical	200 Amp. service; romex wiring; incandescent lighting fixtures, switches, receptacles.	.042	1.07	1.23	2.30
10	Overhead	Contractor's overhead and profit and plans.		6.18	5.37	11.55
		Total		42.60	37.05	**79.65**

RESIDENTIAL Average | Bi-Level

- Simple design from standard plans
- Single family — 1 full bath, 1 kitchen
- No basement
- Asphalt shingles on roof
- Hot air heat
- Gypsum wallboard interior finishes
- Materials and workmanship are average

Note: The illustration shown may contain some optional components (for example: garages and/or fireplaces) whose costs are shown in the modifications, adjustments, & alternatives below or at the end of the square foot section.

Base cost per square foot of living area

Exterior Wall	\multicolumn{11}{c}{Living Area}
	1000	1200	1400	1600	1800	2000	2200	2600	3000	3400	3800
Wood Siding - Wood Frame	103.45	93.75	89.45	86.50	83.60	80.15	78.05	74.00	69.90	68.10	66.55
Brick Veneer - Wood Frame	110.15	99.90	95.25	92.00	88.80	85.20	82.85	78.30	73.85	71.90	70.15
Stucco on Wood Frame	102.35	92.75	88.50	85.55	82.75	79.35	77.30	73.25	69.25	67.45	65.95
Solid Masonry	120.55	109.55	104.20	100.65	96.90	92.95	90.25	85.00	80.05	77.75	75.75
Finished Basement, Add	18.20	18.00	17.40	17.00	16.55	16.25	15.95	15.40	15.00	14.65	14.45
Unfinished Basement, Add	7.15	6.70	6.30	6.05	5.80	5.60	5.40	5.05	4.80	4.65	4.45

Modifications

Add to the total cost

Upgrade Kitchen Cabinets	$ + 3150
Solid Surface Countertops	
Full Bath - including plumbing, wall and floor finishes	+ 4997
Half Bath - including plumbing, wall and floor finishes	+ 3011
One Car Attached Garage	+ 10,672
One Car Detached Garage	+ 14,035
Fireplace & Chimney	+ 4635

Adjustments

For multi family - add to total cost

Additional Kitchen	$ + 5695
Additional Bath	+ 4997
Additional Entry & Exit	+ 1512
Separate Heating	+ 1313
Separate Electric	+ 1718

For Townhouse/Rowhouse - Multiply cost per square foot by

Inner Unit	.91
End Unit	.96

Alternatives

Add to or deduct from the cost per square foot of living area

Cedar Shake Roof	+ 1.30
Clay Tile Roof	+ 2.25
Slate Roof	+ 2.95
Upgrade Walls to Skim Coat Plaster	+ .36
Upgrade Ceilings to Textured Finish	+ .44
Air Conditioning, in Heating Ductwork	+ 1.84
In Separate Ductwork	+ 4.62
Heating Systems, Hot Water	+ 1.46
Heat Pump	+ 2.26
Electric Heat	– .58
Not Heated	– 2.58

Additional upgrades or components

Kitchen Cabinets & Countertops	Page 77
Bathroom Vanities	78
Fireplaces & Chimneys	78
Windows, Skylights & Dormers	78
Appliances	79
Breezeways & Porches	79
Finished Attic	79
Garages	80
Site Improvements	80
Wings & Ells	44

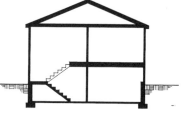

Average Bi-Level

Living Area - 2000 S.F.
Perimeter - 135 L.F.

		Labor-Hours	Cost Per Square Foot Of Living Area		
			Mat.	Labor	Total
1 Site Work	Excavation for lower level, 4' deep. Site preparation for slab.	.029		.66	.66
2 Foundation	Continuous reinforced concrete footing 8" deep x 18" wide, dampproofed and insulated reinforced concrete foundation wall, 8" thick, 4' deep; 4" concrete slab on 4" crushed stone base and polyethylene vapor barrier, trowel finish.	.066	2.91	3.35	6.26
3 Framing	Exterior walls - 2" x 4" wood studs, 16" O.C.; 1/2" plywood sheathing; 2" x 6" rafters 16" O.C. with 1/2" plywood sheathing, 4 in 12 pitch; 2" x 6" ceiling joists 16" O.C.; 2" x 8" floor joists 16" O.C. with 5/8" plywood subfloor; 1/2" plywood subfloor on 1" x 2" wood sleepers 16" O.C.	.118	6.06	7.00	13.06
4 Exterior Walls	Horizontal beveled wood siding; building paper; 3-1/2" batt insulation; wood double hung windows; 3 flush solid core wood exterior doors; storms and screens.	.091	7.84	3.87	11.71
5 Roofing	25 year asphalt shingles; #15 felt building paper; aluminum gutters, downspouts, drip edge and flashings.	.024	.67	1.01	1.68
6 Interiors	Walls and ceilings, 1/2" taped and finished gypsum wallboard, primed and painted with 2 coats; painted baseboard and trim, finished hardwood floor 40%, carpet with 1/2" underlayment 40%, vinyl tile with 1/2" underlayment 15%, ceramic tile with 1/2" underlayment 5%; hollow core and louvered in	.217	12.99	11.86	24.85
7 Specialties	Average grade kitchen cabinets - 14 L.F. wall and base with solid surface counter top and kitchen sink; 40 gallon electric water heater.	.021	1.73	.74	2.47
8 Mechanical	1 lavatory, white, wall hung; 1 water closet, white; 1 bathtub with shower, enameled steel, white; gas fired warm air heat.	.061	2.84	2.41	5.25
9 Electrical	200 Amp. service; romex wiring; incandescent lighting fixtures, switches, receptacles.	.039	1.20	1.38	2.58
10 Overhead	Contractor's overhead and profit and plans.		6.16	5.47	11.63
	Total		42.40	37.75	80.15

37

RESIDENTIAL — Average — Tri-Level

- Simple design from standard plans
- Single family — 1 full bath, 1 kitchen
- No basement
- Asphalt shingles on roof
- Hot air heat
- Gypsum wallboard interior finishes
- Materials and workmanship are average

Note: The illustration shown may contain some optional components (for example: garages and/or fireplaces) whose costs are shown in the modifications, adjustments, & alternatives below or at the end of the square foot section.

Base cost per square foot of living area

Exterior Wall	\multicolumn{11}{c}{Living Area}
	1200	1500	1800	2100	2400	2700	3000	3400	3800	4200	4600
Wood Siding - Wood Frame	98.20	90.75	85.25	80.40	77.30	75.60	73.70	71.75	68.70	66.20	64.95
Brick Veneer - Wood Frame	104.40	96.30	90.25	84.95	81.60	79.75	77.55	75.55	72.20	69.45	68.15
Stucco on Wood Frame	97.20	89.80	84.40	79.65	76.60	74.95	73.00	71.20	68.10	65.60	64.40
Solid Masonry	113.95	105.00	98.05	92.10	88.25	86.15	83.60	81.40	77.55	74.50	73.05
Finished Basement, Add*	20.80	20.45	19.55	18.90	18.45	18.15	17.80	17.60	17.20	16.85	16.65
Unfinished Basement, Add*	7.95	7.35	6.80	6.40	6.15	5.95	5.75	5.60	5.40	5.20	5.10

*Basement under middle level only.

Modifications

Add to the total cost

Upgrade Kitchen Cabinets	$ + 3150
Solid Surface Countertops	
Full Bath - including plumbing, wall and floor finishes	+ 4997
Half Bath - including plumbing, wall and floor finishes	+ 3011
One Car Attached Garage	+ 10,672
One Car Detached Garage	+ 14,035
Fireplace & Chimney	+ 4635

Adjustments

For multi family - add to total cost

Additional Kitchen	$ + 5695
Additional Bath	+ 4997
Additional Entry & Exit	+ 1512
Separate Heating	+ 1313
Separate Electric	+ 1718

For Townhouse/Rowhouse - Multiply cost per square foot by

Inner Unit	.90
End Unit	.95

Alternatives

Add to or deduct from the cost per square foot of living area

Cedar Shake Roof	+ 1.90
Clay Tile Roof	+ 3.25
Slate Roof	+ 4.25
Upgrade Walls to Skim Coat Plaster	+ .31
Upgrade Ceilings to Textured Finish	+ .44
Air Conditioning, in Heating Ductwork	+ 1.55
In Separate Ductwork	+ 4.35
Heating Systems, Hot Water	+ 1.41
Heat Pump	+ 2.34
Electric Heat	– .47
Not Heated	– 2.49

Additional upgrades or components

Kitchen Cabinets & Countertops	Page 77
Bathroom Vanities	78
Fireplaces & Chimneys	78
Windows, Skylights & Dormers	78
Appliances	79
Breezeways & Porches	79
Finished Attic	79
Garages	80
Site Improvements	80
Wings & Ells	44

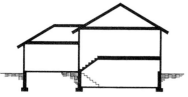

Average Tri-Level

Living Area - 2400 S.F.
Perimeter - 163 L.F.

			Labor-Hours	Cost Per Square Foot Of Living Area		
				Mat.	Labor	Total
1	Site Work	Site preparation for slab; 4' deep trench excavation for foundation wall, excavation for lower level, 4' deep.	.029		.54	.54
2	Foundation	Continuous reinforced concrete footing 8" deep x 18" wide; dampproofed and insulated reinforced concrete foundation wall, 8" thick, 4' deep; 4" concrete slab on 4" crushed stone base and polyethylene vapor barrier, trowel finish.	.080	3.30	3.64	6.94
3	Framing	Exterior walls - 2" x 4" wood studs, 16" O.C.; 1/2" plywood sheathing; 2" x 6" rafters 16" O.C. with 1/2" plywood sheathing, 4 in 12 pitch; 2" x 6" ceiling joists 16" O.C.; 2" x 8" floor joists 16" O.C. with 5/8" plywood subfloor; 1/2" plywood subfloor on 1" x 2" wood sleepers 16" O.C.	.124	5.60	6.61	12.21
4	Exterior Walls	Horizontal beveled wood siding: building paper; 3-1/2" batt insulation; wood double hung windows; 3 flush solid core wood exterior doors; storms and screens.	.083	6.77	3.34	10.11
5	Roofing	25 year asphalt shingles; #15 felt building paper; aluminum gutters, downspouts, drip edge and flashings.	.032	.89	1.35	2.24
6	Interiors	Walls and ceilings, 1/2" taped and finished gypsum wallboard, primed and painted with 2 coats; painted baseboard and trim, finished hardwood floor 40%, carpet with 1/2" underlayment 40%, vinyl tile with 1/2" underlayment 15%, ceramic tile with 1/2" underlayment 5%; hollow core and louvered in	.186	13.20	11.55	24.75
7	Specialties	Average grade kitchen cabinets - 14 L.F. wall and base with solid surface counter top and kitchen sink; 40 gallon electric water heater.	.012	1.45	.62	2.07
8	Mechanical	1 lavatory, white, wall hung; 1 water closet, white; 1 bathtub with shower, enameled steel, white; gas fired warm air heat.	.059	2.51	2.29	4.80
9	Electrical	200 Amp. service; romex wiring; incandescent lighting fixtures, switches, receptacles.	.036	1.13	1.30	2.43
10	Overhead	Contractor's overhead and profit and plans.		5.90	5.31	11.21
		Total		40.75	36.55	**77.30**

RESIDENTIAL — Solid Wall — 1 Story

- Post and beam frame
- Log exterior walls
- Simple design from standard plans
- Single family — 1 full bath, 1 kitchen
- No basement
- Asphalt shingles on roof
- Hot air heat
- Gypsum wallboard interior finishes
- Materials and workmanship are average

Note: The illustration shown may contain some optional components (for example: garages and/or fireplaces) whose costs are shown in the modifications, adjustments, & alternatives below or at the end of the square foot section.

Base cost per square foot of living area

Exterior Wall	Living Area										
	600	800	1000	1200	1400	1600	1800	2000	2400	2800	3200
6" Log - Solid Wall	143.45	130.90	121.40	113.60	106.95	102.70	100.50	97.65	91.90	87.70	84.75
8" Log - Solid Wall	143.60	131.00	121.50	113.75	107.10	102.80	100.55	97.70	91.95	87.80	84.85
Finished Basement, Add	31.35	30.30	28.90	27.60	26.55	25.85	25.45	24.90	24.15	23.50	23.00
Unfinished Basement, Add	13.05	11.80	10.90	10.10	9.45	9.05	8.85	8.50	8.00	7.65	7.35

Modifications

Add to the total cost

Upgrade Kitchen Cabinets	$ + 3150
Solid Surface Countertops	
Full Bath - including plumbing, wall and floor finishes	+ 4997
Half Bath - including plumbing, wall and floor finishes	+ 3011
One Car Attached Garage	+ 10,672
One Car Detached Garage	+ 14,035
Fireplace & Chimney	+ 4635

Adjustments

For multi family - add to total cost

Additional Kitchen	$ + 5695
Additional Bath	+ 4997
Additional Entry & Exit	+ 1512
Separate Heating	+ 1313
Separate Electric	+ 1718

For Townhouse/Rowhouse - Multiply cost per square foot by

Inner Unit	.92
End Unit	.96

Alternatives

Add to or deduct from the cost per square foot of living area

Cedar Shake Roof	+ 1.30
Air Conditioning, in Heating Ductwork	+ 3.04
In Separate Ductwork	+ 5.81
Heating Systems, Hot Water	+ 1.57
Heat Pump	+ 1.92
Electric Heat	− .76
Not Heated	− 2.76

Additional upgrades or components

Kitchen Cabinets & Countertops	Page 77
Bathroom Vanities	78
Fireplaces & Chimneys	78
Windows, Skylights & Dormers	78
Appliances	79
Breezeways & Porches	79
Finished Attic	79
Garages	80
Site Improvements	80
Wings & Ells	44

Solid Wall 1 Story

Living Area - 1600 S.F.
Perimeter - 163 L.F.

#	Category	Description	Labor-Hours	Mat.	Labor	Total
1	Site Work	Site preparation for slab; 4' deep trench excavation for foundation wall.	.048		.82	.82
2	Foundation	Continuous reinforced concrete footing 8" deep x 18" wide; dampproofed and insulated reinforced concrete foundation wall, 8" thick, 4' deep; 4" concrete slab on 4" crushed stone base and polyethylene vapor barrier, trowel finish.	.113	4.95	5.47	10.42
3	Framing	Exterior walls - Precut traditional log home. Handcrafted white cedar or pine logs. Delivery included.	.136	21.78	11.77	33.55
4	Exterior Walls	Wood double-hung windows, solid wood exterior doors	.098	4.96	2.37	7.33
5	Roofing	25 year asphalt shingles; #15 felt building paper; aluminum gutters, downspouts, drip edge and flashings.	.047	1.34	2.01	3.35
6	Interiors	Walls and ceilings, 1/2" taped and finished gypsum wallboard, primed and painted with 2 coats; painted baseboard and trim, finished hardwood floor 40%, carpet with 1/2" underlayment 40%, vinyl tile with 1/2" underlayment 15%, ceramic tile with 1/2" underlayment 5%; hollow core and louvered	.251	10.94	9.68	20.62
7	Specialties	Average grade kitchen cabinets - 14 L.F. wall and base with solid surface counter top and kitchen sink; 40 gallon electric water heater.	.009	2.15	.91	3.06
8	Mechanical	1 lavatory, white, wall hung; 1 water closet, white; 1 bathtub with shower, enameled steel, white; gas fired warm air heat.	.098	3.24	2.60	5.84
9	Electrical	200 Amp. service; romex wiring; incandescent lighting fixtures, switches, receptacles.	.041	1.30	1.51	2.81
10	Overhead	Contractor's overhead and profit and plans.		8.59	6.31	14.90
		Total		59.25	43.45	**102.70**

RESIDENTIAL — Solid Wall — 2 Story

- Post and beam frame
- Log exterior walls
- Simple design from standard plans
- Single family — 1 full bath, 1 kitchen
- No basement
- Asphalt shingles on roof
- Hot air heat
- Gypsum wallboard interior finishes
- Materials and workmanship are average

Note: The illustration shown may contain some optional components (for example: garages and/or fireplaces) whose costs are shown in the modifications, adjustments, & alternatives below or at the end of the square foot section.

Base cost per square foot of living area

Exterior Wall	Living Area										
	1000	1200	1400	1600	1800	2000	2200	2600	3000	3400	3800
6" Log-Solid	128.80	117.70	112.20	108.50	104.65	100.70	97.80	92.40	87.35	85.05	82.95
8" Log-Solid	110.40	100.20	95.50	92.30	89.05	85.40	83.05	78.45	74.05	72.05	70.30
Finished Basement, Add	18.20	18.00	17.40	17.00	16.55	16.25	15.95	15.40	15.00	14.65	14.45
Unfinished Basement, Add	7.15	6.70	6.30	6.05	5.80	5.60	5.40	5.05	4.80	4.65	4.45

Modifications

Add to the total cost

Upgrade Kitchen Cabinets	$ + 3150
Solid Surface Countertops	
Full Bath - including plumbing, wall and floor finishes	+ 4997
Half Bath - including plumbing, wall and floor finishes	+ 3011
One Car Attached Garage	+ 10,672
One Car Detached Garage	+ 14,035
Fireplace & Chimney	+ 5170

Adjustments

For multi family - add to total cost

Additional Kitchen	$ + 5695
Additional Bath	+ 4997
Additional Entry & Exit	+ 1512
Separate Heating	+ 1313
Separate Electric	+ 1718

For Townhouse/Rowhouse - Multiply cost per square foot by

Inner Unit	.92
End Unit	.96

Alternatives

Add to or deduct from the cost per square foot of living area

Cedar Shake Roof	+ 1.90
Air Conditioning, in Heating Ductwork	+ 1.84
In Separate Ductwork	+ 4.62
Heating Systems, Hot Water	+ 1.46
Heat Pump	+ 2.26
Electric Heat	- .58
Not Heated	- 2.58

Additional upgrades or components

Kitchen Cabinets & Countertops	Page 77
Bathroom Vanities	78
Fireplaces & Chimneys	78
Windows, Skylights & Dormers	78
Appliances	79
Breezeways & Porches	79
Finished Attic	79
Garages	80
Site Improvements	80
Wings & Ells	44

Solid Wall 2 Story

Living Area - 2000 S.F.
Perimeter - 135 L.F.

		Labor-Hours	Cost Per Square Foot Of Living Area		
			Mat.	Labor	Total
1 Site Work	Site preparation for slab; 4' deep trench excavation for foundation wall.	.034		.66	.66
2 Foundation	Continuous reinforced concrete footing 8" deep x 18" wide; dampproofed and insulated reinforced concrete foundation wall, 8" thick, 4' deep, 4" concrete slab on 4" crushed stone base and polyethylene vapor barrier, trowel finish.	.066	2.91	3.35	6.26
3 Framing	Exterior walls - Precut traditional log home. Handcrafted white cedar or pine logs. Delivery included.	.131	23.86	12.22	36.08
4 Exterior Walls	Wood double-hung windows, solid wood exterior doors.	.111	5.42	2.55	7.97
5 Roofing	25 year asphalt shingles; #15 felt building paper; aluminum gutters, downspouts, drip edge and flashings.	.024	.67	1.01	1.68
6 Interiors	Walls and ceilings, 1/2" taped and finished gypsum wallboard, primed and painted with 2 coats; painted baseboard and trim, finished hardwood floor 40%, carpet with 1/2" underlayment 40%, vinyl tile with 1/2" underlayment 15%, ceramic tile with 1/2" underlayment 5%; hollow core and louvered	.232	12.31	10.80	23.11
7 Specialties	Average grade kitchen cabinets - 14 L.F. wall and base with solid surface counter top and kitchen sink; 40 gallon electric water heater.	.021	1.73	.74	2.47
8 Mechanical	1 lavatory, white, wall hung; 1 water closet, white; 1 bathtub with shower; enameled steel, white; gas fired warm air heat.	.060	2.84	2.41	5.25
9 Electrical	200 Amp. service; romex wiring; incandescent lighting fixtures, switches, receptacles.	.039	1.20	1.38	2.58
10 Overhead	Contractor's overhead and profit and plans.		8.66	5.98	14.64
	Total		59.60	41.10	**100.70**

RESIDENTIAL | Average | Wings & Ells

1 Story — Base cost per square foot of living area

Exterior Wall	Living Area							
	50	100	200	300	400	500	600	700
Wood Siding - Wood Frame	163.30	126.30	110.15	93.75	88.45	85.35	83.20	84.05
Brick Veneer - Wood Frame	173.95	130.90	112.30	91.85	85.70	82.05	79.55	80.15
Stucco on Wood Frame	159.70	123.60	107.85	92.20	87.10	84.00	81.95	82.80
Solid Masonry	216.55	164.25	141.80	114.85	107.50	103.10	101.60	101.65
Finished Basement, Add	50.35	41.65	37.50	30.50	29.10	28.25	27.70	27.30
Unfinished Basement, Add	23.60	17.90	15.50	11.55	10.75	10.25	9.90	9.70

1-1/2 Story — Base cost per square foot of living area

Exterior Wall	Living Area							
	100	200	300	400	500	600	700	800
Wood Siding - Wood Frame	131.30	105.85	90.70	82.25	77.60	75.50	72.75	71.85
Brick Veneer - Wood Frame	185.95	138.75	115.10	101.00	93.75	89.95	85.85	84.30
Stucco on Wood Frame	164.25	121.35	100.60	89.70	83.35	80.10	76.60	75.05
Solid Masonry	214.85	161.80	134.35	116.00	107.65	103.00	98.20	96.60
Finished Basement, Add	33.95	30.70	27.90	24.80	24.00	23.40	22.90	22.80
Unfinished Basement, Add	14.85	12.45	10.85	9.10	8.60	8.25	8.00	7.95

2 Story — Base cost per square foot of living area

Exterior Wall	Living Area							
	100	200	400	600	800	1000	1200	1400
Wood Siding - Wood Frame	130.75	98.45	84.25	71.20	66.55	63.75	61.95	62.95
Brick Veneer - Wood Frame	189.45	132.30	106.10	85.75	78.70	74.50	71.70	72.05
Stucco on Wood Frame	165.15	114.90	91.60	76.10	70.00	66.40	64.00	64.60
Solid Masonry	221.80	155.35	125.35	98.60	90.25	85.30	81.95	81.95
Finished Basement, Add	26.95	22.70	20.60	17.10	16.40	16.00	15.65	15.50
Unfinished Basement, Add	11.95	9.10	7.90	5.95	5.55	5.30	5.10	5.00

Base costs do not include bathroom or kitchen facilities. Use Modifications/Adjustments/Alternatives on pages 77-80 where appropriate.

RESIDENTIAL | Custom | Illustrations

1 Story

1-1/2 Story

2 Story

2-1/2 Story

Bi-Level

Tri-Level

RESIDENTIAL — Custom — 1 Story

- A distinct residence from designer's plans
- Single family — 1 full bath, 1 half bath, 1 kitchen
- No basement
- Asphalt shingles on roof
- Forced hot air heat/air conditioning
- Gypsum wallboard interior finishes
- Materials and workmanship are above average

Note: The illustration shown may contain some optional components (for example: garages and/or fireplaces) whose costs are shown in the modifications, adjustments, & alternatives below or at the end of the square foot section.

Base cost per square foot of living area

Exterior Wall	800	1000	1200	1400	1600	1800	2000	2400	2800	3200	3600
Wood Siding - Wood Frame	147.80	134.30	123.50	114.75	108.80	105.65	101.75	94.40	89.10	85.45	81.50
Brick Veneer - Wood Frame	167.25	152.90	141.40	131.95	125.60	122.25	117.95	110.10	104.50	100.50	96.25
Stone Veneer - Wood Frame	171.70	156.95	145.00	135.30	128.70	125.25	120.80	112.70	106.85	102.70	98.30
Solid Masonry	174.80	159.75	147.55	137.60	130.85	127.30	122.75	114.50	108.45	104.25	99.70
Finished Basement, Add	47.95	47.70	45.65	44.00	42.85	42.25	41.40	40.15	39.15	38.30	37.60
Unfinished Basement, Add	20.30	19.15	18.20	17.30	16.85	16.50	16.05	15.45	14.95	14.50	14.20

Modifications

Add to the total cost

Upgrade Kitchen Cabinets	$ + 950
Solid Surface Countertops	
Full Bath - including plumbing, wall and floor finishes	+ 5922
Half Bath - including plumbing, wall and floor finishes	+ 3568
Two Car Attached Garage	+ 21,296
Two Car Detached Garage	+ 24,140
Fireplace & Chimney	+ 4825

Adjustments

For multi family - add to total cost

Additional Kitchen	$ + 12,270
Additional Full Bath & Half Bath	+ 9490
Additional Entry & Exit	+ 1512
Separate Heating & Air Conditioning	+ 5440
Separate Electric	+ 1718

For Townhouse/Rowhouse - Multiply cost per square foot by

Inner Unit	.90
End Unit	.95

Alternatives

Add to or deduct from the cost per square foot of living area

Cedar Shake Roof	+ 2.20
Clay Tile Roof	+ 4.05
Slate Roof	+ 5.45
Upgrade Ceilings to Textured Finish	+ .44
Air Conditioning, in Heating Ductwork	Base System
Heating Systems, Hot Water	+ 1.60
Heat Pump	+ 1.91
Electric Heat	– 2.13
Not Heated	– 3.46

Additional upgrades or components

Kitchen Cabinets & Countertops	Page 77
Bathroom Vanities	78
Fireplaces & Chimneys	78
Windows, Skylights & Dormers	78
Appliances	79
Breezeways & Porches	79
Finished Attic	79
Garages	80
Site Improvements	80
Wings & Ells	60

Custom 1 Story

Living Area - 2400 S.F.
Perimeter - 207 L.F.

			Labor-Hours	Cost Per Square Foot Of Living Area		
				Mat.	Labor	Total
1	Site Work	Site preparation for slab; 4' deep trench excavation for foundation wall.	.028		.64	.64
2	Foundation	Continuous reinforced concrete footing 8" deep x 18" wide; dampproofed and insulated reinforced concrete foundation wall, 8" thick, 4' deep; 4" concrete slab on 4" crushed stone base and polyethylene vapor barrier, trowel finish.	.113	5.52	5.99	11.51
3	Framing	Exterior walls - 2" x 6" wood studs, 16" O.C.; 1/2" plywood sheathing; 2" x 8" rafters 16" O.C. with 1/2" plywood sheathing, 4 in 12 pitch; 2" x 6" ceiling joists 16" O.C.; 5/8" plywood subfloor on 1" x 3" wood sleepers 16" O.C.	.190	3.69	5.54	9.23
4	Exterior Walls	Horizontal beveled wood siding; building paper; 6" batt insulation; wood double hung windows; 3 solid core wood exterior doors; storms and screens.	.085	7.83	2.82	10.65
5	Roofing	30 year asphalt shingles; #15 felt building paper; aluminum gutters, downspouts and drip edge; copper flashings.	.082	3.17	2.92	6.09
6	Interiors	Walls and ceilings - 5/8" gypsum wallboard, skim coat plaster, painted with primer and 2 coats; hardwood baseboard and trim, sanded and finished; hardwood floor 70%, ceramic tile with underlayment 20%, vinyl tile with underlayment 10%; wood panel interior doors, primed and painted with 2 coat	.292	13.93	10.54	24.47
7	Specialties	Custom grade kitchen cabinets - 20 L.F. wall and base with solid surface counter top and kitchen sink; 4 L.F. bathroom vanity; 75 gallon electric water heater, medicine cabinet.	.019	4.26	1.04	5.30
8	Mechanical	Gas fired warm air heat/air conditioning; one full bath including: bathtub, corner shower, built in lavatory and water closet; one 1/2 bath including: built in lavatory and water closet.	.092	5.37	2.65	8.02
9	Electrical	200 Amp. service; romex wiring; fluorescent and incandescent lighting fixtures, switches, receptacles.	.039	1.27	1.48	2.75
10	Overhead	Contractor's overhead and profit and design.		9.01	6.73	15.74
		Total		54.05	40.35	**94.40**

RESIDENTIAL — Custom — 1-1/2 Story

- A distinct residence from designer's plans
- Single family — 1 full bath, 1 half bath, 1 kitchen
- No basement
- Asphalt shingles on roof
- Forced hot air heat/air conditioning
- Gypsum wallboard interior finishes
- Materials and workmanship are above average

Note: The illustration shown may contain some optional components (for example: garages and/or fireplaces) whose costs are shown in the modifications, adjustments, & alternatives below or at the end of the square foot section.

©Donald A. Gardner Architects, Inc.

Base cost per square foot of living area

Exterior Wall	\multicolumn{11}{c}{Living Area}
	1000	1200	1400	1600	1800	2000	2400	2800	3200	3600	4000
Wood Siding - Wood Frame	134.20	125.20	118.75	110.90	106.50	102.10	93.60	89.90	86.50	83.65	79.75
Brick Veneer - Wood Frame	143.45	133.80	127.00	118.40	113.55	108.85	99.50	95.55	91.65	88.75	84.45
Stone Veneer - Wood Frame	148.15	138.25	131.15	122.20	117.25	112.30	102.60	98.40	94.35	91.25	86.80
Solid Masonry	151.35	141.30	134.10	124.80	119.70	114.65	104.70	100.40	96.15	93.05	88.50
Finished Basement, Add	31.80	32.05	31.10	29.80	29.10	28.50	27.15	26.55	25.85	25.50	24.95
Unfinished Basement, Add	13.60	13.05	12.65	11.95	11.65	11.35	10.65	10.35	10.00	9.90	9.55

Modifications

Add to the total cost

Upgrade Kitchen Cabinets	$ + 950
Solid Surface Countertops	
Full Bath - including plumbing, wall and floor finishes	+ 5922
Half Bath - including plumbing, wall and floor finishes	+ 3568
Two Car Attached Garage	+ 21,296
Two Car Detached Garage	+ 24,140
Fireplace & Chimney	+ 4825

Adjustments

For multi family - add to total cost

Additional Kitchen	$ + 12,270
Additional Full Bath & Half Bath	+ 9490
Additional Entry & Exit	+ 1512
Separate Heating & Air Conditioning	+ 5440
Separate Electric	+ 1718

For Townhouse/Rowhouse - Multiply cost per square foot by

Inner Unit	.90
End Unit	.95

Alternatives

Add to or deduct from the cost per square foot of living area

Cedar Shake Roof	+ 1.55
Clay Tile Roof	+ 2.95
Slate Roof	+ 3.95
Upgrade Ceilings to Textured Finish	+ .44
Air Conditioning, in Heating Ductwork	Base System
Heating Systems, Hot Water	+ 1.52
Heat Pump	+ 2.01
Electric Heat	– 1.89
Not Heated	– 3.19

Additional upgrades or components

Kitchen Cabinets & Countertops	Page 77
Bathroom Vanities	78
Fireplaces & Chimneys	78
Windows, Skylights & Dormers	78
Appliances	79
Breezeways & Porches	79
Finished Attic	79
Garages	80
Site Improvements	80
Wings & Ells	60

Custom 1-1/2 Story

Living Area - 2800 S.F.
Perimeter - 175 L.F.

#	Category	Description	Labor-Hours	Mat.	Labor	Total
				\multicolumn{3}{c}{Cost Per Square Foot Of Living Area}
1	Site Work	Site preparation for slab; 4' deep trench excavation for foundation wall.	.028		.55	.55
2	Foundation	Continuous reinforced concrete footing 8" deep x 18" wide; dampproofed and insulated reinforced concrete foundation wall, 8" thick, 4' deep; 4" concrete slab on 4" crushed stone base and polyethylene vapor barrier, trowel finish.	.065	3.85	4.26	8.11
3	Framing	Exterior walls - 2" x 6" wood studs, 16" O.C.; 1/2" plywood sheathing; 2" x 8" rafters 16" O.C. with 1/2" plywood sheathing, 8 in 12 pitch; 2" x 10" floor joists 16" O.C. with 5/8" plywood subfloor; 5/8" plywood subfloor on 1" x 3" wood sleepers 16" O.C.	.192	4.85	5.84	10.69
4	Exterior Walls	Horizontal beveled wood siding; building paper; 6" batt insulation; wood double hung windows; 3 solid core wood exterior doors; storms and screens.	.064	7.78	2.84	10.62
5	Roofing	30 year asphalt shingles; #15 felt building paper; aluminum gutters, downspouts and drip edge; copper flashings.	.048	1.98	1.82	3.80
6	Interiors	Walls and ceilings - 5/8" gypsum wallboard, skim coat plaster, painted with primer and 2 coats; hardwood baseboard and trim, sanded and finished; hardwood floor 70%, ceramic tile with underlayment 20%, vinyl tile with underlayment 10%; wood panel interior doors, primed and painted with 2 coat	.259	15.18	11.59	26.77
7	Specialties	Custom grade kitchen cabinets - 20 L.F. wall and base with solid surface counter top and kitchen sink; 4 L.F. bathroom vanity; 75 gallon electric water heater, medicine cabinet.	.030	3.66	.89	4.55
8	Mechanical	Gas fired warm air heat/air conditioning; one full bath including: bathtub, corner shower, built in lavatory and water closet; one 1/2 bath including: built in lavatory and water closet.	.084	4.69	2.48	7.17
9	Electrical	200 Amp. service; romex wiring; fluorescent and incandescent lighting fixtures, switches, receptacles.	.038	1.23	1.42	2.65
10	Overhead	Contractor's overhead and profit and design.		8.63	6.36	14.99
	Total			51.85	38.05	89.90

RESIDENTIAL | Custom | 2 Story

- A distinct residence from designer's plans
- Single family — 1 full bath, 1 half bath, 1 kitchen
- No basement
- Asphalt shingles on roof
- Forced hot air heat/air conditioning
- Gypsum wallboard interior finishes
- Materials and workmanship are above average

Note: The illustration shown may contain some optional components (for example: garages and/or fireplaces) whose costs are shown in the modifications, adjustments, & alternatives below or at the end of the square foot section.

Base cost per square foot of living area

Exterior Wall	Living Area										
	1200	1400	1600	1800	2000	2400	2800	3200	3600	4000	4400
Wood Siding - Wood Frame	126.05	118.85	113.70	109.20	104.20	97.05	91.00	87.00	84.65	82.15	80.05
Brick Veneer - Wood Frame	135.95	128.10	122.50	117.45	112.15	104.25	97.55	93.15	90.55	87.70	85.45
Stone Veneer - Wood Frame	140.95	132.75	126.95	121.70	116.25	107.95	100.85	96.25	93.60	90.60	88.20
Solid Masonry	144.35	136.00	130.05	124.60	119.00	110.45	103.15	98.40	95.70	92.50	90.05
Finished Basement, Add	25.65	25.75	25.10	24.40	23.95	22.90	22.05	21.50	21.15	20.75	20.45
Unfinished Basement, Add	10.95	10.45	10.15	9.85	9.60	9.10	8.65	8.40	8.25	8.05	7.90

Modifications

Add to the total cost

Upgrade Kitchen Cabinets	$ + 950
Solid Surface Countertops	
Full Bath - including plumbing, wall and floor finishes	+ 5922
Half Bath - including plumbing, wall and floor finishes	+ 3568
Two Car Attached Garage	+ 21,296
Two Car Detached Garage	+ 24,140
Fireplace & Chimney	+ 5445

Adjustments

For multi family - add to total cost

Additional Kitchen	$ + 12,270
Additional Full Bath & Half Bath	+ 9490
Additional Entry & Exit	+ 1512
Separate Heating & Air Conditioning	+ 5440
Separate Electric	+ 1718

For Townhouse/Rowhouse - Multiply cost per square foot by

Inner Unit	.87
End Unit	.93

Alternatives

Add to or deduct from the cost per square foot of living area

Cedar Shake Roof	+ 1.10
Clay Tile Roof	+ 2.05
Slate Roof	+ 2.70
Upgrade Ceilings to Textured Finish	+ .44
Air Conditioning, in Heating Ductwork	Base System
Heating Systems, Hot Water	+ 1.49
Heat Pump	+ 2.24
Electric Heat	– 1.89
Not Heated	– 3.02

Additional upgrades or components

Kitchen Cabinets & Countertops	Page 77
Bathroom Vanities	78
Fireplaces & Chimneys	78
Windows, Skylights & Dormers	78
Appliances	79
Breezeways & Porches	79
Finished Attic	79
Garages	80
Site Improvements	80
Wings & Ells	60

Custom 2 Story

Living Area - 2800 S.F.
Perimeter - 156 L.F.

#	Category	Description	Labor-Hours	Cost Per Square Foot Of Living Area		
				Mat.	Labor	Total
1	Site Work	Site preparation for slab; 4' deep trench excavation for foundation wall.	.024		.55	.55
2	Foundation	Continuous reinforced concrete footing 8" deep x 18" wide; dampproofed and insulated reinforced concrete foundation wall, 8" thick, 4' deep; 4" concrete slab on 4" crushed stone base and polyethylene vapor barrier, trowel finish.	.058	3.26	3.67	6.93
3	Framing	Exterior walls - 2" x 6" wood studs, 16" O.C.; 1/2" plywood sheathing; 2" x 8" rafters 16" O.C. with 1/2" plywood sheathing, 6 in 12 pitch; 2" x 8" ceiling joists 16" O.C.; 2" x 10" floor joists 16" O.C. with 5/8" plywood subfloor; 5/8" plywood subfloor on 1" x 3" wood sleepers 16" O.C.	.159	5.35	6.06	11.41
4	Exterior Walls	Horizontal beveled wood siding; building paper; 6" batt insulation; wood double hung windows; 3 solid core wood exterior doors; storms and screens.	.091	8.82	3.25	12.07
5	Roofing	30 year asphalt shingles; #15 felt building paper; aluminum gutters, downspouts and drip edge; copper flashings.	.042	1.59	1.46	3.05
6	Interiors	Walls and ceilings - 5/8" gypsum wallboard, skim coat plaster, painted with primer and 2 coats; hardwood baseboard and trim, sanded and finished; hardwood floor 70%, ceramic tile with underlayment 20%, vinyl tile with underlayment 10%; wood panel interior doors, primed and painted with 2 coat	.271	15.44	11.84	27.28
7	Specialties	Custom grade kitchen cabinets - 20 L.F. wall and base with solid surface counter top and kitchen sink; 4 L.F. bathroom vanity; 75 gallon electric water heater, medicine cabinet.	.028	3.66	.89	4.55
8	Mechanical	Gas fired warm air heat/air conditioning; one full bath including: bathtub, corner shower; built in lavatory and water closet; one 1/2 bath including: built in lavatory and water closet.	.078	4.82	2.52	7.34
9	Electrical	200 Amp. service; romex wiring; fluorescent and incandescent lighting fixtures, switches, receptacles.	.038	1.23	1.42	2.65
10	Overhead	Contractor's overhead and profit and design.		8.83	6.34	15.17
		Total		53.00	38.00	**91.00**

RESIDENTIAL | Custom | 2-1/2 Story

- A distinct residence from designer's plans
- Single family — 1 full bath, 1 half bath, 1 kitchen
- No basement
- Asphalt shingles on roof
- Forced hot air heat/air conditioning
- Gypsum wallboard interior finishes
- Materials and workmanship are above average

Note: The illustration shown may contain some optional components (for example: garages and/or fireplaces) whose costs are shown in the modifications, adjustments, & alternatives below or at the end of the square foot section.

Base cost per square foot of living area

Exterior Wall	\multicolumn{11}{c}{Living Area}
	1500	1800	2100	2400	2800	3200	3600	4000	4500	5000	5500
Wood Siding - Wood Frame	124.55	112.35	105.35	101.20	95.55	90.35	87.45	82.80	80.30	78.30	76.15
Brick Veneer - Wood Frame	134.80	121.75	113.80	109.20	103.20	97.30	94.00	88.90	86.10	83.75	81.40
Stone Veneer - Wood Frame	140.00	126.60	118.10	113.30	107.10	100.80	97.35	92.00	89.05	86.55	84.00
Solid Masonry	143.60	129.95	121.05	116.05	109.80	103.25	99.65	94.15	91.10	88.45	85.90
Finished Basement, Add	20.30	20.30	19.25	18.75	18.25	17.55	17.15	16.65	16.40	16.05	15.80
Unfinished Basement, Add	8.80	8.30	7.75	7.50	7.30	6.95	6.75	6.55	6.40	6.20	6.10

Modifications

Add to the total cost

Upgrade Kitchen Cabinets	$ + 950
Solid Surface Countertops	
Full Bath - including plumbing, wall and floor finishes	+ 5922
Half Bath - including plumbing, wall and floor finishes	+ 3568
Two Car Attached Garage	+ 21,296
Two Car Detached Garage	+ 24,140
Fireplace & Chimney	+ 6150

Adjustments

For multi family - add to total cost

Additional Kitchen	$ + 12,270
Additional Full Bath & Half Bath	+ 9490
Additional Entry & Exit	+ 1512
Separate Heating & Air Conditioning	+ 5440
Separate Electric	+ 1718

For Townhouse/Rowhouse - Multiply cost per square foot by

Inner Unit	.87
End Unit	.94

Alternatives

Add to or deduct from the cost per square foot of living area

Cedar Shake Roof	+ .95
Clay Tile Roof	+ 1.75
Slate Roof	+ 2.35
Upgrade Ceilings to Textured Finish	+ .44
Air Conditioning, in Heating Ductwork	Base System
Heating Systems, Hot Water	+ 1.34
Heat Pump	+ 2.31
Electric Heat	– 3.30
Not Heated	– 3.02

Additional upgrades or components

Kitchen Cabinets & Countertops	Page 77
Bathroom Vanities	78
Fireplaces & Chimneys	78
Windows, Skylights & Dormers	78
Appliances	79
Breezeways & Porches	79
Finished Attic	79
Garages	80
Site Improvements	80
Wings & Ells	60

Custom 2-1/2 Story

Living Area - 3200 S.F.
Perimeter - 150 L.F.

		Labor-Hours	Cost Per Square Foot Of Living Area		
			Mat.	Labor	Total
1 Site Work	Site preparation for slab; 4' deep trench excavation for foundation wall.	.048		.48	.48
2 Foundation	Continuous reinforced concrete footing 8" deep x 18" wide; dampproofed and insulated reinforced concrete foundation wall, 8" thick, 4' deep; 4" concrete slab on 4" crushed stone base and polyethylene vapor barrier, trowel finish.	.063	2.67	3.05	5.72
3 Framing	Exterior walls - 2" x 6" wood studs, 16" O.C.; 1/2" plywood sheathing; 2" x 8" rafters 16" O.C. with 1/2" plywood sheathing, 6 in 12 pitch; 2" x 8" ceiling joists 16" O.C.; 2" x 10" floor joists 16" O.C. with 5/8" plywood subfloor; 5/8" plywood subfloor on 1" x 3" wood sleepers 16" O.C.	.177	5.79	6.30	12.09
4 Exterior Walls	Horizontal beveled wood siding; building paper; 6" batt insulation; wood double hung windows; 3 solid core wood exterior doors; storms and screens.	.134	9.11	3.37	12.48
5 Roofing	30 year asphalt shingles; #15 felt building paper; aluminum gutters, downspouts and drip edge; copper flashings.	.032	1.21	1.12	2.33
6 Interiors	Walls and ceilings - 5/8" gypsum wallboard, skim coat plaster, painted with primer and 2 coats; hardwood baseboard and trim, sanded and finished; hardwood floor 70%, ceramic tile with underlayment 20%, vinyl tile with underlayment 10%; wood panel interior doors, primed and painted with 2 coat	.354	16.22	12.60	28.82
7 Specialties	Custom grade kitchen cabinets - 20 L.F. wall and base with solid surface counter top and kitchen sink; 4 L.F. bathroom vanity; 75 gallon electric water heater, medicine cabinet.	.053	3.18	.78	3.96
8 Mechanical	Gas fired warm air heat/air conditioning; one full bath including: bathtub, corner shower; built in lavatory and water closet; one 1/2 bath including: built in lavatory and water closet.	.104	4.41	2.44	6.85
9 Electrical	200 Amp. service; romex wiring; fluorescent and incandescent lighting fixtures, switches, receptacles.	.048	1.19	1.38	2.57
10 Overhead	Contractor's overhead and profit and design.		8.77	6.28	15.05
	Total		52.55	37.80	**90.35**

RESIDENTIAL | Custom | 3 Story

- A distinct residence from designer's plans
- Single family — 1 full bath, 1 half bath, 1 kitchen
- No basement
- Asphalt shingles on roof
- Forced hot air heat/air conditioning
- Gypsum wallboard interior finishes
- Materials and workmanship are above average

Note: The illustration shown may contain some optional components (for example: garages and/or fireplaces) whose costs are shown in the modifications, adjustments, & alternatives below or at the end of the square foot section.

Base cost per square foot of living area

Exterior Wall	\multicolumn{11}{c}{Living Area}
	1500	1800	2100	2500	3000	3500	4000	4500	5000	5500	6000
Wood Siding - Wood Frame	124.00	112.00	106.25	101.55	94.10	90.40	85.85	81.00	79.30	77.55	75.60
Brick Veneer - Wood Frame	134.70	121.85	115.50	110.35	102.10	97.90	92.75	87.35	85.45	83.45	81.10
Stone Veneer - Wood Frame	140.10	126.90	120.20	114.80	106.10	101.80	96.25	90.55	88.55	86.45	84.00
Solid Masonry	143.85	130.30	123.40	117.90	108.90	104.45	98.65	92.75	90.70	88.55	85.90
Finished Basement, Add	17.85	17.85	17.20	16.65	15.95	15.45	14.95	14.45	14.30	14.05	13.80
Unfinished Basement, Add	7.70	7.30	6.95	6.75	6.40	6.20	5.90	5.70	5.55	5.50	5.30

Modifications

Add to the total cost

Upgrade Kitchen Cabinets	$ + 950
Solid Surface Countertops	
Full Bath - including plumbing, wall and floor finishes	+ 5922
Half Bath - including plumbing, wall and floor finishes	+ 3568
Two Car Attached Garage	+ 21,296
Two Car Detached Garage	+ 24,140
Fireplace & Chimney	+ 6150

Adjustments

For multi family - add to total cost

Additional Kitchen	$ + 12,270
Additional Full Bath & Half Bath	+ 9490
Additional Entry & Exit	+ 1512
Separate Heating & Air Conditioning	+ 5440
Separate Electric	+ 1718

For Townhouse/Rowhouse - Multiply cost per square foot by

Inner Unit	.85
End Unit	.93

Alternatives

Add to or deduct from the cost per square foot of living area

Cedar Shake Roof	+ .75
Clay Tile Roof	+ 1.35
Slate Roof	+ 1.80
Upgrade Ceilings to Textured Finish	+ .44
Air Conditioning, in Heating Ductwork	Base System
Heating Systems, Hot Water	+ 1.34
Heat Pump	+ 2.31
Electric Heat	– 3.30
Not Heated	– 2.92

Additional upgrades or components

Kitchen Cabinets & Countertops	Page 77
Bathroom Vanities	78
Fireplaces & Chimneys	78
Windows, Skylights & Dormers	78
Appliances	79
Breezeways & Porches	79
Finished Attic	79
Garages	80
Site Improvements	80
Wings & Ells	60

Important: See the Reference Section for Location Factors (to adjust for your city) and Estimating Forms

Custom 3 Story

Living Area - 3000 S.F.
Perimeter - 135 L.F.

		Labor-Hours	Cost Per Square Foot Of Living Area		
			Mat.	Labor	Total
1 Site Work	Site preparation for slab; 4' deep trench excavation for foundation wall.	.048		.51	.51
2 Foundation	Continuous reinforced concrete footing 8" deep x 18" wide; dampproofed and insulated reinforced concrete foundation wall, 8" thick, 4' deep; 4" concrete slab on 4" crushed stone base and polyethylene vapor barrier, trowel finish.	.060	2.50	2.88	5.38
3 Framing	Exterior walls - 2" x 6" wood studs, 16" O.C.; 1/2" plywood sheathing; 2" x 8" rafters 16" O.C. with 1/2" plywood sheathing, 6 in 12 pitch; 2" x 8" ceiling joists 16" O.C.; 2" x 10" floor joists 16" O.C. with 5/8" plywood subfloor; 5/8" plywood subfloor on 1" x 3" wood sleepers 16" O.C.	.191	6.10	6.48	12.58
4 Exterior Walls	Horizontal beveled wood siding; building paper; 6" batt insulation; wood double hung windows; 3 solid core wood exterior doors; storms and screens.	.150	10.40	3.86	14.26
5 Roofing	30 year asphalt shingles; #15 felt building paper; aluminum gutters, downspouts and drip edge; copper flashings.	.028	1.06	.97	2.03
6 Interiors	Walls and ceilings - 5/8" gypsum wallboard, skim coat plaster, painted with primer and 2 coats; hardwood baseboard and trim, sanded and finished; hardwood floor 70%, ceramic tile with underlayment 20%, vinyl tile with underlayment 10%; wood panel interior doors, primed and painted with 2 coat	.409	16.68	13.06	29.74
7 Specialties	Custon grade kitchen cabinets - 20 L.F. wall and base with solid surface counter top and kitchen sink; 4 L.F. bathroom vanity; 75 gallon electric water heater, medicine cabinet.	.053	3.40	.83	4.23
8 Mechanical	Gas fired warm air heat/air conditioning; one full bath including: bathtub, corner shower; built in lavatory and water closet; one 1/2 bath including: built in lavatory and water closet.	.105	4.61	2.47	7.08
9 Electrical	200 Amp. service; romex wiring; fluorescent and incandescent lighting fixtures, switches, receptacles.	.048	1.21	1.40	2.61
10 Overhead	Contractor's overhead and profit and design.		9.19	6.49	15.68
	Total		55.15	38.95	**94.10**

RESIDENTIAL | Custom | Bi-Level

- A distinct residence from designer's plans
- Single family — 1 full bath, 1 half bath, 1 kitchen
- No basement
- Asphalt shingles on roof
- Forced hot air heat/air conditioning
- Gypsum wallboard interior finishes
- Materials and workmanship are above average

Note: The illustration shown may contain some optional components (for example: garages and/or fireplaces) whose costs are shown in the modifications, adjustments, & alternatives below or at the end of the square foot section.

Base cost per square foot of living area

Exterior Wall	Living Area										
	1200	1400	1600	1800	2000	2400	2800	3200	3600	4000	4400
Wood Siding - Wood Frame	119.50	112.75	107.80	103.65	98.85	92.20	86.60	82.85	80.65	78.45	76.45
Brick Veneer - Wood Frame	126.90	119.60	114.40	109.85	104.80	97.60	91.50	87.45	85.05	82.60	80.45
Stone Veneer - Wood Frame	130.65	123.15	117.75	113.05	107.85	100.40	94.00	89.80	87.35	84.75	82.55
Solid Masonry	133.25	125.60	120.05	115.20	109.95	102.25	95.75	91.45	88.90	86.20	83.95
Finished Basement, Add	25.65	25.75	25.10	24.40	23.95	22.90	22.05	21.50	21.15	20.75	20.45
Unfinished Basement, Add	10.95	10.45	10.15	9.85	9.60	9.10	8.65	8.40	8.25	8.05	7.90

Modifications

Add to the total cost

Upgrade Kitchen Cabinets	$ + 950
Solid Surface Countertops	
Full Bath - including plumbing, wall and floor finishes	+ 5922
Half Bath - including plumbing, wall and floor finishes	+ 3568
Two Car Attached Garage	+ 21,296
Two Car Detached Garage	+ 24,140
Fireplace & Chimney	+ 4825

Adjustments

For multi family - add to total cost

Additional Kitchen	$ + 12,270
Additional Full Bath & Half Bath	+ 9490
Additional Entry & Exit	+ 1512
Separate Heating & Air Conditioning	+ 5440
Separate Electric	+ 1718

For Townhouse/Rowhouse - Multiply cost per square foot by

Inner Unit	.89
End Unit	.95

Alternatives

Add to or deduct from the cost per square foot of living area

Cedar Shake Roof	+ 1.10
Clay Tile Roof	+ 2.05
Slate Roof	+ 2.70
Upgrade Ceilings to Textured Finish	+ .44
Air Conditioning, in Heating Ductwork	Base System
Heating Systems, Hot Water	+ 1.49
Heat Pump	+ 2.24
Electric Heat	– 1.89
Not Heated	– 2.92

Additional upgrades or components

Kitchen Cabinets & Countertops	Page 77
Bathroom Vanities	78
Fireplaces & Chimneys	78
Windows, Skylights & Dormers	78
Appliances	79
Breezeways & Porches	79
Finished Attic	79
Garages	80
Site Improvements	80
Wings & Ells	60

Custom Bi-Level

Living Area - 2800 S.F.
Perimeter - 156 L.F.

		Labor-Hours	Cost Per Square Foot Of Living Area		
			Mat.	Labor	Total
1 Site Work	Excavation for lower level, 4' deep. Site preparation for slab.	.024		.55	.55
2 Foundation	Continuous reinforced concrete footing 8" deep x 18" wide; dampproofed and insulated reinforced concrete foundation wall, 8" thick, 4' deep; 4" concrete slab on 4" crushed stone base and polyethylene vapor barrier, trowel finish.	.058	3.26	3.67	6.93
3 Framing	Exterior walls - 2" x 6" wood studs, 16" O.C.; 1/2" plywood sheathing; 2" x 8" rafters 16" O.C. with 1/2" plywood sheathing, 6 in 12 pitch; 2" x 8" ceiling joists 16" O.C.; 2" x 10" floor joists 16" O.C. with 5/8" plywood subfloor; 5/8" plywood subfloor on 1" x 3" wood sleepers 16" O.C.	.147	5.08	5.78	10.86
4 Exterior Walls	Horizontal beveled wood siding; building paper; 6" batt insulation; wood double hung windows; 3 solid core wood exterior doors; storms and screens.	.079	6.93	2.52	9.45
5 Roofing	30 year asphalt shingles; #15 felt building paper; aluminum gutters, downspouts and drip edge; copper flashings.	.033	1.59	1.46	3.05
6 Interiors	Walls and ceilings - 5/8" gypsum wallboard, skim coat plaster, painted with primer and 2 coats; hardwood baseboard and trim, sanded and finished; hardwood floor 70%, ceramic tile with underlayment 20%, vinyl tile with underlayment 10%; wood panel interior doors, primed and painted with 2 coat	.257	15.25	11.53	26.78
7 Specialties	Custom grade kitchen cabinets - 20 L.F. wall and base with solid surface counter top and kitchen sink; 4 L.F. bathroom vanity; 75 gallon electric water heater, medicine cabinet.	.028	3.66	.89	4.55
8 Mechanical	Gas fired warm air heat/air conditioning; one full bath including: bathtub, corner shower, built in lavatory and water closet; one 1/2 bath including: built in lavatory and water closet.	.078	4.82	2.52	7.34
9 Electrical	200 Amp. service; romex wiring; fluorescent and incandescent lighting fixtures, switches, receptacles.	.038	1.23	1.42	2.65
10 Overhead	Contractor's overhead and profit and design.		8.38	6.06	14.44
	Total		50.20	36.40	**86.60**

RESIDENTIAL Custom Tri-Level

- A distinct residence from designer's plans
- Single family — 1 full bath, 1 half bath, 1 kitchen
- No basement
- Asphalt shingles on roof
- Forced hot air heat/air conditioning
- Gypsum wallboard interior finishes
- Materials and workmanship are above average

Note: The illustration shown may contain some optional components (for example: garages and/or fireplaces) whose costs are shown in the modifications, adjustments, & alternatives below or at the end of the square foot section.

Base cost per square foot of living area

Exterior Wall	\multicolumn{11}{c}{Living Area}
	1200	1500	1800	2100	2400	2800	3200	3600	4000	4500	5000
Wood Siding - Wood Frame	122.80	111.85	103.75	96.85	92.30	89.10	85.10	80.80	79.20	75.15	73.05
Brick Veneer - Wood Frame	130.10	118.50	109.75	102.30	97.45	94.00	89.70	85.05	83.25	78.95	76.60
Stone Veneer - Wood Frame	133.85	121.85	112.80	105.10	100.00	96.45	92.00	87.15	85.35	80.90	78.45
Solid Masonry	136.40	124.20	114.85	106.95	101.80	98.25	93.60	88.65	86.80	82.15	79.65
Finished Basement, Add*	31.95	31.85	30.45	29.30	28.60	28.05	27.35	26.75	26.45	25.85	25.40
Unfinished Basement, Add*	13.50	12.80	12.10	11.55	11.20	10.95	10.60	10.25	10.15	9.85	9.65

*Basement under middle level only.

Modifications
Add to the total cost

Upgrade Kitchen Cabinets	$ + 950
Solid Surface Countertops	
Full Bath - including plumbing, wall and floor finishes	+ 5922
Half Bath - including plumbing, wall and floor finishes	+ 3568
Two Car Attached Garage	+ 21,296
Two Car Detached Garage	+ 24,140
Fireplace & Chimney	+ 4825

Adjustments
For multi family - add to total cost

Additional Kitchen	$ + 12,270
Additional Full Bath & Half Bath	+ 9490
Additional Entry & Exit	+ 1512
Separate Heating & Air Conditioning	+ 5440
Separate Electric	+ 1718

For Townhouse/Rowhouse - Multiply cost per square foot by

Inner Unit	.87
End Unit	.94

Alternatives
Add to or deduct from the cost per square foot of living area

Cedar Shake Roof	+ 1.55
Clay Tile Roof	+ 2.95
Slate Roof	+ 3.95
Upgrade Ceilings to Textured Finish	+ .44
Air Conditioning, in Heating Ductwork	Base System
Heating Systems, Hot Water	+ 1.44
Heat Pump	+ 2.33
Electric Heat	– 1.68
Not Heated	– 2.92

Additional upgrades or components

Kitchen Cabinets & Countertops	Page 77
Bathroom Vanities	78
Fireplaces & Chimneys	78
Windows, Skylights & Dormers	78
Appliances	79
Breezeways & Porches	79
Finished Attic	79
Garages	80
Site Improvements	80
Wings & Ells	60

Custom Tri-Level

Living Area - 3200 S.F.
Perimeter - 198 L.F.

		Labor-Hours	Cost Per Square Foot Of Living Area		
			Mat.	Labor	Total
1 Site Work	Site preparation for slab; 4' deep trench excavation for foundation wall, excavation for lower level, 4' deep.	.023		.48	.48
2 Foundation	Continuous reinforced concrete footing 8" deep x 18" wide; dampproofed and insulated reinforced concrete foundation wall, 8" thick, 4' deep; 4" concrete slab on 4" crushed stone base and polyethylene vapor barrier, trowel finish.	.073	3.86	4.22	8.08
3 Framing	Exterior walls - 2" x 6" wood studs, 16" O.C.; 1/2" plywood sheathing; 2" x 8" rafters 16" O.C. with 1/2" plywood sheathing, 6 in 12 pitch; 2" x 8" ceiling joists 16" O.C.; 2" x 10" floor joists 16" O.C. with 5/8" plywood subfloor; 5/8" plywood subfloor on 1" x 3" wood sleepers 16" O.C.	.162	4.56	5.66	10.22
4 Exterior Walls	Horizontal beveled wood siding; building paper; 6" batt insulation; wood double hung windows; 3 solid core wood exterior doors; storms and screens.	.076	6.73	2.42	9.15
5 Roofing	30 year asphalt shingles; #15 felt building paper; aluminum gutters, downspouts and drip edge; copper flashings.	.045	2.11	1.94	4.05
6 Interiors	Walls and ceilings - 5/8" gypsum wallboard, skim coat plaster, painted with primer and 2 coats; hardwood baseboard and trim, sanded and finished; hardwood floor 70%, ceramic tile with underlayment 20%, vinyl tile with underlayment 10%; wood panel interior doors, primed and painted with 2 coat	.242	14.55	11.03	25.58
7 Specialties	Custom grade kitchen cabinets - 20 L.F. wall and base with solid surface counter top and kitchen sink; 4 L.F. bathroom vanity; 75 gallon electric water heater, medicine cabinet.	.026	3.18	.78	3.96
8 Mechanical	Gas fired warm air heat/air conditioning; one full bath including: bathtub, corner shower, built in lavatory and water closet; one 1/2 bath including: built in lavatory and water closet.	.073	4.41	2.44	6.85
9 Electrical	200 Amp. service; romex wiring; fluorescent and incandescent lighting fixtures, switches, receptacles.	.036	1.19	1.38	2.57
10 Overhead	Contractor's overhead and profit and design.		8.11	6.05	14.16
	Total		48.70	36.40	**85.10**

RESIDENTIAL Custom Wings & Ells

1 Story — Base cost per square foot of living area

Exterior Wall	Living Area							
	50	100	200	300	400	500	600	700
Wood Siding - Wood Frame	191.75	149.55	131.20	112.00	106.05	102.40	100.00	100.80
Brick Veneer - Wood Frame	216.55	167.35	146.00	121.80	114.90	110.65	107.90	108.45
Stone Veneer - Wood Frame	229.25	176.35	153.50	126.85	119.40	114.90	111.90	112.25
Solid Masonry	237.95	182.60	158.70	130.30	122.50	117.80	114.65	114.95
Finished Basement, Add	75.80	64.80	58.70	48.70	46.65	45.50	44.65	44.10
Unfinished Basement, Add	55.05	38.10	30.30	23.40	21.55	20.40	19.65	19.15

1-1/2 Story — Base cost per square foot of living area

Exterior Wall	Living Area							
	100	200	300	400	500	600	700	800
Wood Siding - Wood Frame	152.75	125.55	109.05	99.60	94.60	92.20	89.20	88.30
Brick Veneer - Wood Frame	174.85	143.30	123.80	111.15	105.25	102.25	98.70	97.75
Stone Veneer - Wood Frame	186.20	152.30	131.35	117.05	110.65	107.40	103.50	102.55
Solid Masonry	194.00	158.60	136.55	121.05	114.40	110.95	106.85	105.85
Finished Basement, Add	50.75	47.00	43.00	38.55	37.30	36.55	35.75	35.75
Unfinished Basement, Add	32.95	25.15	21.45	18.30	17.20	16.50	15.85	15.65

2 Story — Base cost per square foot of living area

Exterior Wall	Living Area							
	100	200	400	600	800	1000	1200	1400
Wood Siding - Wood Frame	152.35	116.95	101.40	86.85	81.80	78.75	76.75	77.80
Brick Veneer - Wood Frame	177.15	134.65	116.15	96.65	90.65	87.00	84.60	85.40
Stone Veneer - Wood Frame	189.85	143.70	123.70	101.70	95.15	91.20	88.60	89.25
Solid Masonry	198.50	150.00	128.90	105.10	98.25	94.15	91.40	91.90
Finished Basement, Add	37.95	32.45	29.40	24.45	23.40	22.80	22.40	22.10
Unfinished Basement, Add	27.55	19.10	15.20	11.70	10.80	10.20	9.85	9.60

Base costs do not include bathroom or kitchen facilities. Use Modifications/Adjustments/Alternatives on pages 77-80 where appropriate.

RESIDENTIAL | Luxury | Illustrations

1 Story

1-1/2 Story

2 Story

2-1/2 Story

Bi-Level

Tri-Level

RESIDENTIAL | Luxury | 1 Story

- Unique residence built from an architect's plan
- Single family — 1 full bath, 1 half bath, 1 kitchen
- No basement
- Cedar shakes on roof
- Forced hot air heat/air conditioning
- Gypsum wallboard interior finishes
- Many special features
- Extraordinary materials and workmanship

Note: The illustration shown may contain some optional components (for example: garages and/or fireplaces) whose costs are shown in the modifications, adjustments, & alternatives below or at the end of the square foot section.

Base cost per square foot of living area

Exterior Wall	Living Area										
	1000	1200	1400	1600	1800	2000	2400	2800	3200	3600	4000
Wood Siding - Wood Frame	169.95	157.10	146.70	139.65	135.80	131.10	122.45	116.30	112.05	107.45	103.65
Brick Veneer - Wood Frame	179.00	165.25	154.15	146.60	142.50	137.40	128.25	121.60	116.95	112.10	108.00
Solid Brick	192.60	177.55	165.30	157.10	152.55	146.95	136.90	129.65	124.35	118.95	114.40
Solid Stone	190.60	175.70	163.55	155.50	151.05	145.50	135.60	128.45	123.25	117.90	113.45
Finished Basement, Add	46.15	49.55	47.50	46.15	45.35	44.25	42.80	41.55	40.55	39.60	38.85
Unfinished Basement, Add	20.20	19.00	18.00	17.35	16.95	16.45	15.65	15.10	14.60	14.15	13.75

Modifications

Add to the total cost

Upgrade Kitchen Cabinets	$ + 1254
Solid Surface Countertops	
Full Bath - including plumbing, wall and floor finishes	+ 6869
Half Bath - including plumbing, wall and floor finishes	+ 4139
Two Car Attached Garage	+ 24,327
Two Car Detached Garage	+ 27,407
Fireplace & Chimney	+ 6760

Adjustments

For multi family - add to total cost

Additional Kitchen	$ + 15,820
Additional Full Bath & Half Bath	+ 11,008
Additional Entry & Exit	+ 2172
Separate Heating & Air Conditioning	+ 5440
Separate Electric	+ 1718

For Townhouse/Rowhouse - Multiply cost per square foot by

Inner Unit	.90
End Unit	.95

Alternatives

Add to or deduct from the cost per square foot of living area

Heavyweight Asphalt Shingles	– 2.20
Clay Tile Roof	+ 1.90
Slate Roof	+ 3.25
Upgrade Ceilings to Textured Finish	+ .44
Air Conditioning, in Heating Ductwork	Base System
Heating Systems, Hot Water	+ 1.71
Heat Pump	+ 2.06
Electric Heat	– 1.87
Not Heated	– 3.74

Additional upgrades or components

Kitchen Cabinets & Countertops	Page 77
Bathroom Vanities	78
Fireplaces & Chimneys	78
Windows, Skylights & Dormers	78
Appliances	79
Breezeways & Porches	79
Finished Attic	79
Garages	80
Site Improvements	80
Wings & Ells	76

Luxury 1 Story

Living Area - 2800 S.F.
Perimeter - 219 L.F.

			Labor-Hours	Cost Per Square Foot Of Living Area		
				Mat.	Labor	Total
1	Site Work	Site preparation for slab; 4' deep trench excavation for foundation wall.	.028		.60	.60
2	Foundation	Continuous reinforced concrete footing 8" deep x 18" wide; dampproofed and insulated reinforced concrete foundation wall, 12" thick, 4' deep; 4" concrete slab on 4" crushed stone base and polyethylene vapor barrier, trowel finish.	.098	6.60	6.20	12.80
3	Framing	Exterior walls - 2" x 6" wood studs, 16" O.C.; 5/8" plywood sheathing; 2" x 10" rafters 16" O.C. with 5/8" plywood sheathing, 6 in 12 pitch; 2" x 8" ceiling joists 16" O.C.; 5/8" plywood subfloor on 1" x 3" wood sleepers 16" O.C.	.260	10.16	11.39	21.55
4	Exterior Walls	Horizontal beveled wood siding; building paper; 6" batt insulation; wood double hung windows; 3 solid core wood exterior doors; storms and screens.	.204	7.75	2.75	10.50
5	Roofing	Red cedar shingles; #15 felt building paper; aluminum gutters, downspouts and drip edge; copper flashings.	.082	3.70	3.40	7.10
6	Interiors	Walls and ceilings - 5/8" gypsum wallboard, skim coat plaster, painted with primer and 2 coats; hardwood baseboard and trim, sanded and finished; hardwood floor 70%, ceramic tile with underlayment 20%, vinyl tile with underlayment 10%; wood panel interior doors, primed and painted with 2 coat	.287	12.36	11.58	23.94
7	Specialties	Luxury grade kitchen cabinets - 25 L.F. wall and base with solid surface counter top and kitchen sink; 6 L.F. bathroom vanity; 75 gallon electric water heater; medicine cabinet.	.052	4.75	1.13	5.88
8	Mechanical	Gas fired warm air heat/air conditioning; one full bath including: bathtub, corner shower; built in lavatory and water closet; one 1/2 bath including: built in lavatory and water closet.	.078	5.47	2.79	8.26
9	Electrical	200 Amp. service; romex wiring; fluorescent and incandescent lighting fixtures; intercom, switches, receptacles.	.044	1.48	1.70	3.18
10	Overhead	Contractor's overhead and profit and architect's fees.		12.53	9.96	22.49
		Total		64.80	51.50	**116.30**

RESIDENTIAL | Luxury | 1-1/2 Story

- Unique residence built from an architect's plan
- Single family — 1 full bath, 1 half bath, 1 kitchen
- No basement
- Cedar shakes on roof
- Forced hot air heat/air conditioning
- Gypsum wallboard interior finishes
- Many special features
- Extraordinary materials and workmanship

Note: The illustration shown may contain some optional components (for example: garages and/or fireplaces) whose costs are shown in the modifications, adjustments, & alternatives below or at the end of the square foot section.

©Larry E. Belk Designs

Base cost per square foot of living area

Exterior Wall	Living Area										
	1000	1200	1400	1600	1800	2000	2400	2800	3200	3600	4000
Wood Siding - Wood Frame	157.80	146.85	138.95	129.65	124.40	119.10	109.15	104.70	100.55	97.35	92.75
Brick Veneer - Wood Frame	168.30	156.65	148.30	138.20	132.45	126.75	115.85	111.10	106.50	103.05	98.05
Solid Brick	184.00	171.40	162.35	150.90	144.50	138.20	126.05	120.70	115.40	111.60	106.00
Solid Stone	181.60	169.15	160.25	149.05	142.75	136.50	124.55	119.25	114.00	110.30	104.85
Finished Basement, Add	32.60	35.30	34.20	32.65	31.75	30.95	29.30	28.60	27.70	27.25	26.55
Unfinished Basement, Add	14.55	13.90	13.40	12.60	12.20	11.85	11.05	10.65	10.25	10.05	9.70

Modifications

Add to the total cost

Upgrade Kitchen Cabinets	$ + 1254
Solid Surface Countertops	
Full Bath - including plumbing, wall and floor finishes	+ 6869
Half Bath - including plumbing, wall and floor finishes	+ 4139
Two Car Attached Garage	+ 24,327
Two Car Detached Garage	+ 27,407
Fireplace & Chimney	+ 6760

Adjustments

For multi family - add to total cost

Additional Kitchen	$ + 15,820
Additional Full Bath & Half Bath	+ 11,008
Additional Entry & Exit	+ 2172
Separate Heating & Air Conditioning	+ 5440
Separate Electric	+ 1718

For Townhouse/Rowhouse - Multiply cost per square foot by

Inner Unit	.90
End Unit	.95

Alternatives

Add to or deduct from the cost per square foot of living area

Heavyweight Asphalt Shingles	– 1.55
Clay Tile Roof	+ 1.35
Slate Roof	+ 2.35
Upgrade Ceilings to Textured Finish	+ .44
Air Conditioning, in Heating Ductwork	Base System
Heating Systems, Hot Water	+ 1.64
Heat Pump	+ 2.28
Electric Heat	– 1.87
Not Heated	– 3.45

Additional upgrades or components

Kitchen Cabinets & Countertops	Page 77
Bathroom Vanities	78
Fireplaces & Chimneys	78
Windows, Skylights & Dormers	78
Appliances	79
Breezeways & Porches	79
Finished Attic	79
Garages	80
Site Improvements	80
Wings & Ells	76

Luxury 1-1/2 Story

Living Area - 2800 S.F.
Perimeter - 175 L.F.

		Labor-Hours	Cost Per Square Foot Of Living Area		
			Mat.	Labor	Total
1	Site Work — Site preparation for slab; 4' deep trench excavation for foundation wall.	.025		.60	.60
2	Foundation — Continuous reinforced concrete footing 8" deep x 18" wide; dampproofed and insulated reinforced concrete foundation wall, 12" thick, 4' deep; 4" concrete slab on 4" crushed stone base and polyethylene vapor barrier, trowel finish.	.066	4.72	4.69	9.41
3	Framing — Exterior walls - 2" x 6" wood studs, 16" O.C.; 5/8" plywood sheathing; 2" x 10" rafters 16" O.C. with 5/8" plywood sheathing, 8 in 12 pitch; 2" x 8" ceiling joists 16" O.C.; 2" x 12" floor joists 16" O.C. with 5/8" plywood subfloor; 5/8" plywood subfloor on 1" x 3" wood sleepers 16" O.C.	.189	6.27	7.40	13.67
4	Exterior Walls — Horizontal beveled wood siding; building paper; 6" batt insulation; wood double hung windows; 3 solid core wood exterior doors; storms and screens.	.174	8.52	3.05	11.57
5	Roofing — Red cedar shingles; #15 felt building paper; aluminum gutters, downspouts and drip edge; copper flashings.	.065	2.32	2.13	4.45
6	Interiors — Walls and ceilings - 5/8" gypsum wallboard, skim coat plaster, painted with primer and 2 coats; hardwood baseboard and trim, sanded and finished; hardwood floor 70%, ceramic tile with underlayment 20%, vinyl tile with underlayment 10%; wood panel interior doors, primed and painted with 2 coa	.260	14.16	13.23	27.39
7	Specialties — Luxury grade kitchen cabinets - 25 L.F. wall and base with solid surface counter top and kitchen sink; 6 L.F. bathroom vanity; 75 gallon electric water heater; medicine cabinet.	.062	4.75	1.13	5.88
8	Mechanical — Gas fired warm air heat/air conditioning; one full bath including: bathtub, corner shower; built in lavatory and water closet; one 1/2 bath including: built in lavatory and water closet.	.080	5.47	2.79	8.26
9	Electrical — 200 Amp. service; romex wiring; fluorescent and incandescent lighting fixtures; intercom, switches, receptacles.	.044	1.48	1.70	3.18
10	Overhead — Contractor's overhead and profit and architect's fees.		11.46	8.83	20.29
	Total		59.15	45.55	**104.70**

RESIDENTIAL | Luxury | 2 Story

- Unique residence built from an architect's plan
- Single family — 1 full bath, 1 half bath, 1 kitchen
- No basement
- Cedar shakes on roof
- Forced hot air heat/air conditioning
- Gypsum wallboard interior finishes
- Many special features
- Extraordinary materials and workmanship

Note: The illustration shown may contain some optional components (for example: garages and/or fireplaces) whose costs are shown in the modifications, adjustments, & alternatives below or at the end of the square foot section.

Base cost per square foot of living area

Exterior Wall	Living Area										
	1200	1400	1600	1800	2000	2400	2800	3200	3600	4000	4400
Wood Siding - Wood Frame	146.30	137.60	131.35	126.00	120.10	111.65	104.65	99.90	97.15	94.15	91.75
Brick Veneer - Wood Frame	157.50	148.05	141.40	135.40	129.20	119.85	112.10	106.85	103.85	100.45	97.85
Solid Brick	174.30	163.70	156.35	149.50	142.75	132.10	123.15	117.40	113.95	109.95	107.00
Solid Stone	171.80	161.35	154.10	147.40	140.75	130.30	121.50	115.80	112.45	108.50	105.65
Finished Basement, Add	26.20	28.40	27.60	26.75	26.15	24.90	23.85	23.10	22.70	22.25	21.90
Unfinished Basement, Add	11.75	11.15	10.80	10.40	10.10	9.50	9.00	8.65	8.50	8.20	8.10

Modifications

Add to the total cost

Upgrade Kitchen Cabinets	$ + 1254
Solid Surface Countertops	
Full Bath - including plumbing, wall and floor finishes	+ 6869
Half Bath - including plumbing, wall and floor finishes	+ 4139
Two Car Attached Garage	+ 24,327
Two Car Detached Garage	+ 27,407
Fireplace & Chimney	+ 7415

Adjustments

For multi family - add to total cost

Additional Kitchen	$ + 15,820
Additional Full Bath & Half Bath	+ 11,008
Additional Entry & Exit	+ 2172
Separate Heating & Air Conditioning	+ 5440
Separate Electric	+ 1718

For Townhouse/Rowhouse - Multiply cost per square foot by

Inner Unit	.86
End Unit	.93

Alternatives

Add to or deduct from the cost per square foot of living area

Heavyweight Asphalt Shingles	– 1.10
Clay Tile Roof	+ .95
Slate Roof	+ 1.60
Upgrade Ceilings to Textured Finish	+ .44
Air Conditioning, in Heating Ductwork	Base System
Heating Systems, Hot Water	+ 1.59
Heat Pump	+ 2.40
Electric Heat	– 1.68
Not Heated	– 3.26

Additional upgrades or components

Kitchen Cabinets & Countertops	Page 77
Bathroom Vanities	78
Fireplaces & Chimneys	78
Windows, Skylights & Dormers	78
Appliances	79
Breezeways & Porches	79
Finished Attic	79
Garages	80
Site Improvements	80
Wings & Ells	76

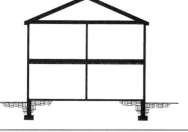

Luxury 2 Story

Living Area - 3200 S.F.
Perimeter - 163 L.F.

		Labor-Hours	Cost Per Square Foot Of Living Area		
			Mat.	Labor	Total
1 Site Work	Site preparation for slab; 4' deep trench excavation for foundation wall.	.024		.52	.52
2 Foundation	Continuous reinforced concrete footing 8" deep x 18" wide; dampproofed and insulated reinforced concrete foundation wall, 12" thick, 4' deep; 4" concrete slab on 4" crushed stone base and polyethylene vapor barrier, trowel finish.	.058	3.82	3.83	7.65
3 Framing	Exterior walls - 2" x 6" wood studs, 16" O.C.; 5/8" plywood sheathing; 2" x 10" rafters 16" O.C. with 5/8" plywood sheathing, 6 in 12 pitch; 2" x 8" ceiling joists 16" O.C.; 2" x 12" floor joists 16" O.C. with 5/8" plywood subfloor; 5/8" plywood subfloor on 1" x 3" wood sleepers 16" O.C.	.193	6.27	7.32	13.59
4 Exterior Walls	Horizontal beveled wood siding, building paper; 6" batt insulation; wood double hung windows; 3 solid core wood exterior doors; storms and screens.	.247	8.96	3.26	12.22
5 Roofing	Red cedar shingles; #15 felt building paper; aluminum gutters, downspouts and drip edge; copper flashings.	.049	1.85	1.71	3.56
6 Interiors	Walls and ceilings - 5/8" gypsum wallboard, skim coat plaster, painted with primer and 2 coats; hardwood baseboard and trim, sanded and finished; hardwood floor 70%, ceramic tile with underlayment 20%, vinyl tile with underlayment 10%; wood panel interior doors, primed and painted with 2 coat	.252	13.98	13.15	27.13
7 Specialties	Luxury grade kitchen cabinets - 25 L.F. wall and base with solid surface counter top and kitchen sink; 6 L.F. bathroom vanity; 75 gallon electric water heater; medicine cabinet.	.057	4.14	.99	5.13
8 Mechanical	Gas fired warm air heat/air conditioning; one full bath including: bathtub, corner shower; built in lavatory and water closet; one 1/2 bath including: built in lavatory and water closet.	.071	5.00	2.69	7.69
9 Electrical	200 Amp. service; romex wiring; fluorescent and incandescent lighting fixtures; intercom, switches, receptacles.	.042	1.43	1.66	3.09
10 Overhead	Contractor's overhead and profit and architect's fee.		10.90	8.42	19.32
	Total		56.35	43.55	99.90

RESIDENTIAL | Luxury | 2-1/2 Story

- Unique residence built from an architect's plan
- Single family — 1 full bath, 1 half bath, 1 kitchen
- No basement
- Cedar shakes on roof
- Forced hot air heat/air conditioning
- Gypsum wallboard interior finishes
- Many special features
- Extraordinary materials and workmanship

Note: The illustration shown may contain some optional components (for example: garages and/or fireplaces) whose costs are shown in the modifications, adjustments, & alternatives below or at the end of the square foot section.

©Larry W. Garnett & Associates, Inc

Base cost per square foot of living area

Exterior Wall	Living Area										
	1500	1800	2100	2500	3000	3500	4000	4500	5000	5500	6000
Wood Siding - Wood Frame	142.85	128.50	120.25	114.25	105.75	99.80	93.80	91.00	88.45	85.90	83.15
Brick Veneer - Wood Frame	154.45	139.30	129.90	123.30	113.95	107.30	100.70	97.55	94.70	91.85	88.85
Solid Brick	171.90	155.35	144.30	136.90	126.35	118.45	111.10	107.40	104.10	100.80	97.45
Solid Stone	169.25	152.90	142.10	134.85	124.50	116.80	109.55	105.85	102.65	99.50	96.10
Finished Basement, Add	20.90	22.40	21.10	20.40	19.40	18.60	18.00	17.55	17.20	16.90	16.60
Unfinished Basement, Add	9.45	8.90	8.25	7.95	7.50	7.05	6.75	6.60	6.40	6.25	6.15

Modifications

Add to the total cost

Upgrade Kitchen Cabinets	$ + 1254
Solid Surface Countertops	
Full Bath - including plumbing, wall and floor finishes	+ 6869
Half Bath - including plumbing, wall and floor finishes	+ 4139
Two Car Attached Garage	+ 24,327
Two Car Detached Garage	+ 27,407
Fireplace & Chimney	+ 8110

Adjustments

For multi family - add to total cost

Additional Kitchen	$ + 15,820
Additional Full Bath & Half Bath	+ 11,008
Additional Entry & Exit	+ 2172
Separate Heating & Air Conditioning	+ 5440
Separate Electric	+ 1718

*For Townhouse/Rowhouse -
Multiply cost per square foot by*

Inner Unit	.86
End Unit	.93

Alternatives

Add to or deduct from the cost per square foot of living area

Heavyweight Asphalt Shingles	– .95
Clay Tile Roof	+ .80
Slate Roof	+ 1.40
Upgrade Ceilings to Textured Finish	+ .44
Air Conditioning, in Heating Ductwork	Base System
Heating Systems, Hot Water	+ 1.44
Heat Pump	+ 2.48
Electric Heat	– 3.30
Not Heated	– 3.26

Additional upgrades or components

Kitchen Cabinets & Countertops	Page 77
Bathroom Vanities	78
Fireplaces & Chimneys	78
Windows, Skylights & Dormers	78
Appliances	79
Breezeways & Porches	79
Finished Attic	79
Garages	80
Site Improvements	80
Wings & Ells	76

Luxury 2-1/2 Story

Living Area - 3000 S.F.
Perimeter - 148 L.F.

			Labor-Hours	Cost Per Square Foot Of Living Area		
				Mat.	Labor	Total
1	Site Work	Site preparation for slab; 4' deep trench excavation for foundation wall.	.055		.55	.55
2	Foundation	Continuous reinforced concrete footing 8" deep x 18" wide; dampproofed and insulated reinforced concrete foundation wall, 12" thick, 4' deep; 4" concrete slab on 4" crushed stone base and polyethylene vapor barrier, trowel finish.	.067	3.34	3.45	6.79
3	Framing	Exterior walls - 2" x 6" wood studs, 16" O.C.; 5/8" plywood sheathing; 2" x 10" rafters 16" O.C. with 5/8" plywood sheathing, 6 in 12 pitch; 2" x 8" ceiling joists 16" O.C.; 2" x 12" floor joists 16" O.C. with 5/8" plywood subfloor; 5/8" plywood subfloor on 1" x 3" wood sleepers 16" O.C.	.209	6.45	7.50	13.95
4	Exterior Walls	Horizontal beveled wood siding; building paper; 6" batt insulation; wood double hung windows; 3 solid core wood exterior doors; storms and screens.	.405	10.47	3.76	14.23
5	Roofing	Red cedar shingles; #15 felt building paper; aluminum gutters, downspouts and drip edge; copper flashings.	.039	1.42	1.31	2.73
6	Interiors	Walls and ceilings - 5/8" gypsum wallboard, skim coat plaster, painted with primer and 2 coats; hardwood baseboard and trim, sanded and finished; hardwood floor 70%, ceramic tile with underlayment 20%, vinyl tile with underlayment 10%; wood panel interior doors, primed and painted with 2 coat	.341	15.75	14.70	30.45
7	Specialties	Luxury grade kitchen cabinets - 25 L.F. wall and base with solid surface counter top and kitchen sink; 6 L.F. bathroom vanity; 75 gallon electric water heater; medicine cabinet.	.119	4.43	1.06	5.49
8	Mechanical	Gas fired warm air heat/air conditioning; one full bath including: bathtub, corner shower; built in lavatory and water closet; one 1/2 bath including: built in lavatory and water closet.	.103	5.23	2.72	7.95
9	Electrical	200 Amp. service; romex wiring; fluorescent and incandescent lighting fixtures; intercom, switches, receptacles.	.054	1.46	1.68	3.14
10	Overhead	Contractor's overhead and profit and architect's fee.		11.65	8.82	20.47
		Total		60.20	45.55	**105.75**

RESIDENTIAL | Luxury | 3 Story

- Unique residence built from an architect's plan
- Single family — 1 full bath, 1 half bath, 1 kitchen
- No basement
- Cedar shakes on roof
- Forced hot air heat/air conditioning
- Gypsum wallboard interior finishes
- Many special features
- Extraordinary materials and workmanship

Note: The illustration shown may contain some optional components (for example: garages and/or fireplaces) whose costs are shown in the modifications, adjustments, & alternatives below or at the end of the square foot section.

Base cost per square foot of living area

Exterior Wall	Living Area										
	1500	1800	2100	2500	3000	3500	4000	4500	5000	5500	6000
Wood Siding - Wood Frame	141.80	127.70	120.95	115.30	106.50	102.15	96.85	91.30	89.20	87.10	84.80
Brick Veneer - Wood Frame	153.90	138.85	131.40	125.25	115.60	110.75	104.70	98.45	96.20	93.85	91.15
Solid Brick	172.05	155.65	147.05	140.25	129.20	123.65	116.40	109.25	106.70	103.90	100.70
Solid Stone	169.30	153.15	144.70	137.95	127.15	121.65	114.60	107.70	105.05	102.40	99.20
Finished Basement, Add	18.40	19.70	18.85	18.30	17.35	16.80	16.10	15.60	15.35	15.10	14.75
Unfinished Basement, Add	8.30	7.80	7.40	7.20	6.75	6.50	6.15	5.90	5.75	5.65	5.50

Modifications

Add to the total cost

Upgrade Kitchen Cabinets	$ + 1254
Solid Surface Countertops	
Full Bath - including plumbing, wall and floor finishes	+ 6869
Half Bath - including plumbing, wall and floor finishes	+ 4139
Two Car Attached Garage	+ 24,327
Two Car Detached Garage	+ 27,407
Fireplace & Chimney	+ 8110

Adjustments

For multi family - add to total cost

Additional Kitchen	$ + 15,820
Additional Full Bath & Half Bath	+ 11,008
Additional Entry & Exit	+ 2172
Separate Heating & Air Conditioning	+ 5440
Separate Electric	+ 1718

For Townhouse/Rowhouse - Multiply cost per square foot by

Inner Unit	.84
End Unit	.92

Alternatives

Add to or deduct from the cost per square foot of living area

Heavyweight Asphalt Shingles	– .75
Clay Tile Roof	+ .65
Slate Roof	+ 1.10
Upgrade Ceilings to Textured Finish	+ .44
Air Conditioning, in Heating Ductwork	Base System
Heating Systems, Hot Water	+ 1.44
Heat Pump	+ 2.48
Electric Heat	– 3.30
Not Heated	– 3.16

Additional upgrades or components

Kitchen Cabinets & Countertops	Page 77
Bathroom Vanities	78
Fireplaces & Chimneys	78
Windows, Skylights & Dormers	78
Appliances	79
Breezeways & Porches	79
Finished Attic	79
Garages	80
Site Improvements	80
Wings & Ells	76

Luxury 3 Story

Living Area - 3000 S.F.
Perimeter - 135 L.F.

		Labor-Hours	Cost Per Square Foot Of Living Area		
			Mat.	Labor	Total
1 Site Work	Site preparation for slab; 4' deep trench excavation for foundation wall.	.055		.55	.55
2 Foundation	Continuous reinforced concrete footing 8" deep x 18" wide; dampproofed and insulated reinforced concrete foundation wall, 12" thick, 4' deep; 4" concrete slab on 4" crushed stone base and polyethylene vapor barrier, trowel finish.	.063	3.01	3.14	6.15
3 Framing	Exterior walls - 2" x 6" wood studs, 16" O.C.; 5/8" plywood sheathing; 2" x 10" rafters 16" O.C. with 5/8" plywood sheathing, 6 in 12 pitch; 2" x 8" ceiling joists 16" O.C.; 2" x 12" floor joists 16" O.C. with 5/8" plywood subfloor; 5/8" plywood subfloor on 1" x 3" wood sleepers 16" O.C.	.225	6.54	7.60	14.14
4 Exterior Walls	Horizontal beveled wood siding; building paper; 6" batt insulation; wood double hung windows; 3 solid core wood exterior doors; storms and screens.	.454	11.40	4.13	15.53
5 Roofing	Red cedar shingles; #15 felt building paper; aluminum gutters, downspouts and drip edge; copper flashings.	.034	1.23	1.13	2.36
6 Interiors	Walls and ceilings - 5/8" gypsum wallboard, skim coat plaster, painted with primer and 2 coats; hardwood baseboard and trim, sanded and finished; hardwood floor 70%, ceramic tile with underlayment 20%, vinyl tile with underlayment 10%; wood panel interior doors, primed and painted with 2 coat	.390	15.78	14.80	30.58
7 Specialties	Luxury grade kitchen cabinets - 25 L.F. wall and base with solid surface counter top and kitchen sink; 6 L.F. bathroom vanity; 75 gallon electric water heater; medicine cabinet.	.119	4.43	1.06	5.49
8 Mechanical	Gas fired warm air heat/air conditioning; one full bath including: bathtub, corner shower; built in lavatory and water closet; one 1/2 bath including: built in lavatory and water closet.	.103	5.23	2.72	7.95
9 Electrical	200 Amp. service; romex wiring; fluorescent and incandescent lighting fixtures; intercom, switches, receptacles.	.053	1.46	1.68	3.14
10 Overhead	Contractor's overhead and profit and architect's fees.		11.77	8.84	20.61
	Total		60.85	45.65	**106.50**

RESIDENTIAL　　Luxury　|　Bi-Level

- Unique residence built from an architect's plan
- Single family — 1 full bath, 1 half bath, 1 kitchen
- No basement
- Cedar shakes on roof
- Forced hot air heat/air conditioning
- Gypsum wallboard interior finishes
- Many special features
- Extraordinary materials and workmanship

Note: The illustration shown may contain some optional components (for example: garages and/or fireplaces) whose costs are shown in the modifications, adjustments, & alternatives below or at the end of the square foot section.

Base cost per square foot of living area

Exterior Wall	\multicolumn{11}{c}{Living Area}
	1200	1400	1600	1800	2000	2400	2800	3200	3600	4000	4400
Wood Siding - Wood Frame	138.70	130.40	124.60	119.60	113.90	106.10	99.50	95.20	92.55	89.85	87.60
Brick Veneer - Wood Frame	147.05	138.25	132.10	126.70	120.75	112.25	105.10	100.40	97.55	94.60	92.20
Solid Brick	159.65	150.05	143.30	137.20	130.90	121.40	113.45	108.25	105.10	101.70	99.10
Solid Stone	157.80	148.30	141.65	135.65	129.45	120.00	112.20	107.00	104.00	100.65	98.00
Finished Basement, Add	26.20	28.40	27.60	26.75	26.15	24.90	23.85	23.10	22.70	22.25	21.90
Unfinished Basement, Add	11.75	11.15	10.80	10.40	10.10	9.50	9.00	8.65	8.50	8.20	8.10

Modifications

Add to the total cost

Upgrade Kitchen Cabinets	$ + 1254
Solid Surface Countertops	
Full Bath - including plumbing, wall and floor finishes	+ 6869
Half Bath - including plumbing, wall and floor finishes	+ 4139
Two Car Attached Garage	+ 24,327
Two Car Detached Garage	+ 27,407
Fireplace & Chimney	+ 6760

Adjustments

For multi family - add to total cost

Additional Kitchen	$ + 15,820
Additional Full Bath & Half Bath	+ 11,008
Additional Entry & Exit	+ 2172
Separate Heating & Air Conditioning	+ 5440
Separate Electric	+ 1718

For Townhouse/Rowhouse - Multiply cost per square foot by

Inner Unit	.89
End Unit	.94

Alternatives

Add to or deduct from the cost per square foot of living area

Heavyweight Asphalt Shingles	– 1.10
Clay Tile Roof	+ .95
Slate Roof	+ 1.60
Upgrade Ceilings to Textured Finish	+ .44
Air Conditioning, in Heating Ductwork	Base System
Heating Systems, Hot Water	+ 1.59
Heat Pump	+ 2.40
Electric Heat	– 1.68
Not Heated	– 3.26

Additional upgrades or components

Kitchen Cabinets & Countertops	Page 77
Bathroom Vanities	78
Fireplaces & Chimneys	78
Windows, Skylights & Dormers	78
Appliances	79
Breezeways & Porches	79
Finished Attic	79
Garages	80
Site Improvements	80
Wings & Ells	76

Luxury Bi-Level

Living Area - 3200 S.F.
Perimeter - 163 L.F.

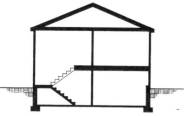

		Labor-Hours	Cost Per Square Foot Of Living Area		
			Mat.	Labor	Total
1 Site Work	Excavation for lower level, 4' deep. Site preparation for slab.	.024		.52	.52
2 Foundation	Continuous reinforced concrete footing 8" deep x 18" wide; dampproofed and insulated reinforced concrete foundation wall, 12" thick, 4' deep; 4" concrete slab on 4" crushed stone base and polyethylene vapor barrier, trowel finish.	.058	3.82	3.83	7.65
3 Framing	Exterior walls - 2" x 6" wood studs, 16" O.C.; 5/8" plywood sheathing; 2" x 10" rafters 16" O.C. with 5/8" plywood sheathing, 6 in 12 pitch; 2" x 8" ceiling joists 16" O.C.; 2" x 12" floor joists 16" O.C. with 5/8" plywood subfloor; 5/8" plywood subfloor on 1" x 3" wood sleepers 16" O.C.	.232	5.97	6.97	12.94
4 Exterior Walls	Horizontal beveled wood siding; building paper; 6" batt insulation; wood double hung windows; 3 solid core wood exterior doors; storms and screens.	.185	7.03	2.53	9.56
5 Roofing	Red cedar shingles: #15 felt building paper; aluminum gutters, downspouts and drip edge; copper flashings.	.042	1.85	1.71	3.56
6 Interiors	Walls and ceilings - 5/8" gypsum wallboard, skim coat plaster, painted with primer and 2 coats; hardwood baseboard and trim, sanded and finished; hardwood floor 70%, ceramic tile with underlayment 20%; vinyl tile with underlayment 10%; wood panel interior doors, primed and painted with 2 coat	.238	13.78	12.83	26.61
7 Specialties	Luxury grade kitchen cabinets - 25 L.F. wall and base with solid surface counter top and kitchen sink; 6 L.F. bathroom vanity; 75 gallon electric water heater; medicine cabinet.	.056	4.14	.99	5.13
8 Mechanical	Gas fired warm air heat/air conditioning; one full bath including: bathtub, corner shower; built in lavatory and water closet; one 1/2 bath including: built in lavatory and water closet.	.071	5.00	2.69	7.69
9 Electrical	200 Amp. service; romex wiring; fluorescent and incandescent lighting fixtures; intercom, switches, receptacles.	.042	1.43	1.66	3.09
10 Overhead	Contractor's overhead and profit and architect's fees.		10.33	8.12	18.45
	Total		53.35	41.85	95.20

RESIDENTIAL | Luxury | Tri-Level

- Unique residence built from an architect's plan
- Single family — 1 full bath, 1 half bath, 1 kitchen
- No basement
- Cedar shakes on roof
- Forced hot air heat/air conditioning
- Gypsum wallboard interior finishes
- Many special features
- Extraordinary materials and workmanship

Note: The illustration shown may contain some optional components (for example: garages and/or fireplaces) whose costs are shown in the modifications, adjustments, & alternatives below or at the end of the square foot section.

©Home Planners, Inc.

Base cost per square foot of living area

Exterior Wall	Living Area										
	1500	1800	2100	2400	2800	3200	3600	4000	4500	5000	5500
Wood Siding - Wood Frame	130.75	121.10	113.00	107.60	103.70	99.10	94.15	92.10	87.50	85.00	82.15
Brick Veneer - Wood Frame	138.30	127.90	119.15	113.40	109.35	104.30	98.95	96.75	91.80	89.00	86.00
Solid Brick	149.65	138.15	128.45	122.15	117.70	112.00	106.20	103.75	98.25	95.10	91.80
Solid Stone	147.90	136.60	127.00	120.85	116.40	110.90	105.10	102.70	97.30	94.20	90.85
Finished Basement, Add*	30.75	33.05	31.70	30.80	30.10	29.30	28.50	28.15	27.35	26.85	26.40
Unfinished Basement, Add*	13.50	12.65	11.95	11.55	11.30	10.85	10.45	10.30	9.95	9.65	9.45

*Basement under middle level only.

Modifications

Add to the total cost

Upgrade Kitchen Cabinets	$ + 1254
Solid Surface Countertops	
Full Bath - including plumbing, wall and floor finishes	+ 6869
Half Bath - including plumbing, wall and floor finishes	+ 4139
Two Car Attached Garage	+ 24,327
Two Car Detached Garage	+ 27,407
Fireplace & Chimney	+ 6760

Adjustments

For multi family - add to total cost

Additional Kitchen	$ + 15,820
Additional Full Bath & Half Bath	+ 11,008
Additional Entry & Exit	+ 2172
Separate Heating & Air Conditioning	+ 5440
Separate Electric	+ 1718

For Townhouse/Rowhouse - Multiply cost per square foot by

Inner Unit	.86
End Unit	.93

Alternatives

Add to or deduct from the cost per square foot of living area

Heavyweight Asphalt Shingles	– 1.55
Clay Tile Roof	+ 1.35
Slate Roof	+ 2.35
Upgrade Ceilings to Textured Finish	+ .44
Air Conditioning, in Heating Ductwork	Base System
Heating Systems, Hot Water	+ 1.54
Heat Pump	+ 2.50
Electric Heat	– 1.49
Not Heated	– 3.16

Additional upgrades or components

Kitchen Cabinets & Countertops	Page 77
Bathroom Vanities	78
Fireplaces & Chimneys	78
Windows, Skylights & Dormers	78
Appliances	79
Breezeways & Porches	79
Finished Attic	79
Garages	80
Site Improvements	80
Wings & Ells	76

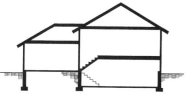

Luxury Tri-Level

Living Area - 3600 S.F.
Perimeter - 207 L.F.

			Labor-Hours	Cost Per Square Foot Of Living Area		
				Mat.	Labor	Total
1	Site Work	Site preparation for slab; 4' deep trench excavation for foundation wall, excavation for lower level, 4' deep.	.021		.46	.46
2	Foundation	Continuous reinforced concrete footing 8" deep x 18" wide; dampproofed and insulated reinforced concrete foundation wall, 12" thick, 4' deep; 4" concrete slab on 4" crushed stone base and polyethylene vapor barrier, trowel finish.	.109	4.58	4.40	8.98
3	Framing	Exterior walls - 2" x 6" wood studs, 16" O.C.; 5/8" plywood sheathing; 2" x 10" rafters 16" O.C. with 5/8" plywood sheathing, 6 in 12 pitch; 2" x 8" ceiling joists 16" O.C.; 2" x 12" floor joists 16" O.C. with 5/8" plywood subfloor; 5/8" plywood subfloor on 1" x 3" wood sleepers 16" O.C.	.204	5.94	6.98	12.92
4	Exterior Walls	Horizontal beveled wood siding; building paper; 6" batt insulation; wood double hung windows; 3 solid core wood exterior doors; storms and screens.	.181	6.77	2.40	9.17
5	Roofing	Red cedar shingles; #15 felt building paper; aluminum gutters, downspouts and drip edge; copper flashings.	.056	2.46	2.27	4.73
6	Interiors	Walls and ceilings - 5/8" gypsum wallboard, skim coat plaster, painted with primer and 2 coats; hardwood baseboard and trim, sanded and finished; hardwood floor 70%, ceramic tile with underlayment 20%, vinyl tile with underlayment 10%; wood panel interior doors, primed and painted with 2 coat	.217	12.91	11.97	24.88
7	Specialties	Luxury grade kitchen cabinets - 25 L.F. wall and base with solid surface counter top and kitchen sink; 6 L.F. bathroom vanity; 75 gallon electric water heater; medicine cabinet.	.048	3.69	.88	4.57
8	Mechanical	Gas fired warm air heat/air conditioning; one full bath including: bathtub, corner shower; built in lavatory and water closet; one 1/2 bath including: built in lavatory and water closet.	.057	4.63	2.57	7.20
9	Electrical	200 Amp. service; romex wiring; fluorescent and incandescent lighting fixtures; intercom, switches, receptacles.	.039	1.40	1.62	3.02
10	Overhead	Contractor's overhead and profit and architect's fees.		10.17	8.05	18.22
		Total		52.55	41.60	**94.15**

RESIDENTIAL | Luxury | Wings & Ells

1 Story — Base cost per square foot of living area

Exterior Wall	Living Area							
	50	100	200	300	400	500	600	700
Wood Siding - Wood Frame	211.40	163.55	142.70	120.70	113.95	109.80	107.05	108.00
Brick Veneer - Wood Frame	239.65	183.70	159.50	131.90	124.00	119.20	116.05	116.65
Solid Brick	281.95	213.95	184.70	148.70	139.05	133.30	129.50	129.55
Solid Stone	275.60	209.35	180.95	146.20	136.75	131.25	127.45	127.60
Finished Basement, Add	82.05	74.25	66.65	53.80	51.25	49.75	48.70	48.00
Unfinished Basement, Add	42.05	32.45	28.45	21.75	20.40	19.60	19.10	18.70

1-1/2 Story — Base cost per square foot of living area

Exterior Wall	Living Area							
	100	200	300	400	500	600	700	800
Wood Siding - Wood Frame	168.20	137.35	118.55	107.70	102.00	99.25	95.80	94.80
Brick Veneer - Wood Frame	193.40	157.50	135.35	120.85	114.10	110.70	106.60	105.55
Solid Brick	231.15	187.75	160.50	140.45	132.20	127.85	122.85	121.55
Solid Stone	225.50	183.20	156.80	137.50	129.50	125.30	120.35	119.15
Finished Basement, Add	53.95	53.35	48.20	42.55	41.00	40.00	39.10	38.95
Unfinished Basement, Add	26.75	22.80	20.05	17.15	16.30	15.80	15.30	15.25

2 Story — Base cost per square foot of living area

Exterior Wall	Living Area							
	100	200	400	600	800	1000	1200	1400
Wood Siding - Wood Frame	165.35	125.00	107.30	90.55	84.75	81.30	78.95	80.20
Brick Veneer - Wood Frame	193.55	145.15	124.10	101.75	94.85	90.70	87.95	88.85
Solid Brick	235.85	175.40	149.25	118.50	109.95	104.80	101.45	101.80
Solid Stone	229.50	170.85	145.45	116.00	107.60	102.70	99.40	99.80
Finished Basement, Add	41.05	37.25	33.40	27.00	25.70	24.95	24.45	24.10
Unfinished Basement, Add	21.05	16.25	14.25	10.90	10.25	9.85	9.55	9.35

Base costs do not include bathroom or kitchen facilities. Use Modifications/Adjustments/Alternatives on pages 77-80 where appropriate.

RESIDENTIAL — Modifications/Adjustments/Alternatives

Kitchen cabinets - Base units, hardwood *(Cost per Unit)*

	Economy	Average	Custom	Luxury
24" deep, 35" high,				
One top drawer,				
One door below				
12" wide	$182	$242	$320	$425
15" wide	191	254	340	445
18" wide	206	275	365	480
21" wide	224	298	395	520
24" wide	244	325	430	570
Four drawers				
12" wide	311	415	550	725
15" wide	259	345	460	605
18" wide	285	380	505	665
24" wide	308	410	545	720
Two top drawers,				
Two doors below				
27" wide	266	355	470	620
30" wide	289	385	510	675
33" wide	319	425	565	745
36" wide	330	440	585	770
42" wide	349	465	620	815
48" wide	364	485	645	850
Range or sink base				
(Cost per unit)				
Two doors below				
30" wide	266	355	470	620
33" wide	281	375	500	655
36" wide	289	385	510	675
42" wide	308	410	545	720
48" wide	319	425	565	745
Corner Base Cabinet				
(Cost per unit)				
36" wide	375	500	665	875
Lazy Susan *(Cost per unit)*				
With revolving door	375	500	665	875

Kitchen cabinets - Wall cabinets, hardwood *(Cost per Unit)*

	Economy	Average	Custom	Luxury
12" deep, 2 doors				
12" high				
30" wide	$150	$200	$ 265	$ 350
36" wide	173	230	305	405
15" high				
30" wide	160	213	285	375
33" wide	171	228	305	400
36" wide	179	239	320	420
24" high				
30" wide	194	258	345	450
36" wide	212	283	375	495
42" wide	229	305	405	535
30" high, 1 door				
12" wide	140	186	245	325
15" wide	155	207	275	360
18" wide	169	225	300	395
24" wide	188	250	335	440
30" high, 2 doors				
27" wide	219	292	390	510
30" wide	220	293	390	515
36" wide	248	330	440	580
42" wide	270	360	480	630
48" wide	300	400	530	700
Corner wall, 30" high				
24" wide	161	214	285	375
30" wide	190	253	335	445
36" wide	205	273	365	480
Broom closet				
84" high, 24" deep				
18" wide	405	540	720	945
Oven Cabinet				
84" high, 24" deep				
27" wide	585	780	1035	1365

Kitchen countertops *(Cost per L.F.)*

	Economy	Average	Custom	Luxury
Solid Surface				
24" wide, no backsplash	89	119	160	210
with backsplash	97	129	170	225
Stock plastic laminate, 24" wide				
with backsplash	17	22	30	40
Custom plastic laminate, no splash				
7/8" thick, alum. molding	24	33	45	55
1-1/4" thick, no splash	33	44	60	75
Marble				
1/2" - 3/4" thick w/splash	45	60	80	105
Maple, laminated				
1-1/2" thick w/splash	65	87	115	150
Stainless steel				
(per S.F.)	119	159	210	280
Cutting blocks, recessed				
16" x 20" x 1" (each)	88	117	155	205

RESIDENTIAL: Modifications/Adjustments/Alternatives

Vanity bases (Cost per Unit)

	Economy	Average	Custom	Luxury
2 door, 30" high, 21" deep				
24" wide	188	251	335	440
30" wide	220	293	390	515
36" wide	289	385	510	675
48" wide	341	455	605	795

Solid surface vanity tops (Cost Each)

	Economy	Average	Custom	Luxury
Center bowl				
22" x 25"	$251	$271	$293	$316
22" x 31"	287	310	335	362
22" x 37"	330	356	385	416
22" x 49"	405	437	472	510

Fireplaces & Chimneys (Cost per Unit)

	1-1/2 Story	2 Story	3 Story
Economy (prefab metal)			
Exterior chimney & 1 fireplace	$4650	$5140	$5635
Interior chimney & 1 fireplace	4455	4955	5185
Average (masonry)			
Exterior chimney & 1 fireplace	4635	5170	5875
Interior chimney & 1 fireplace	4345	4875	5310
For more than 1 flue, add	335	570	950
For more than 1 fireplace, add	3285	3285	3285
Custom (masonry)			
Exterior chimney & 1 fireplace	4825	5445	6150
Interior chimney & 1 fireplace	4525	5120	5520
For more than 1 flue, add	380	655	890
For more than 1 fireplace, add	3465	3465	3465
Luxury (masonry)			
Exterior chimney & 1 fireplace	6760	7415	8110
Interior chimney & 1 fireplace	6450	7055	7465
For more than 1 flue, add	560	930	1300
For more than 1 fireplace, add	5330	5330	5330

Windows and Skylights (Cost Each)

	Economy	Average	Custom	Luxury
Fixed Picture Windows				
3'-6" x 4'-0"	$ 514	$ 556	$ 600	$ 648
4'-0" x 6'-0"	879	949	1025	1107
5'-0" x 6'-0"	986	1065	1150	1242
6'-0" x 6'-0"	986	1065	1150	1242
Bay/Bow Windows				
8'-0" x 5'-0"	1265	1366	1475	1593
10'-0" x 5'-0"	1436	1551	1675	1809
10'-0" x 6'-0"	2315	2500	2700	2916
12'-0" x 6'-0"	2936	3171	3425	3699
Palladian Windows				
3'-2" x 6'-4"		1782	1925	2079
4'-0" x 6'-0"		2106	2275	2457
5'-5" x 6'-10"		2523	2725	2943
8'-0" x 6'-0"		3032	3275	3537
Skylights				
46" x 21-1/2"	401	433	573	619
46" x 28"	433	468	603	651
57" x 44"	530	572	712	769

Dormers (Cost/S.F. of plan area)

	Economy	Average	Custom	Luxury
Framing and Roofing Only				
Gable dormer, 2" x 6" roof frame	$24	$27	$30	$47
2" x 8" roof frame	25	28	31	50
Shed dormer, 2" x 6" roof frame	15	17	19	31
2" x 8" roof frame	17	18	20	32
2" x 10" roof frame	18	20	22	33

RESIDENTIAL — Modifications/Adjustments/Alternatives

Appliances *(Cost per Unit)*

	Economy	Average	Custom	Luxury
Range				
30" free standing, 1 oven	$350	$1088	$1456	$1825
2 oven	1800	1875	1913	1950
30" built-in, 1 oven	580	1265	1608	1950
2 oven	1600	1875	2013	2150
21" free standing				
1 oven	390	438	461	485
Counter Top Ranges				
4 burner standard	300	538	656	775
As above with griddle	635	855	965	1075
Microwave Oven	190	420	535	650
Combination Range, Refrigerator, Sink				
30" wide	1175	2300	2863	3425
60" wide	2775	3192	3400	3608
72" wide	3825	4399	4686	4973
Comb. Range, Refrig., Sink, Microwave Oven & Ice Maker	6153	7076	7538	7999
Compactor				
4 to 1 compaction	555	623	656	690
Deep Freeze				
15 to 23 C.F.	575	668	714	760
30 C.F.	950	1063	1119	1175
Dehumidifier, portable, auto.				
15 pint	175	202	215	228
30 pint	213	245	261	277
Washing Machine, automatic	470	1048	1336	1625
Water Heater				
Electric, glass lined				
30 gal.	475	583	636	690
80 gal.	930	1190	1320	1450
Water Heater, Gas, glass lined				
30 gal.	775	950	1038	1125
50 gal.	965	1195	1310	1425
Water Softener, automatic				
30 grains/gal.	665	765	815	865
100 grains/gal.	905	1041	1109	1177
Dishwasher, built-in				
2 cycles	415	498	539	580
4 or more cycles	435	630	728	825
Dryer, automatic	495	760	893	1025
Garage Door Opener	375	453	491	530
Garbage Disposal	104	164	194	224
Heater, Electric, built-in				
1250 watt ceiling type	178	220	241	262
1250 watt wall type	212	247	265	282
Wall type w/blower				
1500 watt	240	330	375	420
3000 watt	420	462	483	504
Hood For Range, 2 speed				
30" wide	134	467	634	800
42" wide	355	590	708	825
Humidifier, portable				
7 gal. per day	176	203	216	229
15 gal. per day	212	244	260	276
Ice Maker, automatic				
13 lb. per day	990	1139	1213	1287
51 lb. per day	1375	1582	1685	1788
Refrigerator, no frost				
10-12 C.F.	500	568	601	635
14-16 C.F.	545	563	571	580
18-20 C.F.	645	823	911	1000
21-29 C.F.	850	1750	2200	2650
Sump Pump, 1/3 H.P.	236	326	370	415

Breezeway *(Cost per S.F.)*

Class	Type	Area (S.F.)			
		50	100	150	200
Economy	Open	$21.15	$18.00	$15.10	$15.10
	Enclosed	101.50	78.40	65.10	65.10
Average	Open	26.15	23.05	20.15	20.15
	Enclosed	111.20	82.85	67.80	67.80
Custom	Open	37.35	32.85	28.60	28.60
	Enclosed	153.35	114.10	93.25	93.25
Luxury	Open	38.55	33.80	30.45	30.45
	Enclosed	155.20	115.10	93.10	83.35

Porches *(Cost per S.F.)*

Class	Type	Area (S.F.)				
		25	50	100	200	300
Economy	Open	$60.65	$40.60	$31.70	$26.85	$22.90
	Enclosed	121.20	84.45	63.85	49.80	42.65
Average	Open	73.60	46.85	35.95	29.90	29.90
	Enclosed	145.45	98.55	74.55	57.70	48.90
Custom	Open	95.35	63.25	47.60	41.60	37.30
	Enclosed	189.55	129.50	98.55	76.75	66.10
Luxury	Open	102.05	66.70	49.25	44.25	39.45
	Enclosed	199.75	140.30	104.10	80.85	69.60

Finished attic *(Cost per S.F.)*

Class	Area (S.F.)				
	400	500	600	800	1000
Economy	$16.20	$15.65	$15.00	$14.75	$14.20
Average	24.90	24.35	23.75	23.45	22.80
Custom	30.40	29.70	29.05	28.60	28.00
Luxury	38.30	37.40	36.50	35.65	35.05

Alarm system *(Cost per System)*

	Burglar Alarm	Smoke Detector
Economy	$375	$65
Average	430	89
Custom	731	149
Luxury	1075	178

Sauna, prefabricated
(Cost per unit, including heater and controls — 7' high)

Size	Cost
6' x 4'	$4675
6' x 5'	5225
6' x 6'	5600
6' x 9'	6950
8' x 10'	9125
8' x 12'	10,700
10' x 12'	11,400

RESIDENTIAL — Modifications/Adjustments/Alternatives

Garages *

(Costs include exterior wall systems comparable with the quality of the residence. Included in the cost is an allowance for one personnel door, manual overhead door(s) and electrical fixture.)

Class	Detached			Attached			Built-in		Basement	
	One Car	Two Car	Three Car	One Car	Two Car	Three Car	One Car	Two Car	One Car	Two Car
Economy										
Wood	$12,594	$19,230	$25,867	$9770	$16,838	$23,474	$-1733	$-3465	$1354	$1768
Masonry	17,811	25,759	33,708	13,034	21,414	29,363	-2373	-4747		
Average										
Wood	14,035	21,034	28,033	10,672	18,102	25,101	-1910	-3819	1534	2127
Masonry	17,811	25,759	33,708	13,034	21,414	29,363	-2373	-4067		
Custom										
Wood	15,764	24,140	32,516	12,313	21,296	29,673	-3255	-3402	2322	3703
Masonry	19,899	29,315	38,731	14,901	24,923	34,339	-3763	-4418		
Luxury										
Wood	17,633	27,407	37,180	13,947	24,327	34,100	-3332	-3557	3061	4842
Masonry	22,871	33,961	45,052	17,224	28,921	40,012	-3976	-4844		

*See the Introduction to this section for definitions of garage types.

Swimming pools (Cost per S.F.)

Residential	(includes equipment)
In-ground	$22.00 - 56.50
Deck equipment	1.30
Paint pool, preparation & 3 coats (epoxy)	3.28
Rubber base paint	3.03
Pool Cover	.78
Swimming Pool Heaters	(Cost per unit)
(not including wiring, external piping, base or pad)	
Gas	
155 MBH	$2325
190 MBH	2775
500 MBH	9600
Electric	
15 KW 7200 gallon pool	2675
24 KW 9600 gallon pool	3500
54 KW 24,000 gallon pool	5250

Wood and coal stoves

Wood Only	
Free Standing (minimum)	$1725
Fireplace Insert (minimum)	1739
Coal Only	
Free Standing	$1965
Fireplace Insert	2151
Wood and Coal	
Free Standing	$4041
Fireplace Insert	4141

Sidewalks (Cost per S.F.)

Concrete, 3000 psi with wire mesh	4" thick	$3.45
	5" thick	4.23
	6" thick	4.78
Precast concrete patio blocks (natural)	2" thick	9.30
Precast concrete patio blocks (colors)	2" thick	9.85
Flagstone, bluestone	1" thick	13.90
Flagstone, bluestone	1-1/2" thick	18.65
Slate (natural, irregular)	3/4" thick	13.70
Slate (random rectangular)	1/2" thick	20.95
Seeding		
Fine grading & seeding includes lime, fertilizer & seed per S.Y.		2.09
Lawn Sprinkler System per S.F.		.83

Fencing (Cost per L.F.)

Chain Link, 4' high, galvanized	$15.75
Gate, 4' high (each)	162.00
Cedar Picket, 3' high, 2 rail	11.20
Gate (each)	159.00
3 Rail, 4' high	13.50
Gate (each)	170.00
Cedar Stockade, 3 Rail, 6' high	13.70
Gate (each)	169.00
Board & Battens, 2 sides 6' high, pine	19.60
6' high, cedar	27.50
No. 1 Cedar, basketweave, 6' high	15.65
Gate, 6' high (each)	195.00

Carport (Cost per S.F.)

Economy	$7.63
Average	11.56
Custom	17.25
Luxury	19.59

Assemblies Section

Table of Contents

Table No.		Page
1	Site Work	83
1-04	Footing Excavation	84
1-08	Foundation Excavation	86
1-12	Utility Trenching	88
1-16	Sidewalk	90
1-20	Driveway	92
1-24	Septic	94
1-60	Chain Link Fence	96
1-64	Wood Fence	97
2	Foundations	99
2-04	Footing	100
2-08	Block Wall	102
2-12	Concrete Wall	104
2-16	Wood Wall Foundation	106
2-20	Floor Slab	108
3	Framing	111
3-02	Floor (Wood)	112
3-04	Floor (Wood)	114
3-06	Floor (Wood)	116
3-08	Exterior Wall	118
3-12	Gable End Roof	120
3-16	Truss Roof	122
3-20	Hip Roof	124
3-24	Gambrel Roof	126
3-28	Mansard Roof	128
3-32	Shed/Flat Roof	130
3-40	Gable Dormer	132
3-44	Shed Dormer	134
3-48	Partition	136
4	Exterior Walls	139
4-02	Masonry Block Wall	140
4-04	Brick/Stone Veneer	142
4-08	Wood Siding	144
4-12	Shingle Siding	146
4-16	Metal & Plastic Siding	148
4-20	Insulation	150
4-28	Double Hung Window	152
4-32	Casement Window	154
4-36	Awning Window	156
4-40	Sliding Window	158
4-44	Bow/Bay Window	160
4-48	Fixed Window	162
4-52	Entrance Door	164
4-53	Sliding Door	166
4-56	Residential Overhead Door	168
4-58	Aluminum Window	170
4-60	Storm Door & Window	172
4-64	Shutters/Blinds	173
5	Roofing	175
5-04	Gable End Roofing	176
5-08	Hip Roof Roofing	178
5-12	Gambrel Roofing	180
5-16	Mansard Roofing	182
5-20	Shed Roofing	184
5-24	Gable Dormer Roofing	186
5-28	Shed Dormer Roofing	188
5-32	Skylight/Skywindow	190
5-34	Built-up Roofing	192
6	Interiors	195
6-04	Drywall & Thincoat Wall	196
6-08	Drywall & Thincoat Ceiling	198
6-12	Plaster & Stucco Wall	200
6-16	Plaster & Stucco Ceiling	202
6-18	Suspended Ceiling	204
6-20	Interior Door	206
6-24	Closet Door	208
6-60	Carpet	210
6-64	Flooring	211
6-90	Stairways	212
7	Specialties	215
7-08	Kitchen	216
7-12	Appliances	218
7-16	Bath Accessories	219
7-24	Masonry Fireplace	220
7-30	Prefabricated Fireplace	222
7-32	Greenhouse	224
7-36	Swimming Pool	225
7-40	Wood Deck	226
8	Mechanical	229
8-04	Two Fixture Lavatory	230
8-12	Three Fixture Bathroom	232
8-16	Three Fixture Bathroom	234
8-20	Three Fixture Bathroom	236
8-24	Three Fixture Bathroom	238
8-28	Three Fixture Bathroom	240
8-32	Three Fixture Bathroom	242
8-36	Four Fixture Bathroom	244
8-40	Four Fixture Bathroom	246
8-44	Five Fixture Bathroom	248
8-60	Gas Fired Heating/Cooling	250
8-64	Oil Fired Heating/Cooling	252
8-68	Hot Water Heating	254
8-80	Rooftop Heating/Cooling	256
9	Electrical	259
9-10	Electric Service	260
9-20	Electric Heating	261
9-30	Wiring Devices	262
9-40	Light Fixtures	263

How to Use the Assemblies Section

Illustration
Each building assembly system is accompanied by a detailed, illustrated description. Each individual component is labeled. Every element involved in the total system function is shown.

Description
Each page includes a brief outline of any special conditions to be used when pricing a system. All units of measure are defined here.

System Definition
Not only are all components broken down for each system, but alternative components can be found on the opposite page. Simply insert any chosen new element into the chart to develop a custom system.

Labor-hours
Total labor-hours for a system can be found by simply multiplying the quantity of the system required times LABOR-HOURS. The resulting figure is the total labor-hours needed to complete the system.
(QUANTITY OF SYSTEM x LABOR-HOURS = TOTAL SYSTEM LABOR-HOURS)

Materials
This column contains the MATERIAL COST of each element. These cost figures include 10% for profit.

Installation
Labor rates include both the INSTALLATION COST of the contractor and the standard contractor's O&P. On the average, the LABOR COST will be 70.6% over the above BARE LABOR COST.

Totals
This row provides the necessary system cost totals. TOTAL SYSTEM COST can be derived by multiplying the TOTAL times each system's SQUARE FOOT ESTIMATE (TOTAL x SQUARE FEET = TOTAL SYSTEM COST).

Work Sheet
Using the SELECTIVE PRICE SHEET on the page opposite each system, it is possible to create estimates with alternative items for any number of systems.

Note:
Throughout this section, the words assembly and system are used interchangeably.

Quantities
Each material in a system is shown with the quantity required for the system unit. For example, the rafters in this system have 1.170 L.F. per S.F. of ceiling area.

Unit of Measure
In the three right-hand columns, each cost figure is adjusted to agree with the unit of measure for the entire system. In this case, COST PER SQUARE FOOT (S.F.) is the common unit of measure. NOTE: In addition, under the UNIT heading, all the elements of each system are defined in relation to the product as a selling commodity. For example, "fascia board" is defined in linear feet, instead of in board feet.

Total
MATERIAL COST + INSTALLATION COST = TOTAL. Work on the table from left to right across cost columns to derive totals.

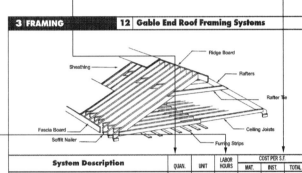

3 | FRAMING — 12 | Gable End Roof Framing Systems

System Description	QUAN.	UNIT	LABOR HOURS	MAT.	INST.	TOTAL
2" X 6" RAFTERS, 16" O.C., 4/12 PITCH						
Rafters, 2" x 6", 16" O.C., 4/12 pitch	1.170	L.F.	.019	.74	.82	1.56
Ceiling joists, 2" x 4", 16" O.C.	1.000	L.F.	.013	.41	.56	.97
Ridge board, 2" x 6"	.050	L.F.	.002	.03	.07	.10
Fascia board, 2" x 6"	.100	L.F.	.005	.07	.23	.30
Rafter tie, 1" x 4", 4' O.C.	.060	L.F.	.001	.03	.05	.08
Soffit nailer (outrigger), 2" x 4", 24" O.C.	.170	L.F.	.004	.07	.19	.26
Sheathing, exterior, plywood, CDX, 1/2" thick	1.170	S.F.	.013	.61	.59	1.20
Furring strips, 1" x 3", 16" O.C.	1.000	L.F.	.023	.25	1	1.25
TOTAL		S.F.	.080	2.21	3.51	5.72
2" X 8" RAFTERS, 16" O.C., 4/12 PITCH						
Rafters, 2" x 8", 16" O.C., 4/12 pitch	1.170	L.F.	.020	1.08	.85	1.93
Ceiling joists, 2" x 6", 16" O.C.	1.000	L.F.	.013	.63	.56	1.19
Ridge board, 2" x 8"	.050	L.F.	.002	.05	.08	.13
Fascia board, 2" x 8"	.100	L.F.	.007	.09	.31	.40
Rafter tie, 1" x 4", 4' O.C.	.060	L.F.	.001	.03	.05	.08
Soffit nailer (outrigger), 2" x 4", 24" O.C.	.170	L.F.	.004	.07	.19	.26
Sheathing, exterior, plywood, CDX, 1/2" thick	1.170	S.F.	.013	.61	.59	1.20
Furring strips, 1" x 3", 16" O.C.	1.000	L.F.	.023	.25	1	1.25
TOTAL		S.F.	.083	2.81	3.63	6.44

The cost of this system is based on the square foot of plan area.
All quantities have been adjusted accordingly.

Description	QUAN.	UNIT	LABOR HOURS	MAT.	INST.	TOTAL

Division 1
Site Work

No part of this publication may be reproduced, stored in a retrieval system, or transmitted in any form or by any means without prior written permission of Reed Construction Data.

1 | SITEWORK — 04 | Footing Excavation Systems

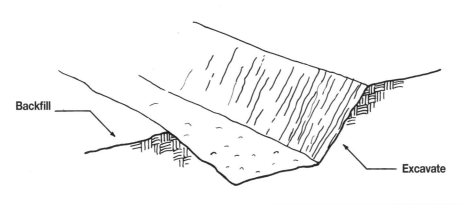

System Description	QUAN.	UNIT	LABOR HOURS	COST EACH MAT.	COST EACH INST.	COST EACH TOTAL
BUILDING, 24' X 38', 4' DEEP						
Cut & chip light trees to 6" diam.	.190	Acre	9.120		541.50	541.50
Excavate, backhoe	174.000	C.Y.	4.641		370.62	370.62
Backfill, dozer, 4" lifts, no compaction	87.000	C.Y.	.580		104.40	104.40
Rough grade, dozer, 30' from building	87.000	C.Y.	.580		104.40	104.40
TOTAL		Ea.	14.921		1,120.92	1,120.92
BUILDING, 26' X 46', 4' DEEP						
Cut & chip light trees to 6" diam.	.210	Acre	10.080		598.50	598.50
Excavate, backhoe	201.000	C.Y.	5.361		428.13	428.13
Backfill, dozer, 4" lifts, no compaction	100.000	C.Y.	.667		120	120
Rough grade, dozer, 30' from building	100.000	C.Y.	.667		120	120
TOTAL		Ea.	16.775		1,266.63	1,266.63
BUILDING, 26' X 60', 4' DEEP						
Cut & chip light trees to 6" diam.	.240	Acre	11.520		684	684
Excavate, backhoe	240.000	C.Y.	6.401		511.20	511.20
Backfill, dozer, 4" lifts, no compaction	120.000	C.Y.	.800		144	144
Rough grade, dozer, 30' from building	120.000	C.Y.	.800		144	144
TOTAL		Ea.	19.521		1,483.20	1,483.20
BUILDING, 30' X 66', 4' DEEP						
Cut & chip light trees to 6" diam.	.260	Acre	12.480		741	741
Excavate, backhoe	268.000	C.Y.	7.148		570.84	570.84
Backfill, dozer, 4" lifts, no compaction	134.000	C.Y.	.894		160.80	160.80
Rough grade, dozer, 30' from building	134.000	C.Y.	.894		160.80	160.80
TOTAL		Ea.	21.416		1,633.44	1,633.44

The costs in this system are on a cost each basis.
Quantities are based on 1'-0" clearance on each side of footing.

Description	QUAN.	UNIT	LABOR HOURS	COST EACH MAT.	COST EACH INST.	COST EACH TOTAL

Since its founding in 1975, Reed Construction Data has developed an online and print portfolio of innovative products and services for the construction, design and manufacturing community. Our products and services are designed specifically to help construction industry professionals advance their businesses with timely, accurate and actionable project, product and cost data. Reed Construction Data is your all-inclusive source of construction information encompassing all phases of the construction process.

Reed Bulletin and Reed Connect™ deliver the most comprehensive, timely and reliable project information to support contractors, distributors and building product manufacturers in identifying, bidding and tracking projects – private and public, general building and civil. Reed Construction Data also offers in-depth construction activity statistics and forecasts covering major project categories, many at the county and metropolitan level.

Reed Bulletin

www.reedbulletin.com

- Project leads targeted by geographic region and formatted by construction stage – available online or in print.
- Locate those hard-to-find jobs that are more profitable to your business.
- Optional automatic e-mail updates sent whenever project details change.
- Download plans and specs online or order print copies.

Reed Research & Analytics

www.buildingteamforecast.com

Reed Construction Forecast

- Delivers timely construction industry activity combining historical data, current year projections and forecasts.
- Modeled at the individual MSA-level to capture changing local market conditions.
- Covers 21 major project categories.

Reed Construction Starts

- Available in a monthly report or as an interactive database.
- Data provided in square footage and dollar value.
- Highly effective and efficient business planning tool.

Market Fundamentals

- Metropolitan area-specific reporting.
- Five-year forecast of industry performance including major projects in development and underway.
- Property types include office, retail, hotel, warehouse and apartment.

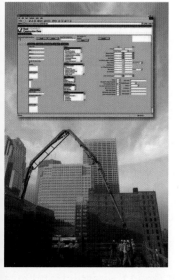

Reed Connect

www.reedconnect.com

- Customized web-based project lead delivery service featuring advanced search capabilities.
- Manage and track actionable leads from planning to quote through winning the job.
- Competitive analysis tool to analyze lost sales opportunities.
- Potential integration with your CRM application.

For more information about Reed Construction Data, please call 877-REED411, visit our website at www.reedconstructiondata.com, or E-mail: marketing@reedbusiness.com

Reed Construction Data
CONNECT • FIRST SOURCE • BULLETIN
RSMEANS • ACP • RESEARCH • PLANSDIRECT

Reed Construction Data®

The design community utilizes the Reed First Source® suite of products to search, select and specify nationally available building products during the formative stages of project design, as well as during other stages of product selection and specification. Reed Design Registry is a detailed database of architecture firms. This tool features sophisticated search and sort capabilities to support the architect selection process and ensure locating the best manufacturers to partner with on your next project.

Reed First Source - The Leading Product Directory to "Find It, Choose It, Use It"

www.reedfirstsource.com

- Comprehensive directory of over 11,000 commercial building product manufacturers classified by MasterFormat™ 2004 categories.

- SPEC-DATA's 10-part format provides performance data along with technical and installation information.

- MANU-SPEC delivers manufacturer guide specifications in the CSI 3-part SectionFormat.

- Search for products, download CAD, research building codes and view catalogs online.

Reed Design Registry – The Premier Source of Architecture Firms

www.reedregistry.com

- Comprehensive website directory contains over 30,000 architect firms.

- Profiles list firm size, firm's specialties and more.

- Official database of AIA member-owned firms.

- Allows architects to share project information with the building community.

- Coming soon – Reed Registry to include engineers and landscape architects.

For more information about Reed Construction Data, please call 877-REED411, visit our website at www.reedconstructiondata.com, or E-mail: marketing@reedbusiness.com

Reed Construction Data
CONNECT • FIRST SOURCE • BULLETIN
RSMEANS • ACP • RESEARCH • PLANSDIRECT

Footing Excavation Price Sheet	QUAN.	UNIT	LABOR HOURS	COST EACH MAT.	COST EACH INST.	COST EACH TOTAL
Clear and grub, medium brush, 30' from building, 24' x 38'	.190	Acre	9.120		545	545
26' x 46'	.210	Acre	10.080		595	595
26' x 60'	.240	Acre	11.520		685	685
30' x 66'	.260	Acre	12.480		740	740
Light trees, to 6" dia. cut & chip, 24' x 38'	.190	Acre	9.120		545	545
26' x 46'	.210	Acre	10.080		595	595
26' x 60'	.240	Acre	11.520		685	685
30' x 66'	.260	Acre	12.480		740	740
Medium trees, to 10" dia. cut & chip, 24' x 38'	.190	Acre	13.029		770	770
26' x 46'	.210	Acre	14.400		850	850
26' x 60'	.240	Acre	16.457		975	975
30' x 66'	.260	Acre	17.829		1,050	1,050
Excavation, footing, 24' x 38', 2' deep	68.000	C.Y.	.906		145	145
4' deep	174.000	C.Y.	2.319		370	370
8' deep	384.000	C.Y.	5.119		815	815
26' x 46', 2' deep	79.000	C.Y.	1.053		169	169
4' deep	201.000	C.Y.	2.679		430	430
8' deep	404.000	C.Y.	5.385		860	860
26' x 60', 2' deep	94.000	C.Y.	1.253		200	200
4' deep	240.000	C.Y.	3.199		510	510
8' deep	483.000	C.Y.	6.438		1,025	1,025
30' x 66', 2' deep	105.000	C.Y.	1.400		224	224
4' deep	268.000	C.Y.	3.572		570	570
8' deep	539.000	C.Y.	7.185		1,150	1,150
Backfill, 24' x 38', 2" lifts, no compaction	34.000	C.Y.	.227		41	41
Compaction, air tamped, add	34.000	C.Y.	2.267		360	360
4" lifts, no compaction	87.000	C.Y.	.580		104	104
Compaction, air tamped, add	87.000	C.Y.	5.800		920	920
8" lifts, no compaction	192.000	C.Y.	1.281		231	231
Compaction, air tamped, add	192.000	C.Y.	12.801		2,025	2,025
26' x 46', 2" lifts, no compaction	40.000	C.Y.	.267		48	48
Compaction, air tamped, add	40.000	C.Y.	2.667		420	420
4" lifts, no compaction	100.000	C.Y.	.667		120	120
Compaction, air tamped, add	100.000	C.Y.	6.667		1,050	1,050
8" lifts, no compaction	202.000	C.Y.	1.347		243	243
Compaction, air tamped, add	202.000	C.Y.	13.467		2,125	2,125
26' x 60', 2" lifts, no compaction	47.000	C.Y.	.313		56.50	56.50
Compaction, air tamped, add	47.000	C.Y.	3.133		495	495
4" lifts, no compaction	120.000	C.Y.	.800		144	144
Compaction, air tamped, add	120.000	C.Y.	8.000		1,275	1,275
8" lifts, no compaction	242.000	C.Y.	1.614		290	290
Compaction, air tamped, add	242.000	C.Y.	16.134		2,550	2,550
30' x 66', 2" lifts, no compaction	53.000	C.Y.	.354		63.50	63.50
Compaction, air tamped, add	53.000	C.Y.	3.534		560	560
4" lifts, no compaction	134.000	C.Y.	.894		161	161
Compaction, air tamped, add	134.000	C.Y.	8.934		1,425	1,425
8" lifts, no compaction	269.000	C.Y.	1.794		325	325
Compaction, air tamped, add	269.000	C.Y.	17.934		2,850	2,850
Rough grade, 30' from building, 24' x 38'	87.000	C.Y.	.580		104	104
26' x 46'	100.000	C.Y.	.667		120	120
26' x 60'	120.000	C.Y.	.800		144	144
30' x 66'	134.000	C.Y.	.894		161	161

1 | SITEWORK   08 | Foundation Excavation Systems

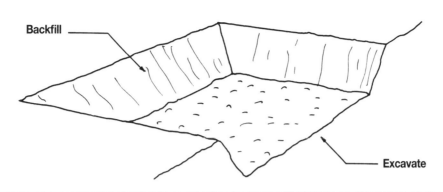

System Description	QUAN.	UNIT	LABOR HOURS	COST EACH MAT.	COST EACH INST.	COST EACH TOTAL
BUILDING, 24' X 38', 8' DEEP						
Clear & grub, dozer, medium brush, 30' from building	.190	Acre	2.027		213.75	213.75
Excavate, track loader, 1-1/2 C.Y. bucket	550.000	C.Y.	7.860		704	704
Backfill, dozer, 8" lifts, no compaction	180.000	C.Y.	1.201		216	216
Rough grade, dozer, 30' from building	280.000	C.Y.	1.868		336	336
TOTAL		Ea.	12.956		1,469.75	1,469.75
BUILDING, 26' X 46', 8' DEEP						
Clear & grub, dozer, medium brush, 30' from building	.210	Acre	2.240		236.25	236.25
Excavate, track loader, 1-1/2 C.Y. bucket	672.000	C.Y.	9.603		860.16	860.16
Backfill, dozer, 8" lifts, no compaction	220.000	C.Y.	1.467		264	264
Rough grade, dozer, 30' from building	340.000	C.Y.	2.268		408	408
TOTAL		Ea.	15.578		1,768.41	1,768.41
BUILDING, 26' X 60', 8' DEEP						
Clear & grub, dozer, medium brush, 30' from building	.240	Acre	2.560		270	270
Excavate, track loader, 1-1/2 C.Y. bucket	829.000	C.Y.	11.846		1,061.12	1,061.12
Backfill, dozer, 8" lifts, no compaction	270.000	C.Y.	1.801		324	324
Rough grade, dozer, 30' from building	420.000	C.Y.	2.801		504	504
TOTAL		Ea.	19.008		2,159.12	2,159.12
BUILDING, 30' X 66', 8' DEEP						
Clear & grub, dozer, medium brush, 30' from building	.260	Acre	2.773		292.50	292.50
Excavate, track loader, 1-1/2 C.Y. bucket	990.000	C.Y.	14.147		1,267.20	1,267.20
Backfill dozer, 8" lifts, no compaction	320.000	C.Y.	2.134		384	384
Rough grade, dozer, 30' from building	500.000	C.Y.	3.335		600	600
TOTAL		Ea.	22.389		2,543.70	2,543.70

The costs in this system are on a cost each basis.
Quantities are based on 1'-0" clearance beyond footing projection.

Description	QUAN.	UNIT	LABOR HOURS	COST EACH MAT.	COST EACH INST.	COST EACH TOTAL

Foundation Excavation Price Sheet

	QUAN.	UNIT	LABOR HOURS	COST EACH MAT.	COST EACH INST.	COST EACH TOTAL
Clear & grub, medium brush, 30' from building, 24' x 38'	.190	Acre	2.027		214	214
26' x 46'	.210	Acre	2.240		236	236
26' x 60'	.240	Acre	2.560		270	270
30' x 66'	.260	Acre	2.773		293	293
Light trees, to 6" dia. cut & chip, 24' x 38'	.190	Acre	9.120		545	545
26' x 46'	.210	Acre	10.080		595	595
26' x 60'	.240	Acre	11.520		685	685
30' x 66'	.260	Acre	12.480		740	740
Medium trees, to 10" dia. cut & chip, 24' x 38'	.190	Acre	13.029		770	770
26' x 46'	.210	Acre	14.400		850	850
26' x 60'	.240	Acre	16.457		975	975
30' x 66'	.260	Acre	17.829		1,050	1,050
Excavation, basement, 24' x 38', 2' deep	98.000	C.Y.	1.400		126	126
4' deep	220.000	C.Y.	3.144		281	281
8' deep	550.000	C.Y.	7.860		705	705
26' x 46', 2' deep	123.000	C.Y.	1.758		158	158
4' deep	274.000	C.Y.	3.915		350	350
8' deep	672.000	C.Y.	9.603		860	860
26' x 60', 2' deep	157.000	C.Y.	2.244		202	202
4' deep	345.000	C.Y.	4.930		440	440
8' deep	829.000	C.Y.	11.846		1,050	1,050
30' x 66', 2' deep	192.000	C.Y.	2.744		246	246
4' deep	419.000	C.Y.	5.988		535	535
8' deep	990.000	C.Y.	14.147		1,275	1,275
Backfill, 24' x 38', 2" lifts, no compaction	32.000	C.Y.	.213		38.50	38.50
Compaction, air tamped, add	32.000	C.Y.	2.133		340	340
4" lifts, no compaction	72.000	C.Y.	.480		86.50	86.50
Compaction, air tamped, add	72.000	C.Y.	4.800		760	760
8" lifts, no compaction	180.000	C.Y.	1.201		216	216
Compaction, air tamped, add	180.000	C.Y.	12.001		1,900	1,900
26' x 46', 2" lifts, no compaction	40.000	C.Y.	.267		48	48
Compaction, air tamped, add	40.000	C.Y.	2.667		420	420
4" lifts, no compaction	90.000	C.Y.	.600		108	108
Compaction, air tamped, add	90.000	C.Y.	6.000		950	950
8" lifts, no compaction	220.000	C.Y.	1.467		264	264
Compacton, air tamped, add	220.000	C.Y.	14.667		2,325	2,325
26' x 60', 2" lifts, no compaction	50.000	C.Y.	.334		60	60
Compaction, air tamped, add	50.000	C.Y.	3.334		530	530
4" lifts, no compaction	110.000	C.Y.	.734		132	132
Compaction, air tamped, add	110.000	C.Y.	7.334		1,175	1,175
8" lifts, no compaction	270.000	C.Y.	1.801		325	325
Compaction, air tamped, add	270.000	C.Y.	18.001		2,850	2,850
30' x 66', 2" lifts, no compaction	60.000	C.Y.	.400		72	72
Compaction, air tamped, add	60.000	C.Y.	4.000		635	635
4" lifts, no compaction	130.000	C.Y.	.867		156	156
Compaction, air tamped, add	130.000	C.Y.	8.667		1,375	1,375
8" lifts, no compaction	320.000	C.Y.	2.134		385	385
Compaction, air tamped, add	320.000	C.Y.	21.334		3,375	3,375
Rough grade, 30' from building, 24' x 38'	280.000	C.Y.	1.868		335	335
26' x 46'	340.000	C.Y.	2.268		410	410
26' x 60'	420.000	C.Y.	2.801		500	500
30' x 66'	500.000	C.Y.	3.335		600	600

1 | SITEWORK — 12 | Utility Trenching Systems

System Description	QUAN.	UNIT	LABOR HOURS	COST PER L.F. MAT.	COST PER L.F. INST.	COST PER L.F. TOTAL
2' DEEP						
Excavation, backhoe	.296	C.Y.	.032		1.72	1.72
Bedding, sand	.111	C.Y.	.044	1.73	1.63	3.36
Utility, sewer, 6" cast iron	1.000	L.F.	.283	15.05	11.91	26.96
Backfill, incl. compaction	.185	C.Y.	.044		1.38	1.38
TOTAL		L.F.	.403	16.78	16.64	33.42
4' DEEP						
Excavation, backhoe	.889	C.Y.	.095		5.15	5.15
Bedding, sand	.111	C.Y.	.044	1.73	1.63	3.36
Utility, sewer, 6" cast iron	1.000	L.F.	.283	15.05	11.91	26.96
Backfill, incl. compaction	.778	C.Y.	.183		5.80	5.80
TOTAL		L.F.	.605	16.78	24.49	41.27
6' DEEP						
Excavation, backhoe	1.770	C.Y.	.189		10.27	10.27
Bedding, sand	.111	C.Y.	.044	1.73	1.63	3.36
Utility, sewer, 6" cast iron	1.000	L.F.	.283	15.05	11.91	26.96
Backfill, incl. compaction	1.660	C.Y.	.391		12.37	12.37
TOTAL		L.F.	.907	16.78	36.18	52.96
8' DEEP						
Excavation, backhoe	2.960	C.Y.	.316		17.17	17.17
Bedding, sand	.111	C.Y.	.044	1.73	1.63	3.36
Utility, sewer, 6" cast iron	1.000	L.F.	.283	15.05	11.91	26.96
Backfill, incl. compaction	2.850	C.Y.	.671		21.23	21.23
TOTAL		L.F.	1.314	16.78	51.94	68.72

The costs in this system are based on a cost per linear foot of trench, and based on 2' wide at bottom of trench up to 6' deep.

Description	QUAN.	UNIT	LABOR HOURS	COST PER L.F. MAT.	COST PER L.F. INST.	COST PER L.F. TOTAL

Utility Trenching Price Sheet	QUAN.	UNIT	LABOR HOURS	MAT.	INST.	TOTAL
Excavation, bottom of trench 2' wide, 2' deep	.296	C.Y.	.032		1.72	1.72
4' deep	.889	C.Y.	.095		5.15	5.15
6' deep	1.770	C.Y.	.142		8.15	8.15
8' deep	2.960	C.Y.	.105		15.55	15.55
Bedding, sand, bottom of trench 2' wide, no compaction, pipe, 2" diameter	.070	C.Y.	.028	1.09	1.03	2.12
4" diameter	.084	C.Y.	.034	1.31	1.23	2.54
6" diameter	.105	C.Y.	.042	1.64	1.55	3.19
8" diameter	.122	C.Y.	.049	1.90	1.80	3.70
Compacted, pipe, 2" diameter	.074	C.Y.	.030	1.15	1.09	2.24
4" diameter	.092	C.Y.	.037	1.44	1.35	2.79
6" diameter	.111	C.Y.	.044	1.73	1.63	3.36
8" diameter	.129	C.Y.	.052	2.01	1.89	3.90
3/4" stone, bottom of trench 2' wide, pipe, 4" diameter	.082	C.Y.	.033	1.28	1.21	2.49
6" diameter	.099	C.Y.	.040	1.54	1.46	3
3/8" stone, bottom of trench 2' wide, pipe, 4" diameter	.084	C.Y.	.034	1.31	1.23	2.54
6" diameter	.102	C.Y.	.041	1.59	1.50	3.09
Utilities, drainage & sewerage, corrugated plastic, 6" diameter	1.000	L.F.	.069	4.79	2.25	7.04
8" diameter	1.000	L.F.	.072	9.95	2.36	12.31
Bituminous fiber, 4" diameter	1.000	L.F.	.064	2.56	2.10	4.66
6" diameter	1.000	L.F.	.069	4.79	2.25	7.04
8" diameter	1.000	L.F.	.072	9.95	2.36	12.31
Concrete, non-reinforced, 6" diameter	1.000	L.F.	.181	5.30	7.15	12.45
8" diameter	1.000	L.F.	.214	5.85	8.50	14.35
PVC, SDR 35, 4" diameter	1.000	L.F.	.064	2.56	2.10	4.66
6" diameter	1.000	L.F.	.069	4.79	2.25	7.04
8" diameter	1.000	L.F.	.072	9.95	2.36	12.31
Vitrified clay, 4" diameter	1.000	L.F.	.091	1.87	2.98	4.85
6" diameter	1.000	L.F.	.120	3.12	3.95	7.07
8" diameter	1.000	L.F.	.140	4.43	5.75	10.18
Gas & service, polyethylene, 1-1/4" diameter	1.000	L.F.	.059	2.06	2.27	4.33
Steel sched.40, 1" diameter	1.000	L.F.	.107	6.30	5.15	11.45
2" diameter	1.000	L.F.	.114	9.85	5.55	15.40
Sub-drainage, PVC, perforated, 3" diameter	1.000	L.F.	.064	2.56	2.10	4.66
4" diameter	1.000	L.F.	.064	2.56	2.10	4.66
5" diameter	1.000	L.F.	.069	4.79	2.25	7.04
6" diameter	1.000	L.F.	.069	4.79	2.25	7.04
Porous wall concrete, 4" diameter	1.000	L.F.	.072	3.16	2.36	5.52
Vitrified clay, perforated, 4" diameter	1.000	L.F.	.120	3.32	4.75	8.07
6" diameter	1.000	L.F.	.152	5.50	6.05	11.55
Water service, copper, type K, 3/4"	1.000	L.F.	.083	6.30	4.04	10.34
1" diameter	1.000	L.F.	.093	8.30	4.54	12.84
PVC, 3/4"	1.000	L.F.	.121	1.02	5.85	6.87
1" diameter	1.000	L.F.	.134	1.25	6.50	7.75
Backfill, bottom of trench 2' wide no compact, 2' deep, pipe, 2" diameter	.226	L.F.	.053		1.68	1.68
4" diameter	.212	L.F.	.050		1.58	1.58
6" diameter	.185	L.F.	.044		1.38	1.38
4' deep, pipe, 2" diameter	.819	C.Y.	.193		6.10	6.10
4" diameter	.805	C.Y.	.189		6	6
6" diameter	.778	C.Y.	.183		5.80	5.80
6' deep, pipe, 2" diameter	1.700	C.Y.	.400		12.65	12.65
4" diameter	1.690	C.Y.	.398		12.60	12.60
6" diameter	1.660	C.Y.	.391		12.35	12.35
8' deep, pipe, 2" diameter	2.890	C.Y.	.680		21.50	21.50
4" diameter	2.870	C.Y.	.675		21.50	21.50
6" diameter	2.850	C.Y.	.671		21	21

1 | SITEWORK 16 | Sidewalk Systems

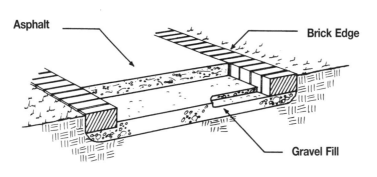

System Description	QUAN.	UNIT	LABOR HOURS	COST PER S.F.		
				MAT.	INST.	TOTAL
ASPHALT SIDEWALK SYSTEM, 3' WIDE WALK						
Gravel fill, 4" deep	1.000	S.F.	.001	.34	.03	.37
Compact fill	.012	C.Y.			.01	.01
Handgrade	1.000	S.F.	.004		.14	.14
Walking surface, bituminous paving, 2" thick	1.000	S.F.	.007	.62	.27	.89
Edging, brick, laid on edge	.670	L.F.	.079	1.61	3.08	4.69
TOTAL		S.F.	.091	2.57	3.53	6.10
CONCRETE SIDEWALK SYSTEM, 3' WIDE WALK						
Gravel fill, 4" deep	1.000	S.F.	.001	.34	.03	.37
Compact fill	.012	C.Y.			.01	.01
Handgrade	1.000	S.F.	.004		.14	.14
Walking surface, concrete, 4" thick	1.000	S.F.	.040	1.91	1.54	3.45
Edging, brick, laid on edge	.670	L.F.	.079	1.61	3.08	4.69
TOTAL		S.F.	.124	3.86	4.80	8.66
PAVERS, BRICK SIDEWALK SYSTEM, 3' WIDE WALK						
Sand base fill, 4" deep	1.000	S.F.	.001	.39	.07	.46
Compact fill	.012	C.Y.			.01	.01
Handgrade	1.000	S.F.	.004		.14	.14
Walking surface, brick pavers	1.000	S.F.	.160	2.94	6.20	9.14
Edging, redwood, untreated, 1" x 4"	.670	L.F.	.032	1.66	1.41	3.07
TOTAL		S.F.	.197	4.99	7.83	12.82

The costs in this system are based on a cost per square foot of sidewalk area. Concrete used is 3000 p.s.i.

Description	QUAN.	UNIT	LABOR HOURS	COST PER S.F.		
				MAT.	INST.	TOTAL

Sidewalk Price Sheet	QUAN.	UNIT	LABOR HOURS	COST PER S.F.		
				MAT.	INST.	TOTAL
Base, crushed stone, 3" deep	1.000	S.F.	.001	.38	.08	.46
6" deep	1.000	S.F.	.001	.76	.08	.84
9" deep	1.000	S.F.	.002	1.10	.11	1.21
12" deep	1.000	S.F.	.002	1.76	.13	1.89
Bank run gravel, 6" deep	1.000	S.F.	.001	.51	.05	.56
9" deep	1.000	S.F.	.001	.74	.08	.82
12" deep	1.000	S.F.	.001	1.01	.10	1.11
Compact base, 3" deep	.009	C.Y.	.001		.01	.01
6" deep	.019	C.Y.	.001		.02	.02
9" deep	.028	C.Y.	.001		.03	.03
Handgrade	1.000	S.F.	.004		.14	.14
Surface, brick, pavers dry joints, laid flat, running bond	1.000	S.F.	.160	2.94	6.20	9.14
Basket weave	1.000	S.F.	.168	3.31	6.55	9.86
Herringbone	1.000	S.F.	.174	3.31	6.75	10.06
Laid on edge, running bond	1.000	S.F.	.229	2.90	8.85	11.75
Mortar jts. laid flat, running bond	1.000	S.F.	.192	3.53	7.45	10.98
Basket weave	1.000	S.F.	.202	3.97	7.85	11.82
Herringbone	1.000	S.F.	.209	3.97	8.10	12.07
Laid on edge, running bond	1.000	S.F.	.274	3.48	10.60	14.08
Bituminous paving, 1-1/2" thick	1.000	S.F.	.006	.46	.21	.67
2" thick	1.000	S.F.	.007	.62	.27	.89
2-1/2" thick	1.000	S.F.	.008	.79	.29	1.08
Sand finish, 3/4" thick	1.000	S.F.	.001	.28	.09	.37
1" thick	1.000	S.F.	.001	.34	.11	.45
Concrete, reinforced, broom finish, 4" thick	1.000	S.F.	.040	1.91	1.54	3.45
5" thick	1.000	S.F.	.044	2.54	1.69	4.23
6" thick	1.000	S.F.	.047	2.97	1.81	4.78
Crushed stone, white marble, 3" thick	1.000	S.F.	.009	.23	.30	.53
Bluestone, 3" thick	1.000	S.F.	.009	.25	.30	.55
Flagging, bluestone, 1"	1.000	S.F.	.198	6.25	7.65	13.90
1-1/2"	1.000	S.F.	.188	11.35	7.30	18.65
Slate, natural cleft, 3/4"	1.000	S.F.	.174	6.95	6.75	13.70
Random rect., 1/2"	1.000	S.F.	.152	15.05	5.90	20.95
Granite blocks	1.000	S.F.	.174	8.45	6.75	15.20
Edging, corrugated aluminum, 4", 3' wide walk	.666	L.F.	.008	.29	.36	.65
4' wide walk	.500	L.F.	.006	.22	.27	.49
6", 3' wide walk	.666	L.F.	.010	.37	.42	.79
4' wide walk	.500	L.F.	.007	.28	.32	.60
Redwood-cedar-cypress, 1" x 4", 3' wide walk	.666	L.F.	.021	.83	.93	1.76
4' wide walk	.500	L.F.	.016	.63	.70	1.33
2" x 4", 3' wide walk	.666	L.F.	.032	1.66	1.41	3.07
4' wide walk	.500	L.F.	.024	1.25	1.06	2.31
Brick, dry joints, 3' wide walk	.666	L.F.	.079	1.61	3.08	4.69
4' wide walk	.500	L.F.	.059	1.20	2.30	3.50
Mortar joints, 3' wide walk	.666	L.F.	.095	1.93	3.70	5.63
4' wide walk	.500	L.F.	.071	1.44	2.76	4.20

1 | SITEWORK 20 | Driveway Systems

System Description	QUAN.	UNIT	LABOR HOURS	COST PER S.F.		
				MAT.	INST.	TOTAL
ASPHALT DRIVEWAY TO 10' WIDE						
Excavation, driveway to 10' wide, 6" deep	.019	C.Y.			.03	.03
Base, 6" crushed stone	1.000	S.F.	.001	.76	.08	.84
Handgrade base	1.000	S.F.	.004		.14	.14
2" thick base	1.000	S.F.	.002	.62	.16	.78
1" topping	1.000	S.F.	.001	.34	.11	.45
Edging, brick pavers	.200	L.F.	.024	.48	.92	1.40
TOTAL		S.F.	.032	2.20	1.44	3.64
CONCRETE DRIVEWAY TO 10' WIDE						
Excavation, driveway to 10' wide, 6" deep	.019	C.Y.			.03	.03
Base, 6" crushed stone	1.000	S.F.	.001	.76	.08	.84
Handgrade base	1.000	S.F.	.004		.14	.14
Surface, concrete, 4" thick	1.000	S.F.	.040	1.91	1.54	3.45
Edging, brick pavers	.200	L.F.	.024	.48	.92	1.40
TOTAL		S.F.	.069	3.15	2.71	5.86
PAVERS, BRICK DRIVEWAY TO 10' WIDE						
Excavation, driveway to 10' wide, 6" deep	.019	C.Y.			.03	.03
Base, 6" sand	1.000	S.F.	.001	.62	.10	.72
Handgrade base	1.000	S.F.	.004		.14	.14
Surface, pavers, brick laid flat, running bond	1.000	S.F.	.160	2.94	6.20	9.14
Edging, redwood, untreated, 2" x 4"	.200	L.F.	.010	.50	.42	.92
TOTAL		S.F.	.175	4.06	6.89	10.95

Description	QUAN.	UNIT	LABOR HOURS	COST PER S.F.		
				MAT.	INST.	TOTAL

Driveway Price Sheet

Driveway Price Sheet	QUAN.	UNIT	LABOR HOURS	COST PER S.F. MAT.	COST PER S.F. INST.	COST PER S.F. TOTAL
Excavation, by machine, 10' wide, 6" deep	.019	C.Y.	.001		.03	.03
12" deep	.037	C.Y.	.001		.06	.06
18" deep	.055	C.Y.	.001		.08	.08
20' wide, 6" deep	.019	C.Y.	.001		.03	.03
12" deep	.037	C.Y.	.001		.06	.06
18" deep	.055	C.Y.	.001		.08	.08
Base, crushed stone, 10' wide, 3" deep	1.000	S.F.	.001	.38	.04	.42
6" deep	1.000	S.F.	.001	.76	.08	.84
9" deep	1.000	S.F.	.002	1.10	.11	1.21
20' wide, 3" deep	1.000	S.F.	.001	.38	.04	.42
6" deep	1.000	S.F.	.001	.76	.08	.84
9" deep	1.000	S.F.	.002	1.10	.11	1.21
Bank run gravel, 10' wide, 3" deep	1.000	S.F.	.001	.26	.03	.29
6" deep	1.000	S.F.	.001	.51	.05	.56
9" deep	1.000	S.F.	.001	.74	.08	.82
20' wide, 3" deep	1.000	S.F.	.001	.26	.03	.29
6" deep	1.000	S.F.	.001	.51	.05	.56
9" deep	1.000	S.F.	.001	.74	.08	.82
Handgrade, 10' wide	1.000	S.F.	.004		.14	.14
20' wide	1.000	S.F.	.004		.14	.14
Surface, asphalt, 10' wide, 3/4" topping, 1" base	1.000	S.F.	.002	.76	.20	.96
2" base	1.000	S.F.	.003	.90	.25	1.15
1" topping, 1" base	1.000	S.F.	.002	.82	.22	1.04
2" base	1.000	S.F.	.003	.96	.27	1.23
20' wide, 3/4" topping, 1" base	1.000	S.F.	.002	.76	.20	.96
2" base	1.000	S.F.	.003	.90	.25	1.15
1" topping, 1" base	1.000	S.F.	.002	.82	.22	1.04
2" base	1.000	S.F.	.003	.96	.27	1.23
Concrete, 10' wide, 4" thick	1.000	S.F.	.040	1.91	1.54	3.45
6" thick	1.000	S.F.	.047	2.97	1.81	4.78
20' wide, 4" thick	1.000	S.F.	.040	1.91	1.54	3.45
6" thick	1.000	S.F.	.047	2.97	1.81	4.78
Paver, brick 10' wide dry joints, running bond, laid flat	1.000	S.F.	.160	2.94	6.20	9.14
Laid on edge	1.000	S.F.	.229	2.90	8.85	11.75
Mortar joints, laid flat	1.000	S.F.	.192	3.53	7.45	10.98
Laid on edge	1.000	S.F.	.274	3.48	10.60	14.08
20' wide, running bond, dry jts., laid flat	1.000	S.F.	.160	2.94	6.20	9.14
Laid on edge	1.000	S.F.	.229	2.90	8.85	11.75
Mortar joints, laid flat	1.000	S.F.	.192	3.53	7.45	10.98
Laid on edge	1.000	S.F.	.274	3.48	10.60	14.08
Crushed stone, 10' wide, white marble, 3"	1.000	S.F.	.009	.23	.30	.53
Bluestone, 3"	1.000	S.F.	.009	.25	.30	.55
20' wide, white marble, 3"	1.000	S.F.	.009	.23	.30	.53
Bluestone, 3"	1.000	S.F.	.009	.25	.30	.55
Soil cement, 10' wide	1.000	S.F.	.007	.28	.76	1.04
20' wide	1.000	S.F.	.007	.28	.76	1.04
Granite blocks, 10' wide	1.000	S.F.	.174	8.45	6.75	15.20
20' wide	1.000	S.F.	.174	8.45	6.75	15.20
Asphalt block, solid 1-1/4" thick	1.000	S.F.	.119	6.40	4.60	11
Solid 3" thick	1.000	S.F.	.123	8.95	4.78	13.73
Edging, brick, 10' wide	.200	L.F.	.024	.48	.92	1.40
20' wide	.100	L.F.	.012	.24	.46	.70
Redwood, untreated 2" x 4", 10' wide	.200	L.F.	.010	.50	.42	.92
20' wide	.100	L.F.	.005	.25	.21	.46
Granite, 4 1/2" x 12" straight, 10' wide	.200	L.F.	.032	1.17	1.77	2.94
20' wide	.100	L.F.	.016	.59	.89	1.48
Finishes, asphalt sealer, 10' wide	1.000	S.F.	.023	.73	.75	1.48
20' wide	1.000	S.F.	.023	.73	.75	1.48
Concrete, exposed aggregate 10' wide	1.000	S.F.	.013	.20	.51	.71
20' wide	1.000	S.F.	.013	.20	.51	.71

1 | SITEWORK   24 | Septic Systems

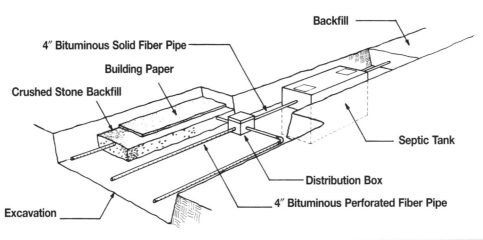

System Description	QUAN.	UNIT	LABOR HOURS	COST EACH MAT.	COST EACH INST.	COST EACH TOTAL
SEPTIC SYSTEM WITH 1000 S.F. LEACHING FIELD, 1000 GALLON TANK						
Tank, 1000 gallon, concrete	1.000	Ea.	3.500	695	143.50	838.50
Distribution box, concrete	1.000	Ea.	1.000	129	32	161
4" PVC pipe	25.000	L.F.	1.600	64	52.50	116.50
Tank and field excavation	119.000	C.Y.	13.130		972.23	972.23
Crushed stone backfill	76.000	C.Y.	12.160	1,938	561.64	2,499.64
Backfill with excavated material	36.000	C.Y.	.240		43.20	43.20
Building paper	125.000	S.Y.	2.430	56.25	101.25	157.50
4" PVC perforated pipe	145.000	L.F.	9.280	371.20	304.50	675.70
4" pipe fittings	2.000	Ea.	1.939	33.50	85	118.50
TOTAL		Ea.	45.279	3,286.95	2,295.82	5,582.77
SEPTIC SYSTEM WITH 2 LEACHING PITS, 1000 GALLON TANK						
Tank, 1000 gallon, concrete	1.000	Ea.	3.500	695	143.50	838.50
Distribution box, concrete	1.000	Ea.	1.000	129	32	161
4" PVC pipe	75.000	L.F.	4.800	192	157.50	349.50
Excavation for tank only	20.000	C.Y.	2.207		163.40	163.40
Crushed stone backfill	10.000	C.Y.	1.600	255	73.90	328.90
Backfill with excavated material	55.000	C.Y.	.367		66	66
Pits, 6' diameter, including excavation and stone backfill	2.000	Ea.		2,150		2,150
TOTAL		Ea.	13.474	3,421	636.30	4,057.30

The costs in this system include all necessary piping and excavation.

Description	QUAN.	UNIT	LABOR HOURS	COST EACH MAT.	COST EACH INST.	COST EACH TOTAL

Septic Systems Price Sheet

	QUAN.	UNIT	LABOR HOURS	COST EACH MAT.	COST EACH INST.	COST EACH TOTAL
Tank, precast concrete, 1000 gallon	1.000	Ea.	3.500	695	144	839
2000 gallon	1.000	Ea.	5.600	2,175	230	2,405
Distribution box, concrete, 5 outlets	1.000	Ea.	1.000	129	32	161
12 outlets	1.000	Ea.	2.000	475	63.50	538.50
4" pipe, PVC, solid	25.000	L.F.	1.600	64	52.50	116.50
Tank and field excavation, 1000 S.F. field	119.000	C.Y.	6.565		970	970
2000 S.F. field	190.000	C.Y.	10.482		1,550	1,550
Tank excavation only, 1000 gallon tank	20.000	C.Y.	1.103		164	164
2000 gallon tank	32.000	C.Y.	1.765		261	261
Backfill, crushed stone 1000 S.F. field	76.000	C.Y.	12.160	1,950	560	2,510
2000 S.F. field	140.000	C.Y.	22.400	3,575	1,025	4,600
Backfill with excavated material, 1000 S.F. field	36.000	C.Y.	.240		43.50	43.50
2000 S.F. field	60.000	C.Y.	.400		72	72
6' diameter pits	55.000	C.Y.	.367		66	66
3' diameter pits	42.000	C.Y.	.280		50	50
Building paper, 1000 S.F. field	125.000	S.Y.	2.376	55	99	154
2000 S.F. field	250.000	S.Y.	4.860	113	203	316
4" pipe, PVC, perforated, 1000 S.F. field	145.000	L.F.	9.280	370	305	675
2000 S.F. field	265.000	L.F.	16.960	680	555	1,235
Pipe fittings, bituminous fiber, 1000 S.F. field	2.000	Ea.	1.939	33.50	85	118.50
2000 S.F. field	4.000	Ea.	3.879	67	170	237
Leaching pit, including excavation and stone backfill, 3' diameter	1.000	Ea.		805		805
6' diameter	1.000	Ea.		1,075		1,075

1 | SITEWORK — 60 | Chain Link Fence Price Sheet

System Description	QUAN.	UNIT	LABOR HOURS	COST PER UNIT		
				MAT.	INST.	TOTAL
Chain link fence						
Galv. 9ga. wire, 1-5/8"post 10'O.C., 1-3/8"top rail, 2"corner post, 3'hi	1.000	L.F.	.130	7.45	4.27	11.72
4' high	1.000	L.F.	.141	11.10	4.64	15.74
6' high	1.000	L.F.	.209	12.55	6.85	19.40
Add for gate 3' wide 1-3/8" frame 3' high	1.000	Ea.	2.000	67	66	133
4' high	1.000	Ea.	2.400	83	79	162
6' high	1.000	Ea.	2.400	149	79	228
Add for gate 4' wide 1-3/8" frame 3' high	1.000	Ea.	2.667	78.50	87.50	166
4' high	1.000	Ea.	2.667	102	87.50	189.50
6' high	1.000	Ea.	3.000	189	98.50	287.50
Alum. 9ga. wire, 1-5/8"post, 10'O.C., 1-3/8"top rail, 2"corner post, 3'hi	1.000	L.F.	.130	9.15	4.27	13.42
4' high	1.000	L.F.	.141	10.40	4.64	15.04
6' high	1.000	L.F.	.209	13.35	6.85	20.20
Add for gate 3' wide 1-3/8" frame 3' high	1.000	Ea.	2.000	89	66	155
4' high	1.000	Ea.	2.400	122	79	201
6' high	1.000	Ea.	2.400	182	79	261
Add for gate 4' wide 1-3/8" frame 3' high	1.000	Ea.	2.400	122	79	201
4' high	1.000	Ea.	2.667	162	87.50	249.50
6' high	1.000	Ea.	3.000	253	98.50	351.50
Vinyl 9ga. wire, 1-5/8"post 10'O.C., 1-3/8"top rail, 2"corner post, 3'hi	1.000	L.F.	.130	8.10	4.27	12.37
4' high	1.000	L.F.	.141	13.30	4.64	17.94
6' high	1.000	L.F.	.209	15.20	6.85	22.05
Add for gate 3' wide 1-3/8" frame 3' high	1.000	Ea.	2.000	99.50	66	165.50
4' high	1.000	Ea.	2.400	129	79	208
6' high	1.000	Ea.	2.400	199	79	278
Add for gate 4' wide 1-3/8" frame 3' high	1.000	Ea.	2.400	135	79	214
4' high	1.000	Ea.	2.667	179	87.50	266.50
6' high	1.000	Ea.	3.000	259	98.50	357.50
Tennis court, chain link fence, 10' high						
Galv. 11ga. wire, 2"post 10'O.C., 1-3/8"top rail, 2-1/2"corner post	1.000	L.F.	.253	19.90	8.30	28.20
Add for gate 3' wide 1-3/8" frame	1.000	Ea.	2.400	249	79	328
Alum. 11ga. wire, 2"post 10'O.C., 1-3/8"top rail, 2-1/2"corner post	1.000	L.F.	.253	28.50	8.30	36.80
Add for gate 3' wide 1-3/8" frame	1.000	Ea.	2.400	325	79	404
Vinyl 11ga. wire, 2"post 10' O.C., 1-3/8"top rail, 2-1/2"corner post	1.000	L.F.	.253	24	8.30	32.30
Add for gate 3' wide 1-3/8" frame	1.000	Ea.	2.400	360	79	439
Railings, commercial						
Aluminum balcony rail, 1-1/2" posts with pickets	1.000	L.F.	.164	55.50	9.70	65.20
With expanded metal panels	1.000	L.F.	.164	71.50	9.70	81.20
With porcelain enamel panel inserts	1.000	L.F.	.164	64	9.70	73.70
Mild steel, ornamental rounded top rail	1.000	L.F.	.164	62	9.70	71.70
As above, but pitch down stairs	1.000	L.F.	.183	67.50	10.75	78.25
Steel pipe, welded, 1-1/2" round, painted	1.000	L.F.	.160	23.50	9.45	32.95
Galvanized	1.000	L.F.	.160	33	9.45	42.45
Residential, stock units, mild steel, deluxe	1.000	L.F.	.102	14.20	6	20.20
Economy	1.000	L.F.	.102	10.65	6	16.65

1 | SITEWORK — 64 | Wood Fence Price Sheet

System Description	QUAN.	UNIT	LABOR HOURS	COST PER UNIT MAT.	COST PER UNIT INST.	COST PER UNIT TOTAL
Basketweave, 3/8"x4" boards, 2"x4" stringers on spreaders, 4"x4" posts						
No. 1 cedar, 6' high	1.000	L.F.	.150	9.75	5.90	15.65
Treated pine, 6' high	1.000	L.F.	.160	11.85	6.30	18.15
Board fence, 1"x4" boards, 2"x4" rails, 4"x4" posts						
Preservative treated, 2 rail, 3' high	1.000	L.F.	.166	7.25	6.50	13.75
4' high	1.000	L.F.	.178	7.95	7	14.95
3 rail, 5' high	1.000	L.F.	.185	8.95	7.25	16.20
6' high	1.000	L.F.	.192	10.25	7.55	17.80
Western cedar, No. 1, 2 rail, 3' high	1.000	L.F.	.166	7.90	6.50	14.40
3 rail, 4' high	1.000	L.F.	.178	9.35	7	16.35
5' high	1.000	L.F.	.185	10.80	7.25	18.05
6' high	1.000	L.F.	.192	11.85	7.55	19.40
No. 1 cedar, 2 rail, 3' high	1.000	L.F.	.166	11.85	6.50	18.35
4' high	1.000	L.F.	.178	13.50	7	20.50
3 rail, 5' high	1.000	L.F.	.185	15.60	7.25	22.85
6' high	1.000	L.F.	.192	17.40	7.55	24.95
Shadow box, 1"x6" boards, 2"x4" rails, 4"x4" posts						
Fir, pine or spruce, treated, 3 rail, 6' high	1.000	L.F.	.160	13.30	6.30	19.60
No. 1 cedar, 3 rail, 4' high	1.000	L.F.	.185	16.35	7.25	23.60
6' high	1.000	L.F.	.192	20	7.55	27.55
Open rail, split rails, No. 1 cedar, 2 rail, 3' high	1.000	L.F.	.150	6.55	5.90	12.45
3 rail, 4' high	1.000	L.F.	.160	8.85	6.30	15.15
No. 2 cedar, 2 rail, 3' high	1.000	L.F.	.150	5.10	5.90	11
3 rail, 4' high	1.000	L.F.	.160	5.80	6.30	12.10
Open rail, rustic rails, No. 1 cedar, 2 rail, 3' high	1.000	L.F.	.150	4.09	5.90	9.99
3 rail, 4' high	1.000	L.F.	.160	5.50	6.30	11.80
No. 2 cedar, 2 rail, 3' high	1.000	L.F.	.150	3.93	5.90	9.83
3 rail, 4' high	1.000	L.F.	.160	4.15	6.30	10.45
Rustic picket, molded pine pickets, 2 rail, 3' high	1.000	L.F.	.171	5.80	6.75	12.55
3 rail, 4' high	1.000	L.F.	.197	6.65	7.80	14.45
No. 1 cedar, 2 rail, 3' high	1.000	L.F.	.171	7.90	6.75	14.65
3 rail, 4' high	1.000	L.F.	.197	9.10	7.80	16.90
Picket fence, fir, pine or spruce, preserved, treated						
2 rail, 3' high	1.000	L.F.	.171	5.05	6.75	11.80
3 rail, 4' high	1.000	L.F.	.185	5.95	7.25	13.20
Western cedar, 2 rail, 3' high	1.000	L.F.	.171	6.30	6.75	13.05
3 rail, 4' high	1.000	L.F.	.185	6.45	7.25	13.70
No. 1 cedar, 2 rail, 3' high	1.000	L.F.	.171	12.60	6.75	19.35
3 rail, 4' high	1.000	L.F.	.185	14.70	7.25	21.95
Stockade, No. 1 cedar, 3-1/4" rails, 6' high	1.000	L.F.	.150	11.90	5.90	17.80
8' high	1.000	L.F.	.155	15.40	6.10	21.50
No. 2 cedar, treated rails, 6' high	1.000	L.F.	.150	11.90	5.90	17.80
Treated pine, treated rails, 6' high	1.000	L.F.	.150	11.65	5.90	17.55
Gates, No. 2 cedar, picket, 3'-6" wide 4' high	1.000	Ea.	2.667	63	105	168
No. 2 cedar, rustic round, 3' wide, 3' high	1.000	Ea.	2.667	80.50	105	185.50
No. 2 cedar, stockade screen, 3'-6" wide, 6' high	1.000	Ea.	3.000	70	118	188
General, wood, 3'-6" wide, 4' high	1.000	Ea.	2.400	61	94.50	155.50
6' high	1.000	Ea.	3.000	76.50	118	194.50

Division 2
Foundations

2 | FOUNDATIONS — 04 | Footing Systems

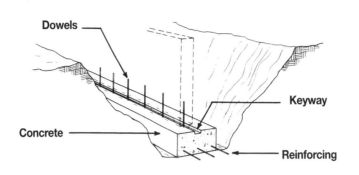

System Description	QUAN.	UNIT	LABOR HOURS	COST PER L.F. MAT.	COST PER L.F. INST.	COST PER L.F. TOTAL
8" THICK BY 18" WIDE FOOTING						
Concrete, 3000 psi	.040	C.Y.		4.56		4.56
Place concrete, direct chute	.040	C.Y.	.016		.56	.56
Forms, footing, 4 uses	1.330	SFCA	.103	.86	3.87	4.73
Reinforcing, 1/2" diameter bars, 2 each	1.380	Lb.	.011	.70	.52	1.22
Keyway, 2" x 4", beveled, 4 uses	1.000	L.F.	.015	.22	.66	.88
Dowels, 1/2" diameter bars, 2' long, 6' O.C.	.166	Ea.	.006	.12	.27	.39
TOTAL		L.F.	.151	6.46	5.88	12.34
12" THICK BY 24" WIDE FOOTING						
Concrete, 3000 psi	.070	C.Y.		7.98		7.98
Place concrete, direct chute	.070	C.Y.	.028		.98	.98
Forms, footing, 4 uses	2.000	SFCA	.155	1.30	5.82	7.12
Reinforcing, 1/2" diameter bars, 2 each	1.380	Lb.	.011	.70	.52	1.22
Keyway, 2" x 4", beveled, 4 uses	1.000	L.F.	.015	.22	.66	.88
Dowels, 1/2" diameter bars, 2' long, 6' O.C.	.166	Ea.	.006	.12	.27	.39
TOTAL		L.F.	.215	10.32	8.25	18.57
12" THICK BY 36" WIDE FOOTING						
Concrete, 3000 psi	.110	C.Y.		12.54		12.54
Place concrete, direct chute	.110	C.Y.	.044		1.53	1.53
Forms, footing, 4 uses	2.000	SFCA	.155	1.30	5.82	7.12
Reinforcing, 1/2" diameter bars, 2 each	1.380	Lb.	.011	.70	.52	1.22
Keyway, 2" x 4", beveled, 4 uses	1.000	L.F.	.015	.22	.66	.88
Dowels, 1/2" diameter bars, 2' long, 6' O.C.	.166	Ea.	.006	.12	.27	.39
TOTAL		L.F.	.231	14.88	8.80	23.68

The footing costs in this system are on a cost per linear foot basis

Description	QUAN.	UNIT	LABOR HOURS	COST PER S.F. MAT.	COST PER S.F. INST.	COST PER S.F. TOTAL

Footing Price Sheet	QUAN.	UNIT	LABOR HOURS	COST PER L.F.		
				MAT.	INST.	TOTAL
Concrete, 8" thick by 18" wide footing						
2000 psi concrete	.040	C.Y.		4.40		4.40
2500 psi concrete	.040	C.Y.		4.44		4.44
3000 psi concrete	.040	C.Y.		4.56		4.56
3500 psi concrete	.040	C.Y.		4.64		4.64
4000 psi concrete	.040	C.Y.		4.76		4.76
12" thick by 24" wide footing						
2000 psi concrete	.070	C.Y.		7.70		7.70
2500 psi concrete	.070	C.Y.		7.75		7.75
3000 psi concrete	.070	C.Y.		8		8
3500 psi concrete	.070	C.Y.		8.10		8.10
4000 psi concrete	.070	C.Y.		8.35		8.35
12" thick by 36" wide footing						
2000 psi concrete	.110	C.Y.		12.10		12.10
2500 psi concrete	.110	C.Y.		12.20		12.20
3000 psi concrete	.110	C.Y.		12.55		12.55
3500 psi concrete	.110	C.Y.		12.75		12.75
4000 psi concrete	.110	C.Y.		13.10		13.10
Place concrete, 8" thick by 18" wide footing, direct chute	.040	C.Y.	.016		.56	.56
Pumped concrete	.040	C.Y.	.017		.81	.81
Crane & bucket	.040	C.Y.	.032		1.67	1.67
12" thick by 24" wide footing, direct chute	.070	C.Y.	.028		.98	.98
Pumped concrete	.070	C.Y.	.030		1.43	1.43
Crane & bucket	.070	C.Y.	.056		2.92	2.92
12" thick by 36" wide footing, direct chute	.110	C.Y.	.044		1.53	1.53
Pumped concrete	.110	C.Y.	.047		2.24	2.24
Crane & bucket	.110	C.Y.	.088		4.59	4.59
Forms, 8" thick footing, 1 use	1.330	SFCA	.140	.25	5.25	5.50
4 uses	1.330	SFCA	.103	.86	3.87	4.73
12" thick footing, 1 use	2.000	SFCA	.211	.38	7.90	8.28
4 uses	2.000	SFCA	.155	1.30	5.80	7.10
Reinforcing, 3/8" diameter bar, 1 each	.400	Lb.	.003	.20	.15	.35
2 each	.800	Lb.	.006	.41	.30	.71
3 each	1.200	Lb.	.009	.61	.46	1.07
1/2" diameter bar, 1 each	.700	Lb.	.005	.36	.27	.63
2 each	1.380	Lb.	.011	.70	.52	1.22
3 each	2.100	Lb.	.016	1.07	.80	1.87
5/8" diameter bar, 1 each	1.040	Lb.	.008	.53	.40	.93
2 each	2.080	Lb.	.016	1.06	.79	1.85
Keyway, beveled, 2" x 4", 1 use	1.000	L.F.	.030	.44	1.32	1.76
2 uses	1.000	L.F.	.023	.33	.99	1.32
2" x 6", 1 use	1.000	L.F.	.032	.64	1.40	2.04
2 uses	1.000	L.F.	.024	.48	1.05	1.53
Dowels, 2 feet long, 6' O.C., 3/8" bar	.166	Ea.	.005	.07	.25	.32
1/2" bar	.166	Ea.	.006	.12	.27	.39
5/8" bar	.166	Ea.	.006	.19	.30	.49
3/4" bar	.166	Ea.	.006	.19	.30	.49

2 | FOUNDATIONS   08 | Block Wall Systems

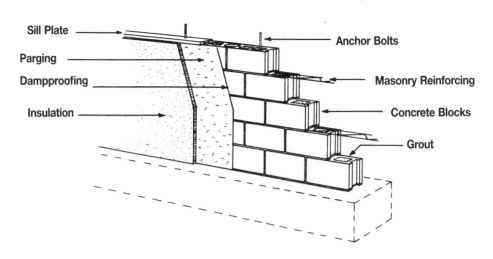

System Description	QUAN.	UNIT	LABOR HOURS	COST PER S.F.		
				MAT.	INST.	TOTAL
8" WALL, GROUTED, FULL HEIGHT						
Concrete block, 8" x 16" x 8"	1.000	S.F.	.094	2.70	3.76	6.46
Masonry reinforcing, every second course	.750	L.F.	.002	.17	.09	.26
Parging, plastering with portland cement plaster, 1 coat	1.000	S.F.	.014	.25	.58	.83
Dampproofing, bituminous coating, 1 coat	1.000	S.F.	.012	.14	.48	.62
Insulation, 1" rigid polystyrene	1.000	S.F.	.010	.52	.44	.96
Grout, solid, pumped	1.000	S.F.	.059	1.20	2.31	3.51
Anchor bolts, 1/2" diameter, 8" long, 4' O.C.	.060	Ea.	.002	.05	.11	.16
Sill plate, 2" x 4", treated	.250	L.F.	.007	.15	.32	.47
TOTAL		S.F.	.200	5.18	8.09	13.27
12" WALL, GROUTED, FULL HEIGHT						
Concrete block, 8" x 16" x 12"	1.000	S.F.	.160	3.77	6.20	9.97
Masonry reinforcing, every second course	.750	L.F.	.003	.19	.14	.33
Parging, plastering with portland cement plaster, 1 coat	1.000	S.F.	.014	.25	.58	.83
Dampproofing, bituminous coating, 1 coat	1.000	S.F.	.012	.14	.48	.62
Insulation, 1" rigid polystyrene	1.000	S.F.	.010	.52	.44	.96
Grout, solid, pumped	1.000	S.F.	.063	1.96	2.46	4.42
Anchor bolts, 1/2" diameter, 8" long, 4' O.C.	.060	Ea.	.002	.05	.11	.16
Sill plate, 2" x 4", treated	.250	L.F.	.007	.15	.32	.47
TOTAL		S.F.	.271	7.03	10.73	17.76

The costs in this system are based on a square foot of wall. Do not subtract for window or door openings.

Description	QUAN.	UNIT	LABOR HOURS	COST PER S.F.		
				MAT.	INST.	TOTAL

Block Wall Systems	QUAN.	UNIT	LABOR HOURS	COST PER S.F. MAT.	COST PER S.F. INST.	COST PER S.F. TOTAL
Concrete, block, 8" x 16" x, 6" thick	1.000	S.F.	.089	2.50	3.51	6.01
8" thick	1.000	S.F.	.093	2.70	3.76	6.46
10" thick	1.000	S.F.	.095	3.51	4.56	8.07
12" thick	1.000	S.F.	.122	3.77	6.20	9.97
Solid block, 8" x 16" x, 6" thick	1.000	S.F.	.091	2.53	3.63	6.16
8" thick	1.000	S.F.	.096	3.72	3.85	7.57
10" thick	1.000	S.F.	.096	3.72	3.85	7.57
12" thick	1.000	S.F.	.126	5.50	5.30	10.80
Masonry reinforcing, wire strips, to 8" wide, every course	1.500	L.F.	.004	.33	.18	.51
Every 2nd course	.750	L.F.	.002	.17	.09	.26
Every 3rd course	.500	L.F.	.001	.11	.06	.17
Every 4th course	.400	L.F.	.001	.09	.05	.14
Wire strips to 12" wide, every course	1.500	L.F.	.006	.38	.27	.65
Every 2nd course	.750	L.F.	.003	.19	.14	.33
Every 3rd course	.500	L.F.	.002	.13	.09	.22
Every 4th course	.400	L.F.	.002	.10	.07	.17
Parging, plastering with portland cement plaster, 1 coat	1.000	S.F.	.014	.25	.58	.83
2 coats	1.000	S.F.	.022	.39	.88	1.27
Dampproofing, bituminous, brushed on, 1 coat	1.000	S.F.	.012	.14	.48	.62
2 coats	1.000	S.F.	.016	.28	.64	.92
Sprayed on, 1 coat	1.000	S.F.	.010	.14	.39	.53
2 coats	1.000	S.F.	.016	.27	.64	.91
Troweled on, 1/16" thick	1.000	S.F.	.016	.26	.64	.90
1/8" thick	1.000	S.F.	.020	.46	.81	1.27
1/2" thick	1.000	S.F.	.023	1.49	.92	2.41
Insulation, rigid, fiberglass, 1.5#/C.F., unfaced						
1-1/2" thick R 6.2	1.000	S.F.	.008	.87	.35	1.22
2" thick R 8.5	1.000	S.F.	.008	.75	.35	1.10
3" thick R 13	1.000	S.F.	.010	.87	.44	1.31
Foamglass, 1-1/2" thick R 2.64	1.000	S.F.	.010	1.44	.44	1.88
2" thick R 5.26	1.000	S.F.	.011	3.48	.48	3.96
Perlite, 1" thick R 2.77	1.000	S.F.	.010	.33	.44	.77
2" thick R 5.55	1.000	S.F.	.011	.65	.48	1.13
Polystyrene, extruded, 1" thick R 5.4	1.000	S.F.	.010	.52	.44	.96
2" thick R 10.8	1.000	S.F.	.011	1.47	.48	1.95
Molded 1" thick R 3.85	1.000	S.F.	.010	.24	.44	.68
2" thick R 7.7	1.000	S.F.	.011	.83	.48	1.31
Grout, concrete block cores, 6" thick	1.000	S.F.	.044	.90	1.73	2.63
8" thick	1.000	S.F.	.059	1.20	2.31	3.51
10" thick	1.000	S.F.	.061	1.58	2.38	3.96
12" thick	1.000	S.F.	.063	1.96	2.46	4.42
Anchor bolts, 2' on center, 1/2" diameter, 8" long	.120	Ea.	.005	.10	.21	.31
12" long	.120	Ea.	.005	.18	.22	.40
3/4" diameter, 8" long	.120	Ea.	.006	.21	.27	.48
12" long	.120	Ea.	.006	.27	.28	.55
4' on center, 1/2" diameter, 8" long	.060	Ea.	.002	.05	.11	.16
12" long	.060	Ea.	.003	.09	.11	.20
3/4" diameter, 8" long	.060	Ea.	.003	.11	.13	.24
12" long	.060	Ea.	.003	.13	.14	.27
Sill plates, treated, 2" x 4"	.250	L.F.	.007	.15	.32	.47
4" x 4"	.250	L.F.	.007	.38	.30	.68

2 | FOUNDATIONS 12 | Concrete Wall Systems

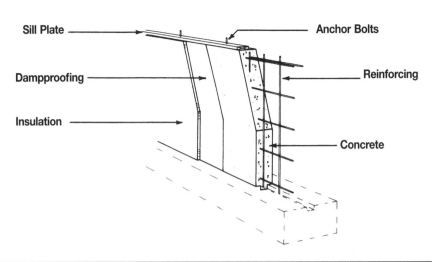

System Description	QUAN.	UNIT	LABOR HOURS	COST PER S.F. MAT.	COST PER S.F. INST.	COST PER S.F. TOTAL
8" THICK, POURED CONCRETE WALL						
Concrete, 8" thick, 3000 psi	.025	C.Y.		2.85		2.85
Forms, prefabricated plywood, 4 uses per month	2.000	SFCA	.076	1.46	2.92	4.38
Reinforcing, light	.670	Lb.	.004	.34	.17	.51
Placing concrete, direct chute	.025	C.Y.	.013		.46	.46
Dampproofing, brushed on, 2 coats	1.000	S.F.	.016	.28	.64	.92
Rigid insulation, 1" polystyrene	1.000	S.F.	.010	.52	.44	.96
Anchor bolts, 1/2" diameter, 12" long, 4' O.C.	.060	Ea.	.003	.09	.11	.20
Sill plates, 2" x 4", treated	.250	L.F.	.007	.15	.32	.47
TOTAL		S.F.	.129	5.69	5.06	10.75
12" THICK, POURED CONCRETE WALL						
Concrete, 12" thick, 3000 psi	.040	C.Y.		4.56		4.56
Forms, prefabricated plywood, 4 uses per month	2.000	SFCA	.076	1.46	2.92	4.38
Reinforcing, light	1.000	Lb.	.005	.51	.26	.77
Placing concrete, direct chute	.040	C.Y.	.019		.67	.67
Dampproofing, brushed on, 2 coats	1.000	S.F.	.016	.28	.64	.92
Rigid insulation, 1" polystyrene	1.000	S.F.	.010	.52	.44	.96
Anchor bolts, 1/2" diameter, 12" long, 4' O.C.	.060	Ea.	.003	.09	.11	.20
Sill plates, 2" x 4" treated	.250	L.F.	.007	.15	.32	.47
TOTAL		S.F.	.136	7.57	5.36	12.93

The costs in this system are based on sq. ft. of wall. Do not subtract for window and door openings. The costs assume a 4' high wall.

Description	QUAN.	UNIT	LABOR HOURS	COST PER S.F. MAT.	COST PER S.F. INST.	COST PER S.F. TOTAL

Concrete Wall Price Sheet	QUAN.	UNIT	LABOR HOURS	COST PER S.F.		
				MAT.	INST.	TOTAL
Formwork, prefabricated plywood, 1 use per month	2.000	SFCA	.081	4.42	3.12	7.54
4 uses per month	2.000	SFCA	.076	1.46	2.92	4.38
Job built forms, 1 use per month	2.000	SFCA	.320	5.50	12.20	17.70
4 uses per month	2.000	SFCA	.221	2.06	8.45	10.51
Reinforcing, 8" wall, light reinforcing	.670	Lb.	.004	.34	.17	.51
Heavy reinforcing	1.500	Lb.	.008	.77	.39	1.16
10" wall, light reinforcing	.850	Lb.	.005	.43	.22	.65
Heavy reinforcing	2.000	Lb.	.011	1.02	.52	1.54
12" wall light reinforcing	1.000	Lb.	.005	.51	.26	.77
Heavy reinforcing	2.250	Lb.	.012	1.15	.59	1.74
Placing concrete, 8" wall, direct chute	.025	C.Y.	.013		.46	.46
Pumped concrete	.025	C.Y.	.016		.76	.76
Crane & bucket	.025	C.Y.	.023		1.18	1.18
10" wall, direct chute	.030	C.Y.	.016		.56	.56
Pumped concrete	.030	C.Y.	.019		.91	.91
Crane & bucket	.030	C.Y.	.027		1.41	1.41
12" wall, direct chute	.040	C.Y.	.019		.67	.67
Pumped concrete	.040	C.Y.	.023		1.10	1.10
Crane & bucket	.040	C.Y.	.032		1.67	1.67
Dampproofing, bituminous, brushed on, 1 coat	1.000	S.F.	.012	.14	.48	.62
2 coats	1.000	S.F.	.016	.28	.64	.92
Sprayed on, 1 coat	1.000	S.F.	.010	.14	.39	.53
2 coats	1.000	S.F.	.016	.27	.64	.91
Troweled on, 1/16" thick	1.000	S.F.	.016	.26	.64	.90
1/8" thick	1.000	S.F.	.020	.46	.81	1.27
1/2" thick	1.000	S.F.	.023	1.49	.92	2.41
Insulation rigid, fiberglass, 1.5#/C.F., unfaced						
1-1/2" thick, R 6.2	1.000	S.F.	.008	.87	.35	1.22
2" thick, R 8.3	1.000	S.F.	.008	.75	.35	1.10
3" thick, R 12.4	1.000	S.F.	.010	.87	.44	1.31
Foamglass, 1-1/2" thick R 2.64	1.000	S.F.	.010	1.44	.44	1.88
2" thick R 5.26	1.000	S.F.	.011	3.48	.48	3.96
Perlite, 1" thick R 2.77	1.000	S.F.	.010	.33	.44	.77
2" thick R 5.55	1.000	S.F.	.011	.65	.48	1.13
Polystyrene, extruded, 1" thick R 5.40	1.000	S.F.	.010	.52	.44	.96
2" thick R 10.8	1.000	S.F.	.011	1.47	.48	1.95
Molded, 1" thick R 3.85	1.000	S.F.	.010	.24	.44	.68
2" thick R 7.70	1.000	S.F.	.011	.83	.48	1.31
Anchor bolts, 2' on center, 1/2" diameter, 8" long	.120	Ea.	.005	.10	.21	.31
12" long	.120	Ea.	.005	.18	.22	.40
3/4" diameter, 8" long	.120	Ea.	.006	.21	.27	.48
12" long	.120	Ea.	.006	.27	.28	.55
Sill plates, treated lumber, 2" x 4"	.250	L.F.	.007	.15	.32	.47
4" x 4"	.250	L.F.	.007	.38	.30	.68

2 | FOUNDATIONS — 16 | Wood Wall Foundation Systems

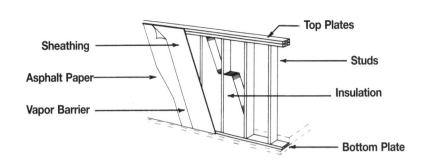

System Description	QUAN.	UNIT	LABOR HOURS	COST PER S.F. MAT.	COST PER S.F. INST.	COST PER S.F. TOTAL
2" X 4" STUDS, 16" O.C., WALL						
Studs, 2" x 4", 16" O.C., treated	1.000	L.F.	.015	.58	.63	1.21
Plates, double top plate, single bottom plate, treated, 2" x 4"	.750	L.F.	.011	.44	.47	.91
Sheathing, 1/2", exterior grade, CDX, treated	1.000	S.F.	.014	.87	.62	1.49
Asphalt paper, 15# roll	1.100	S.F.	.002	.06	.10	.16
Vapor barrier, 4 mil polyethylene	1.000	S.F.	.002	.05	.09	.14
Insulation, batts, fiberglass, 3-1/2" thick, R 11	1.000	S.F.	.005	.41	.22	.63
TOTAL		S.F.	.049	2.41	2.13	4.54
2" X 6" STUDS, 16" O.C., WALL						
Studs, 2" x 6", 16" O.C., treated	1.000	L.F.	.016	.91	.70	1.61
Plates, double top plate, single bottom plate, treated, 2" x 6"	.750	L.F.	.012	.68	.53	1.21
Sheathing, 5/8" exterior grade, CDX, treated	1.000	S.F.	.015	1.31	.66	1.97
Asphalt paper, 15# roll	1.100	S.F.	.002	.06	.10	.16
Vapor barrier, 4 mil polyethylene	1.000	S.F.	.002	.05	.09	.14
Insulation, batts, fiberglass, 6" thick, R 19	1.000	S.F.	.006	.47	.26	.73
TOTAL		S.F.	.053	3.48	2.34	5.82
2" X 8" STUDS, 16" O.C., WALL						
Studs, 2" x 8", 16" O.C. treated	1.000	L.F.	.018	1.26	.78	2.04
Plates, double top plate, single bottom plate, treated, 2" x 8"	.750	L.F.	.013	.95	.59	1.54
Sheathing, 3/4" exterior grade, CDX, treated	1.000	S.F.	.016	1.51	.72	2.23
Asphalt paper, 15# roll	1.100	S.F.	.002	.06	.10	.16
Vapor barrier, 4 mil polyethylene	1.000	S.F.	.002	.05	.09	.14
Insulation, batts, fiberglass, 9" thick, R 30	1.000	S.F.	.006	1.07	.26	1.33
TOTAL		S.F.	.057	4.90	2.54	7.44

The costs in this system are based on a sq. ft. of wall area. Do not subtract for window or door openings. The costs assume a 4' high wall.

Description	QUAN.	UNIT	LABOR HOURS	COST PER S.F. MAT.	COST PER S.F. INST.	COST PER S.F. TOTAL

Wood Wall Foundation Price Sheet	QUAN.	UNIT	LABOR HOURS	COST PER S.F.		
				MAT.	INST.	TOTAL
Studs, treated, 2" x 4", 12" O.C.	1.250	L.F.	.018	.73	.79	1.52
16" O.C.	1.000	L.F.	.015	.58	.63	1.21
2" x 6", 12" O.C.	1.250	L.F.	.020	1.14	.88	2.02
16" O.C.	1.000	L.F.	.016	.91	.70	1.61
2" x 8", 12" O.C.	1.250	L.F.	.022	1.58	.98	2.56
16" O.C.	1.000	L.F.	.018	1.26	.78	2.04
Plates, treated double top single bottom, 2" x 4"	.750	L.F.	.011	.44	.47	.91
2" x 6"	.750	L.F.	.012	.68	.53	1.21
2" x 8"	.750	L.F.	.013	.95	.59	1.54
Sheathing, treated exterior grade CDX, 1/2" thick	1.000	S.F.	.014	.87	.62	1.49
5/8" thick	1.000	S.F.	.015	1.31	.66	1.97
3/4" thick	1.000	S.F.	.016	1.51	.72	2.23
Asphalt paper, 15# roll	1.100	S.F.	.002	.06	.10	.16
Vapor barrier, polyethylene, 4 mil	1.000	S.F.	.002	.03	.09	.12
10 mil	1.000	S.F.	.002	.06	.09	.15
Insulation, rigid, fiberglass, 1.5#/C.F., unfaced	1.000	S.F.	.008	.48	.35	.83
1-1/2" thick, R 6.2	1.000	S.F.	.008	.87	.35	1.22
2" thick, R 8.3	1.000	S.F.	.008	.75	.35	1.10
3" thick, R 12.4	1.000	S.F.	.010	.89	.45	1.34
Foamglass 1 1/2" thick, R 2.64	1.000	S.F.	.010	1.44	.44	1.88
2" thick, R 5.26	1.000	S.F.	.011	3.48	.48	3.96
Perlite 1" thick, R 2.77	1.000	S.F.	.010	.33	.44	.77
2" thick, R 5.55	1.000	S.F.	.011	.65	.48	1.13
Polystyrene, extruded, 1" thick, R 5.40	1.000	S.F.	.010	.52	.44	.96
2" thick, R 10.8	1.000	S.F.	.011	1.47	.48	1.95
Molded 1" thick, R 3.85	1.000	S.F.	.010	.24	.44	.68
2" thick, R 7.7	1.000	S.F.	.011	.83	.48	1.31
Non rigid, batts, fiberglass, paper backed, 3-1/2" thick roll, R 11	1.000	S.F.	.005	.41	.22	.63
6", R 19	1.000	S.F.	.006	.47	.26	.73
9", R 30	1.000	S.F.	.006	1.07	.26	1.33
12", R 38	1.000	S.F.	.006	.99	.26	1.25
Mineral fiber, paper backed, 3-1/2", R 13	1.000	S.F.	.005	.42	.22	.64
6", R 19	1.000	S.F.	.005	.55	.22	.77
10", R 30	1.000	S.F.	.006	.81	.26	1.07

2 | FOUNDATIONS   20 | Floor Slab Systems

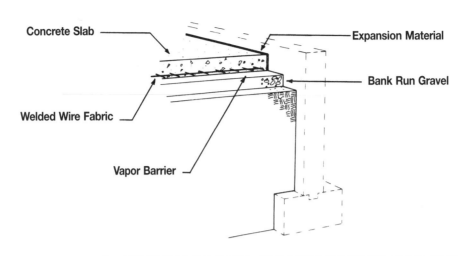

System Description	QUAN.	UNIT	LABOR HOURS	COST PER S.F. MAT.	COST PER S.F. INST.	COST PER S.F. TOTAL
4" THICK SLAB						
Concrete, 4" thick, 3000 psi concrete	.012	C.Y.		1.37		1.37
Place concrete, direct chute	.012	C.Y.	.005		.19	.19
Bank run gravel, 4" deep	1.000	S.F.	.001	.38	.04	.42
Polyethylene vapor barrier, .006" thick	1.000	S.F.	.002	.05	.09	.14
Edge forms, expansion material	.100	L.F.	.005	.03	.20	.23
Welded wire fabric, 6 x 6, 10/10 (W1.4/W1.4)	1.100	S.F.	.005	.15	.25	.40
Steel trowel finish	1.000	S.F.	.015		.58	.58
TOTAL		S.F.	.033	1.98	1.35	3.33
6" THICK SLAB						
Concrete, 6" thick, 3000 psi concrete	.019	C.Y.		2.17		2.17
Place concrete, direct chute	.019	C.Y.	.008		.29	.29
Bank run gravel, 4" deep	1.000	S.F.	.001	.38	.04	.42
Polyethylene vapor barrier, .006" thick	1.000	S.F.	.002	.05	.09	.14
Edge forms, expansion material	.100	L.F.	.005	.03	.20	.23
Welded wire fabric, 6 x 6, 10/10 (W1.4/W1.4)	1.100	S.F.	.005	.15	.25	.40
Steel trowel finish	1.000	S.F.	.015		.58	.58
TOTAL		S.F.	.036	2.78	1.45	4.23

The slab costs in this section are based on a cost per square foot of floor area.

Description	QUAN.	UNIT	LABOR HOURS	COST PER S.F. MAT.	COST PER S.F. INST.	COST PER S.F. TOTAL

Floor Slab Price Sheet	QUAN.	UNIT	LABOR HOURS	COST PER S.F. MAT.	COST PER S.F. INST.	COST PER S.F. TOTAL
Concrete, 4" thick slab, 2000 psi concrete	.012	C.Y.		1.32		1.32
2500 psi concrete	.012	C.Y.		1.33		1.33
3000 psi concrete	.012	C.Y.		1.37		1.37
3500 psi concrete	.012	C.Y.		1.39		1.39
4000 psi concrete	.012	C.Y.		1.43		1.43
4500 psi concrete	.012	C.Y.		1.45		1.45
5" thick slab, 2000 psi concrete	.015	C.Y.		1.65		1.65
2500 psi concrete	.015	C.Y.		1.67		1.67
3000 psi concrete	.015	C.Y.		1.71		1.71
3500 psi concrete	.015	C.Y.		1.74		1.74
4000 psi concrete	.015	C.Y.		1.79		1.79
4500 psi concrete	.015	C.Y.		1.82		1.82
6" thick slab, 2000 psi concrete	.019	C.Y.		2.09		2.09
2500 psi concrete	.019	C.Y.		2.11		2.11
3000 psi concrete	.019	C.Y.		2.17		2.17
3500 psi concrete	.019	C.Y.		2.20		2.20
4000 psi concrete	.019	C.Y.		2.26		2.26
4500 psi concrete	.019	C.Y.		2.30		2.30
Place concrete, 4" slab, direct chute	.012	C.Y.	.005		.19	.19
Pumped concrete	.012	C.Y.	.006		.29	.29
Crane & bucket	.012	C.Y.	.008		.41	.41
5" slab, direct chute	.015	C.Y.	.007		.23	.23
Pumped concrete	.015	C.Y.	.007		.36	.36
Crane & bucket	.015	C.Y.	.010		.52	.52
6" slab, direct chute	.019	C.Y.	.008		.29	.29
Pumped concrete	.019	C.Y.	.009		.44	.44
Crane & bucket	.019	C.Y.	.012		.65	.65
Gravel, bank run, 4" deep	1.000	S.F.	.001	.38	.04	.42
6" deep	1.000	S.F.	.001	.51	.05	.56
9" deep	1.000	S.F.	.001	.74	.08	.82
12" deep	1.000	S.F.	.001	1.01	.10	1.11
3/4" crushed stone, 3" deep	1.000	S.F.	.001	.38	.04	.42
6" deep	1.000	S.F.	.001	.76	.08	.84
9" deep	1.000	S.F.	.002	1.10	.11	1.21
12" deep	1.000	S.F.	.002	1.76	.13	1.89
Vapor barrier polyethylene, .004" thick	1.000	S.F.	.002	.03	.09	.12
.006" thick	1.000	S.F.	.002	.05	.09	.14
Edge forms, expansion material, 4" thick slab	.100	L.F.	.004	.02	.13	.15
6" thick slab	.100	L.F.	.005	.03	.20	.23
Welded wire fabric 6 x 6, 10/10 (W1.4/W1.4)	1.100	S.F.	.005	.15	.25	.40
6 x 6, 6/6 (W2.9/W2.9)	1.100	S.F.	.006	.24	.30	.54
4 x 4, 10/10 (W1.4/W1.4)	1.100	S.F.	.006	.22	.28	.50
Finish concrete, screed finish	1.000	S.F.	.009		.36	.36
Float finish	1.000	S.F.	.011		.44	.44
Steel trowel, for resilient floor	1.000	S.F.	.013		.53	.53
For finished floor	1.000	S.F.	.015		.58	.58

Division 3
Framing

3 | FRAMING — 02 | Floor Framing Systems

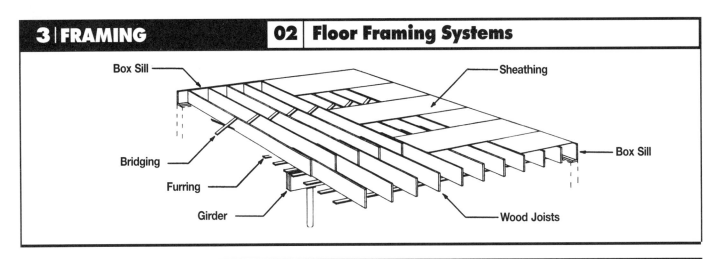

System Description	QUAN.	UNIT	LABOR HOURS	COST PER S.F. MAT.	COST PER S.F. INST.	COST PER S.F. TOTAL
2" X 8", 16" O.C.						
Wood joists, 2" x 8", 16" O.C.	1.000	L.F.	.015	.92	.63	1.55
Bridging, 1" x 3", 6' O.C.	.080	Pr.	.005	.03	.21	.24
Box sills, 2" x 8"	.150	L.F.	.002	.14	.09	.23
Concrete filled steel column, 4" diameter	.125	L.F.	.002	.12	.11	.23
Girder, built up from three 2" x 8"	.125	L.F.	.013	.34	.58	.92
Sheathing, plywood, subfloor, 5/8" CDX	1.000	S.F.	.012	.65	.52	1.17
Furring, 1" x 3", 16" O.C.	1.000	L.F.	.023	.25	1	1.25
TOTAL		S.F.	.072	2.45	3.14	5.59
2" X 10", 16" O.C.						
Wood joists, 2" x 10", 16" OC	1.000	L.F.	.018	1.31	.78	2.09
Bridging, 1" x 3", 6' OC	.080	Pr.	.005	.03	.21	.24
Box sills, 2" x 10"	.150	L.F.	.003	.20	.12	.32
Concrete filled steel column, 4" diameter	.125	L.F.	.002	.12	.11	.23
Girder, built up from three 2" x 8"	.125	L.F.	.014	.49	.62	1.11
Sheathing, plywood, subfloor, 5/8" CDX	1.000	S.F.	.012	.65	.52	1.17
Furring, 1" x 3", 16" OC	1.000	L.F.	.023	.25	1	1.25
TOTAL		S.F.	.077	3.05	3.36	6.41
2" X 12", 16" O.C.						
Wood joists, 2" x 12", 16" O.C.	1.000	L.F.	.018	1.57	.80	2.37
Bridging, 1" x 3", 6' O.C.	.080	Pr.	.005	.03	.21	.24
Box sills, 2" x 12"	.150	L.F.	.003	.24	.12	.36
Concrete filled steel column, 4" diameter	.125	L.F.	.002	.12	.11	.23
Girder, built up from three 2" x 12"	.125	L.F.	.015	.59	.65	1.24
Sheathing, plywood, subfloor, 5/8" CDX	1.000	S.F.	.012	.65	.52	1.17
Furring, 1" x 3", 16" O.C.	1.000	L.F.	.023	.25	1	1.25
TOTAL		S.F.	.078	3.45	3.41	6.86

Floor costs on this page are given on a cost per square foot basis.

Description	QUAN.	UNIT	LABOR HOURS	COST PER S.F. MAT.	COST PER S.F. INST.	COST PER S.F. TOTAL

Floor Framing Price Sheet (Wood)	QUAN.	UNIT	LABOR HOURS	COST PER S.F. MAT.	COST PER S.F. INST.	COST PER S.F. TOTAL
Joists, #2 or better, pine, 2" x 4", 12" O.C.	1.250	L.F.	.016	.51	.70	1.21
16" O.C.	1.000	L.F.	.013	.41	.56	.97
2" x 6", 12" O.C.	1.250	L.F.	.016	.79	.70	1.49
16" O.C.	1.000	L.F.	.013	.63	.56	1.19
2" x 8", 12" O.C.	1.250	L.F.	.018	1.15	.79	1.94
16" O.C.	1.000	L.F.	.015	.92	.63	1.55
2" x 10", 12" O.C.	1.250	L.F.	.022	1.64	.98	2.62
16" O.C.	1.000	L.F.	.018	1.31	.78	2.09
2"x 12", 12" O.C.	1.250	L.F.	.023	1.96	1	2.96
16" O.C.	1.000	L.F.	.018	1.57	.80	2.37
Bridging, wood 1" x 3", joists 12" O.C.	.100	Pr.	.006	.04	.27	.31
16" O.C.	.080	Pr.	.005	.03	.21	.24
Metal, galvanized, joists 12" O.C.	.100	Pr.	.006	.14	.27	.41
16" O.C.	.080	Pr.	.005	.11	.21	.32
Compression type, joists 12" O.C.	.100	Pr.	.004	.17	.17	.34
16" O.C.	.080	Pr.	.003	.13	.14	.27
Box sills, #2 or better pine, 2" x 4"	.150	L.F.	.002	.06	.08	.14
2" x 6"	.150	L.F.	.002	.09	.08	.17
2" x 8"	.150	L.F.	.002	.14	.09	.23
2" x 10"	.150	L.F.	.003	.20	.12	.32
2" x 12"	.150	L.F.	.003	.24	.12	.36
Girders, including lally columns, 3 pieces spiked together, 2" x 8"	.125	L.F.	.015	.46	.69	1.15
2" x 10"	.125	L.F.	.016	.61	.73	1.34
2" x 12"	.125	L.F.	.017	.71	.76	1.47
Solid girders, 3" x 8"	.040	L.F.	.004	.24	.18	.42
3" x 10"	.040	L.F.	.004	.27	.19	.46
3" x 12"	.040	L.F.	.004	.30	.20	.50
4" x 8"	.040	L.F.	.004	.29	.20	.49
4" x 10"	.040	L.F.	.004	.37	.21	.58
4" x 12"	.040	L.F.	.004	.35	.21	.56
Steel girders, bolted & including fabrication, wide flange shapes						
12" deep, 14#/l.f.	.040	L.F.	.003	.82	.23	1.05
10" deep, 15#/l.f.	.040	L.F.	.003	.82	.23	1.05
8" deep, 10#/l.f.	.040	L.F.	.003	.54	.23	.77
6" deep, 9#/l.f.	.040	L.F.	.003	.49	.23	.72
5" deep, 16#/l.f.	.040	L.F.	.003	.82	.23	1.05
Sheathing, plywood exterior grade CDX, 1/2" thick	1.000	S.F.	.011	.52	.50	1.02
5/8" thick	1.000	S.F.	.012	.65	.52	1.17
3/4" thick	1.000	S.F.	.013	.78	.56	1.34
Boards, 1" x 8" laid regular	1.000	S.F.	.016	1.41	.70	2.11
Laid diagonal	1.000	S.F.	.019	1.41	.82	2.23
1" x 10" laid regular	1.000	S.F.	.015	1.17	.63	1.80
Laid diagonal	1.000	S.F.	.018	1.17	.78	1.95
Furring, 1" x 3", 12" O.C.	1.250	L.F.	.029	.31	1.25	1.56
16" O.C.	1.000	L.F.	.023	.25	1	1.25
24" O.C.	.750	L.F.	.017	.19	.75	.94

3 | FRAMING — 04 | Floor Framing Systems

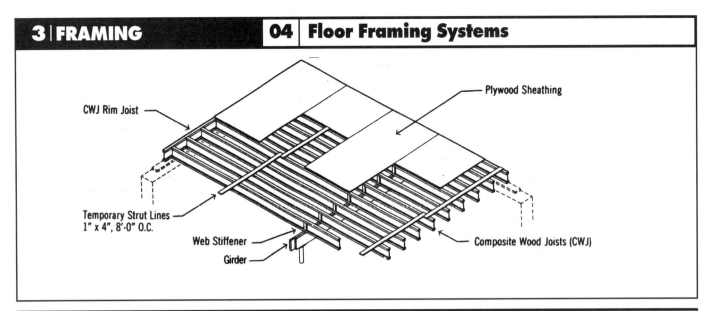

System Description	QUAN.	UNIT	LABOR HOURS	COST PER S.F. MAT.	COST PER S.F. INST.	COST PER S.F. TOTAL
9-1/2" COMPOSITE WOOD JOISTS, 16" O.C.						
CWJ, 9-1/2", 16" O.C., 15' span	1.000	L.F.	.018	2.05	.78	2.83
Temp. strut line, 1" x 4", 8' O.C.	.160	L.F.	.003	.07	.14	.21
CWJ rim joist, 9-1/2"	.150	L.F.	.003	.31	.12	.43
Concrete filled steel column, 4" diameter	.125	L.F.	.002	.12	.11	.23
Girder, built up from three 2" x 8"	.125	L.F.	.013	.34	.58	.92
Sheathing, plywood, subfloor, 5/8" CDX	1.000	S.F.	.012	.65	.52	1.17
TOTAL		S.F.	.051	3.54	2.25	5.79
11-1/2" COMPOSITE WOOD JOISTS, 16" O.C.						
CWJ, 11-1/2", 16" O.C., 18' span	1.000	L.F.	.018	2.18	.80	2.98
Temp. strut line, 1" x 4", 8' O.C.	.160	L.F.	.003	.07	.14	.21
CWJ rim joist, 11-1/2"	.150	L.F.	.003	.33	.12	.45
Concrete filled steel column, 4" diameter	.125	L.F.	.002	.12	.11	.23
Girder, built up from three 2" x 10"	.125	L.F.	.014	.49	.62	1.11
Sheathing, plywood, subfloor, 5/8" CDX	1.000	S.F.	.012	.65	.52	1.17
TOTAL		S.F.	.052	3.84	2.31	6.15
14" COMPOSITE WOOD JOISTS, 16" O.C.						
CWJ, 14", 16" O.C., 22' span	1.000	L.F.	.020	2.55	.85	3.40
Temp. strut line, 1" x 4", 8' O.C.	.160	L.F.	.003	.07	.14	.21
CWJ rim joist, 14"	.150	L.F.	.003	.38	.13	.51
Concrete filled steel column, 4" diameter	.600	L.F.	.002	.12	.11	.23
Girder, built up from three 2" x 12"	.600	L.F.	.015	.59	.65	1.24
Sheathing, plywood, subfloor, 5/8" CDX	1.000	S.F.	.012	.65	.52	1.17
TOTAL		S.F.	.055	4.36	2.40	6.76

Floor costs on this page are given on a cost per square foot basis.

Description	QUAN.	UNIT	LABOR HOURS	COST PER S.F. MAT.	COST PER S.F. INST.	COST PER S.F. TOTAL

Floor Framing Price Sheet (Wood)	QUAN.	UNIT	LABOR HOURS	COST PER S.F.		
				MAT.	INST.	TOTAL
Composite wood joist 9-1/2" deep, 12" O.C.	1.250	L.F.	.022	2.56	.97	3.53
16" O.C.	1.000	L.F.	.018	2.05	.78	2.83
11-1/2" deep, 12" O.C.	1.250	L.F.	.023	2.72	.99	3.71
16" O.C.	1.000	L.F.	.018	2.18	.80	2.98
14" deep, 12" O.C.	1.250	L.F.	.024	3.19	1.06	4.25
16" O.C.	1.000	L.F.	.020	2.55	.85	3.40
16" deep, 12" O.C.	1.250	L.F.	.026	3.56	1.12	4.68
16" O.C.	1.000	L.F.	.021	2.85	.90	3.75
CWJ rim joist, 9-1/2"	.150	L.F.	.003	.31	.12	.43
11-1/2"	.150	L.F.	.003	.33	.12	.45
14"	.150	L.F.	.003	.38	.13	.51
16"	.150	L.F.	.003	.43	.13	.56
Girders, including lally columns, 3 pieces spiked together, 2" x 8"	.125	L.F.	.015	.46	.69	1.15
2" x 10"	.125	L.F.	.016	.61	.73	1.34
2" x 12"	.125	L.F.	.017	.71	.76	1.47
Solid girders, 3" x 8"	.040	L.F.	.004	.24	.18	.42
3" x 10"	.040	L.F.	.004	.27	.19	.46
3" x 12"	.040	L.F.	.004	.30	.20	.50
4" x 8"	.040	L.F.	.004	.29	.20	.49
4" x 10"	.040	L.F.	.004	.37	.21	.58
4" x 12"	.040	L.F.	.004	.35	.21	.56
Steel girders, bolted & including fabrication, wide flange shapes						
12" deep, 14#/l.f.	.040	L.F.	.061	18.75	5.40	24.15
10" deep, 15#/l.f.	.040	L.F.	.067	20.50	5.90	26.40
8" deep, 10#/l.f.	.040	L.F.	.067	13.55	5.90	19.45
6" deep, 9#/l.f.	.040	L.F.	.067	12.20	5.90	18.10
5" deep, 16#/l.f.	.040	L.F.	.064	19.80	5.70	25.50
Sheathing, plywood exterior grade CDX, 1/2" thick	1.000	S.F.	.011	.52	.50	1.02
5/8" thick	1.000	S.F.	.012	.65	.52	1.17
3/4" thick	1.000	S.F.	.013	.78	.56	1.34
Boards, 1" x 8" laid regular	1.000	S.F.	.016	1.41	.70	2.11
Laid diagonal	1.000	S.F.	.019	1.41	.82	2.23
1" x 10" laid regular	1.000	S.F.	.015	1.17	.63	1.80
Laid diagonal	1.000	S.F.	.018	1.17	.78	1.95
Furring, 1" x 3", 12" O.C.	1.250	L.F.	.029	.31	1.25	1.56
16" O.C.	1.000	L.F.	.023	.25	1	1.25
24" O.C.	.750	L.F.	.017	.19	.75	.94

3 | FRAMING 06 | Floor Framing Systems

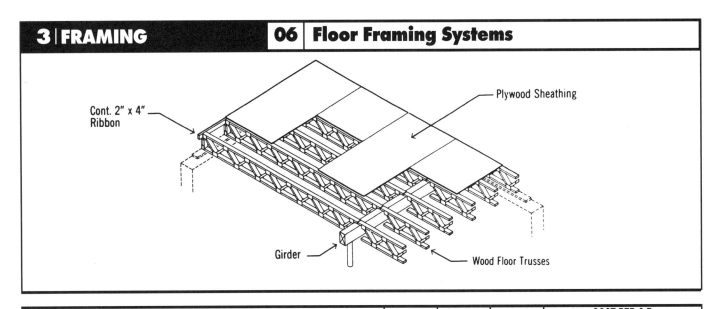

System Description	QUAN.	UNIT	LABOR HOURS	COST PER S.F. MAT.	COST PER S.F. INST.	COST PER S.F. TOTAL
12" OPEN WEB JOISTS, 16" O.C.						
OWJ 12", 16" O.C., 21' span	1.000	L.F.	.018	2	.80	2.80
Continuous ribbing, 2" x 4"	.150	L.F.	.002	.06	.08	.14
Concrete filled steel column, 4" diameter	.125	L.F.	.002	.12	.11	.23
Girder, built up from three 2" x 8"	.125	L.F.	.013	.34	.58	.92
Sheathing, plywood, subfloor, 5/8" CDX	1.000	S.F.	.012	.65	.52	1.17
Furring, 1" x 3", 16" O.C.	1.000	L.F.	.023	.25	1	1.25
TOTAL		S.F.	.070	3.42	3.09	6.51
14" OPEN WEB WOOD JOISTS, 16" O.C.						
OWJ 14", 16" O.C., 22' span	1.000	L.F.	.020	2.33	.85	3.18
Continuous ribbing, 2" x 4"	.150	L.F.	.002	.06	.08	.14
Concrete filled steel column, 4" diameter	.125	L.F.	.002	.12	.11	.23
Girder, built up from three 2" x 10"	.125	L.F.	.014	.49	.62	1.11
Sheathing, plywood, subfloor, 5/8" CDX	1.000	S.F.	.012	.65	.52	1.17
Furring, 1" x 3", 16" O.C.	1.000	L.F.	.023	.25	1	1.25
TOTAL		S.F.	.073	3.90	3.18	7.08
16" OPEN WEB WOOD JOISTS, 16" O.C.						
OWJ 16", 16" O.C., 24' span	1.000	L.F.	.021	2.43	.90	3.33
Continuous ribbing, 2" x 4"	.150	L.F.	.002	.06	.08	.14
Concrete filled steel column, 4" diameter	.125	L.F.	.002	.12	.11	.23
Girder, built up from three 2" x 12"	.125	L.F.	.015	.59	.65	1.24
Sheathing, plywood, subfloor, 5/8" CDX	1.000	S.F.	.012	.65	.52	1.17
Furring, 1" x 3", 16" O.C.	1.000	L.F.	.023	.25	1	1.25
TOTAL		S.F.	.075	4.10	3.26	7.36

Floor costs on this page are given on a cost per square foot basis.

Description	QUAN.	UNIT	LABOR HOURS	COST PER S.F. MAT.	COST PER S.F. INST.	COST PER S.F. TOTAL

Floor Framing Price Sheet (Wood)	QUAN.	UNIT	LABOR HOURS	COST PER S.F.		
				MAT.	INST.	TOTAL
Open web joists, 12" deep, 12" O.C.	1.250	L.F.	.023	2.50	.99	3.49
16" O.C.	1.000	L.F.	.018	2	.80	2.80
14" deep, 12" O.C.	1.250	L.F.	.024	2.91	1.06	3.97
16" O.C.	1.000	L.F.	.020	2.33	.85	3.18
16" deep, 12" O.C.	1.250	L.F.	.026	3.03	1.12	4.15
16" O.C.	1.000	L.F.	.021	2.43	.90	3.33
18" deep, 12" O.C.	1.250	L.F.	.027	3.09	1.18	4.27
16" O.C.	1.000	L.F.	.022	2.48	.95	3.43
Continuous ribbing, 2" x 4"	.150	L.F.	.002	.06	.08	.14
2" x 6"	.150	L.F.	.002	.09	.08	.17
2" x 8"	.150	L.F.	.002	.14	.09	.23
2" x 10"	.150	L.F.	.003	.20	.12	.32
2" x 12"	.150	L.F.	.003	.24	.12	.36
Girders, including lally columns, 3 pieces spiked together, 2" x 8"	.125	L.F.	.015	.46	.69	1.15
2" x 10"	.125	L.F.	.016	.61	.73	1.34
2" x 12"	.125	L.F.	.017	.71	.76	1.47
Solid girders, 3" x 8"	.040	L.F.	.004	.24	.18	.42
3" x 10"	.040	L.F.	.004	.27	.19	.46
3" x 12"	.040	L.F.	.004	.30	.20	.50
4" x 8"	.040	L.F.	.004	.29	.20	.49
4" x 10"	.040	L.F.	.004	.37	.21	.58
4" x 12"	.040	L.F.	.004	.35	.21	.56
Steel girders, bolted & including fabrication, wide flange shapes						
12" deep, 14#/l.f.	.040	L.F.	.061	18.75	5.40	24.15
10" deep, 15#/l.f.	.040	L.F.	.067	20.50	5.90	26.40
8" deep, 10#/l.f.	.040	L.F.	.067	13.55	5.90	19.45
6" deep, 9#/l.f.	.040	L.F.	.067	12.20	5.90	18.10
5" deep, 16#/l.f.	.040	L.F.	.064	19.80	5.70	25.50
Sheathing, plywood exterior grade CDX, 1/2" thick	1.000	S.F.	.011	.52	.50	1.02
5/8" thick	1.000	S.F.	.012	.65	.52	1.17
3/4" thick	1.000	S.F.	.013	.78	.56	1.34
Boards, 1" x 8" laid regular	1.000	S.F.	.016	1.41	.70	2.11
Laid diagonal	1.000	S.F.	.019	1.41	.82	2.23
1" x 10" laid regular	1.000	S.F.	.015	1.17	.63	1.80
Laid diagonal	1.000	S.F.	.018	1.17	.78	1.95
Furring, 1" x 3", 12" O.C.	1.250	L.F.	.029	.31	1.25	1.56
16" O.C.	1.000	L.F.	.023	.25	1	1.25
24" O.C.	.750	L.F.	.017	.19	.75	.94

3 | FRAMING   08 | Exterior Wall Framing Systems

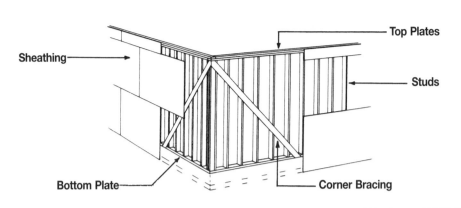

System Description	QUAN.	UNIT	LABOR HOURS	COST PER S.F. MAT.	COST PER S.F. INST.	COST PER S.F. TOTAL
2" X 4", 16" O.C.						
2" x 4" studs, 16" O.C.	1.000	L.F.	.015	.41	.63	1.04
Plates, 2" x 4", double top, single bottom	.375	L.F.	.005	.15	.24	.39
Corner bracing, let-in, 1" x 6"	.063	L.F.	.003	.05	.15	.20
Sheathing, 1/2" plywood, CDX	1.000	S.F.	.011	.52	.50	1.02
TOTAL		S.F.	.034	1.13	1.52	2.65
2" X 4", 24" O.C.						
2" x 4" studs, 24" O.C.	.750	L.F.	.011	.31	.47	.78
Plates, 2" x 4", double top, single bottom	.375	L.F.	.005	.15	.24	.39
Corner bracing, let-in, 1" x 6"	.063	L.F.	.002	.05	.10	.15
Sheathing, 1/2" plywood, CDX	1.000	S.F.	.011	.52	.50	1.02
TOTAL		S.F.	.029	1.03	1.31	2.34
2" X 6", 16" O.C.						
2" x 6" studs, 16" O.C.	1.000	L.F.	.016	.63	.70	1.33
Plates, 2" x 6", double top, single bottom	.375	L.F.	.006	.24	.26	.50
Corner bracing, let-in, 1" x 6"	.063	L.F.	.003	.05	.15	.20
Sheathing, 1/2" plywood, CDX	1.000	S.F.	.014	.52	.62	1.14
TOTAL		S.F.	.039	1.44	1.73	3.17
2" X 6", 24" O.C.						
2" x 6" studs, 24" O.C.	.750	L.F.	.012	.47	.53	1
Plates, 2" x 6", double top, single bottom	.375	L.F.	.006	.24	.26	.50
Corner bracing, let-in, 1" x 6"	.063	L.F.	.002	.05	.10	.15
Sheathing, 1/2" plywood, CDX	1.000	S.F.	.011	.52	.50	1.02
TOTAL		S.F.	.031	1.28	1.39	2.67

The wall costs on this page are given in cost per square foot of wall.
For window and door openings see below.

Description	QUAN.	UNIT	LABOR HOURS	COST PER S.F. MAT.	COST PER S.F. INST.	COST PER S.F. TOTAL

Exterior Wall Framing Price Sheet	QUAN.	UNIT	LABOR HOURS	COST PER S.F.		
				MAT.	INST.	TOTAL
Studs, #2 or better, 2" x 4", 12" O.C.	1.250	L.F.	.018	.51	.79	1.30
16" O.C.	1.000	L.F.	.015	.41	.63	1.04
24" O.C.	.750	L.F.	.011	.31	.47	.78
32" O.C.	.600	L.F.	.009	.25	.38	.63
2" x 6", 12" O.C.	1.250	L.F.	.020	.79	.88	1.67
16" O.C.	1.000	L.F.	.016	.63	.70	1.33
24" O.C.	.750	L.F.	.012	.47	.53	1
32" O.C.	.600	L.F.	.010	.38	.42	.80
2" x 8", 12" O.C.	1.250	L.F.	.025	1.50	1.09	2.59
16" O.C.	1.000	L.F.	.020	1.20	.87	2.07
24" O.C.	.750	L.F.	.015	.90	.65	1.55
32" O.C.	.600	L.F.	.012	.72	.52	1.24
Plates, #2 or better, double top, single bottom, 2" x 4"	.375	L.F.	.005	.15	.24	.39
2" x 6"	.375	L.F.	.006	.24	.26	.50
2" x 8"	.375	L.F.	.008	.45	.33	.78
Corner bracing, let-in 1" x 6" boards, studs, 12" O.C.	.070	L.F.	.004	.05	.16	.21
16" O.C.	.063	L.F.	.003	.05	.15	.20
24" O.C.	.063	L.F.	.002	.05	.10	.15
32" O.C.	.057	L.F.	.002	.04	.09	.13
Let-in steel ("T" shape), studs, 12" O.C.	.070	L.F.	.001	.04	.04	.08
16" O.C.	.063	L.F.	.001	.04	.04	.08
24" O.C.	.063	L.F.	.001	.04	.04	.08
32" O.C.	.057	L.F.	.001	.03	.03	.06
Sheathing, plywood CDX, 3/8" thick	1.000	S.F.	.010	.48	.46	.94
1/2" thick	1.000	S.F.	.011	.52	.50	1.02
5/8" thick	1.000	S.F.	.012	.65	.54	1.19
3/4" thick	1.000	S.F.	.013	.78	.58	1.36
Boards, 1" x 6", laid regular	1.000	S.F.	.025	1.56	1.07	2.63
Laid diagonal	1.000	S.F.	.027	1.56	1.19	2.75
1" x 8", laid regular	1.000	S.F.	.021	1.41	.91	2.32
Laid diagonal	1.000	S.F.	.025	1.41	1.07	2.48
Wood fiber, regular, no vapor barrier, 1/2" thick	1.000	S.F.	.013	.61	.58	1.19
5/8" thick	1.000	S.F.	.013	.79	.58	1.37
Asphalt impregnated 25/32" thick	1.000	S.F.	.013	.31	.58	.89
1/2" thick	1.000	S.F.	.013	.20	.58	.78
Polystyrene, regular, 3/4" thick	1.000	S.F.	.010	.52	.44	.96
2" thick	1.000	S.F.	.011	1.47	.48	1.95
Fiberglass, foil faced, 1" thick	1.000	S.F.	.008	1	.35	1.35
2" thick	1.000	S.F.	.009	1.86	.39	2.25

Window & Door Openings	QUAN.	UNIT	LABOR HOURS	COST EACH		
				MAT.	INST.	TOTAL
The following costs are to be added to the total costs of the wall for each opening. Do not subtract the area of the openings.						
Headers, 2" x 6" double, 2' long	4.000	L.F.	.178	2.52	7.75	10.27
3' long	6.000	L.F.	.267	3.78	11.65	15.43
4' long	8.000	L.F.	.356	5.05	15.50	20.55
5' long	10.000	L.F.	.444	6.30	19.40	25.70
2" x 8" double, 4' long	8.000	L.F.	.376	7.35	16.40	23.75
5' long	10.000	L.F.	.471	9.20	20.50	29.70
6' long	12.000	L.F.	.565	11.05	24.50	35.55
8' long	16.000	L.F.	.753	14.70	33	47.70
2" x 10" double, 4' long	8.000	L.F.	.400	10.50	17.45	27.95
6' long	12.000	L.F.	.600	15.70	26	41.70
8' long	16.000	L.F.	.800	21	35	56
10' long	20.000	L.F.	1.000	26	43.50	69.50
2" x 12" double, 8' long	16.000	L.F.	.853	25	37.50	62.50
12' long	24.000	L.F.	1.280	37.50	56	93.50

3 | FRAMING — 12 | Gable End Roof Framing Systems

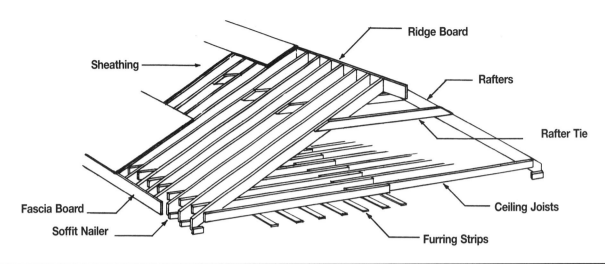

System Description	QUAN.	UNIT	LABOR HOURS	COST PER S.F. MAT.	COST PER S.F. INST.	COST PER S.F. TOTAL
2" X 6" RAFTERS, 16" O.C., 4/12 PITCH						
Rafters, 2" x 6", 16" O.C., 4/12 pitch	1.170	L.F.	.019	.74	.82	1.56
Ceiling joists, 2" x 4", 16" O.C.	1.000	L.F.	.013	.41	.56	.97
Ridge board, 2" x 6"	.050	L.F.	.002	.03	.07	.10
Fascia board, 2" x 6"	.100	L.F.	.005	.07	.23	.30
Rafter tie, 1" x 4", 4' O.C.	.060	L.F.	.001	.03	.05	.08
Soffit nailer (outrigger), 2" x 4", 24" O.C.	.170	L.F.	.004	.07	.19	.26
Sheathing, exterior, plywood, CDX, 1/2" thick	1.170	S.F.	.013	.61	.59	1.20
Furring strips, 1" x 3", 16" O.C.	1.000	L.F.	.023	.25	1	1.25
TOTAL		S.F.	.080	2.21	3.51	5.72
2" X 8" RAFTERS, 16" O.C., 4/12 PITCH						
Rafters, 2" x 8", 16" O.C., 4/12 pitch	1.170	L.F.	.020	1.08	.85	1.93
Ceiling joists, 2" x 6", 16" O.C.	1.000	L.F.	.013	.63	.56	1.19
Ridge board, 2" x 8"	.050	L.F.	.002	.05	.08	.13
Fascia board, 2" x 8"	.100	L.F.	.007	.09	.31	.40
Rafter tie, 1" x 4", 4' O.C.	.060	L.F.	.001	.03	.05	.08
Soffit nailer (outrigger), 2" x 4", 24" O.C.	.170	L.F.	.004	.07	.19	.26
Sheathing, exterior, plywood, CDX, 1/2" thick	1.170	S.F.	.013	.61	.59	1.20
Furring strips, 1" x 3", 16" O.C.	1.000	L.F.	.023	.25	1	1.25
TOTAL		S.F.	.083	2.81	3.63	6.44

The cost of this system is based on the square foot of plan area.
All quantities have been adjusted accordingly.

Description	QUAN.	UNIT	LABOR HOURS	COST PER S.F. MAT.	COST PER S.F. INST.	COST PER S.F. TOTAL

Gable End Roof Framing Price Sheet	QUAN.	UNIT	LABOR HOURS	COST PER S.F.		
				MAT.	INST.	TOTAL
Rafters, #2 or better, 16" O.C., 2" x 6", 4/12 pitch	1.170	L.F.	.019	.74	.82	1.56
8/12 pitch	1.330	L.F.	.027	.84	1.16	2
2" x 8", 4/12 pitch	1.170	L.F.	.020	1.08	.85	1.93
8/12 pitch	1.330	L.F.	.028	1.22	1.24	2.46
2" x 10", 4/12 pitch	1.170	L.F.	.030	1.53	1.30	2.83
8/12 pitch	1.330	L.F.	.043	1.74	1.88	3.62
24" O.C., 2" x 6", 4/12 pitch	.940	L.F.	.015	.59	.66	1.25
8/12 pitch	1.060	L.F.	.021	.67	.92	1.59
2" x 8", 4/12 pitch	.940	L.F.	.016	.86	.69	1.55
8/12 pitch	1.060	L.F.	.023	.98	.99	1.97
2" x 10", 4/12 pitch	.940	L.F.	.024	1.23	1.04	2.27
8/12 pitch	1.060	L.F.	.034	1.39	1.49	2.88
Ceiling joist, #2 or better, 2" x 4", 16" O.C.	1.000	L.F.	.013	.41	.56	.97
24" O.C.	.750	L.F.	.010	.31	.42	.73
2" x 6", 16" O.C.	1.000	L.F.	.013	.63	.56	1.19
24" O.C.	.750	L.F.	.010	.47	.42	.89
2" x 8", 16" O.C.	1.000	L.F.	.015	.92	.63	1.55
24" O.C.	.750	L.F.	.011	.69	.47	1.16
2" x 10", 16" O.C.	1.000	L.F.	.018	1.31	.78	2.09
24" O.C.	.750	L.F.	.013	.98	.59	1.57
Ridge board, #2 or better, 1" x 6"	.050	L.F.	.001	.04	.06	.10
1" x 8"	.050	L.F.	.001	.05	.06	.11
1" x 10"	.050	L.F.	.002	.07	.07	.14
2" x 6"	.050	L.F.	.002	.03	.07	.10
2" x 8"	.050	L.F.	.002	.05	.08	.13
2" x 10"	.050	L.F.	.002	.07	.09	.16
Fascia board, #2 or better, 1" x 6"	.100	L.F.	.004	.05	.17	.22
1" x 8"	.100	L.F.	.005	.06	.20	.26
1" x 10"	.100	L.F.	.005	.07	.22	.29
2" x 6"	.100	L.F.	.006	.07	.25	.32
2" x 8"	.100	L.F.	.007	.09	.31	.40
2" x 10"	.100	L.F.	.004	.26	.16	.42
Rafter tie, #2 or better, 4' O.C., 1" x 4"	.060	L.F.	.001	.03	.05	.08
1" x 6"	.060	L.F.	.001	.03	.06	.09
2" x 4"	.060	L.F.	.002	.03	.07	.10
2" x 6"	.060	L.F.	.002	.04	.09	.13
Soffit nailer (outrigger), 2" x 4", 16" O.C.	.220	L.F.	.006	.09	.25	.34
24" O.C.	.170	L.F.	.004	.07	.19	.26
2" x 6", 16" O.C.	.220	L.F.	.006	.10	.28	.38
24" O.C.	.170	L.F.	.005	.08	.23	.31
Sheathing, plywood CDX, 4/12 pitch, 3/8" thick.	1.170	S.F.	.012	.56	.54	1.10
1/2" thick	1.170	S.F.	.013	.61	.59	1.20
5/8" thick	1.170	S.F.	.014	.76	.63	1.39
8/12 pitch, 3/8"	1.330	S.F.	.014	.64	.61	1.25
1/2" thick	1.330	S.F.	.015	.69	.67	1.36
5/8" thick	1.330	S.F.	.016	.86	.72	1.58
Boards, 4/12 pitch roof, 1" x 6"	1.170	S.F.	.026	1.83	1.12	2.95
1" x 8"	1.170	S.F.	.021	1.65	.94	2.59
8/12 pitch roof, 1" x 6"	1.330	S.F.	.029	2.07	1.28	3.35
1" x 8"	1.330	S.F.	.024	1.88	1.06	2.94
Furring, 1" x 3", 12" O.C.	1.200	L.F.	.027	.30	1.20	1.50
16" O.C.	1.000	L.F.	.023	.25	1	1.25
24" O.C.	.800	L.F.	.018	.20	.80	1

3 | FRAMING — 16 | Truss Roof Framing Systems

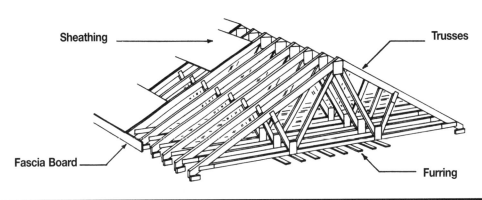

System Description	QUAN.	UNIT	LABOR HOURS	COST PER S.F. MAT.	COST PER S.F. INST.	COST PER S.F. TOTAL
TRUSS, 16" O.C., 4/12 PITCH, 1' OVERHANG, 26' SPAN						
Truss, 40# loading, 16" O.C., 4/12 pitch, 26' span	.030	Ea.	.021	2.57	1.25	3.82
Fascia board, 2" x 6"	.100	L.F.	.005	.07	.23	.30
Sheathing, exterior, plywood, CDX, 1/2" thick	1.170	S.F.	.013	.61	.59	1.20
Furring, 1" x 3", 16" O.C.	1.000	L.F.	.023	.25	1	1.25
TOTAL		S.F.	.062	3.50	3.07	6.57
TRUSS, 16" O.C., 8/12 PITCH, 1' OVERHANG, 26' SPAN						
Truss, 40# loading, 16" O.C., 8/12 pitch, 26' span	.030	Ea.	.023	2.97	1.36	4.33
Fascia board, 2" x 6"	.100	L.F.	.005	.07	.23	.30
Sheathing, exterior, plywood, CDX, 1/2" thick	1.330	S.F.	.015	.69	.67	1.36
Furring, 1" x 3", 16" O.C.	1.000	L.F.	.023	.25	1	1.25
TOTAL		S.F.	.066	3.98	3.26	7.24
TRUSS, 24" O.C., 4/12 PITCH, 1' OVERHANG, 26' SPAN						
Truss, 40# loading, 24" O.C., 4/12 pitch, 26' span	.020	Ea.	.014	1.71	.83	2.54
Fascia board, 2" x 6"	.100	L.F.	.005	.07	.23	.30
Sheathing, exterior, plywood, CDX, 1/2" thick	1.170	S.F.	.013	.61	.59	1.20
Furring, 1" x 3", 16" O.C.	1.000	L.F.	.023	.25	1	1.25
TOTAL		S.F.	.055	2.64	2.65	5.29
TRUSS, 24" O.C., 8/12 PITCH, 1' OVERHANG, 26' SPAN						
Truss, 40# loading, 24" O.C., 8/12 pitch, 26' span	.020	Ea.	.015	1.98	.91	2.89
Fascia board, 2" x 6"	.100	L.F.	.005	.07	.23	.30
Sheathing, exterior, plywood, CDX, 1/2" thick	1.330	S.F.	.015	.69	.67	1.36
Furring, 1" x 3", 16" O.C.	1.000	L.F.	.023	.25	1	1.25
TOTAL		S.F.	.058	2.99	2.81	5.80

The cost of this system is based on the square foot of plan area.
A one foot overhang is included.

Description	QUAN.	UNIT	LABOR HOURS	COST PER S.F. MAT.	COST PER S.F. INST.	COST PER S.F. TOTAL

Truss Roof Framing Price Sheet	QUAN.	UNIT	LABOR HOURS	COST PER S.F.		
				MAT.	INST.	TOTAL
Truss, 40# loading, including 1' overhang, 4/12 pitch, 24' span, 16" O.C.	.033	Ea.	.022	2.72	1.30	4.02
24" O.C.	.022	Ea.	.015	1.82	.86	2.68
26' span, 16" O.C.	.030	Ea.	.021	2.57	1.25	3.82
24" O.C.	.020	Ea.	.014	1.71	.83	2.54
28' span, 16" O.C.	.027	Ea.	.020	2.05	1.21	3.26
24" O.C.	.019	Ea.	.014	1.44	.85	2.29
32' span, 16" O.C.	.024	Ea.	.019	2.64	1.14	3.78
24" O.C.	.016	Ea.	.013	1.76	.76	2.52
36' span, 16" O.C.	.022	Ea.	.019	3.21	1.13	4.34
24" O.C.	.015	Ea.	.013	2.19	.77	2.96
8/12 pitch, 24' span, 16" O.C.	.033	Ea.	.024	3.14	1.42	4.56
24" O.C.	.022	Ea.	.016	2.09	.95	3.04
26' span, 16" O.C.	.030	Ea.	.023	2.97	1.36	4.33
24" O.C.	.020	Ea.	.015	1.98	.91	2.89
28' span, 16" O.C.	.027	Ea.	.022	2.94	1.30	4.24
24" O.C.	.019	Ea.	.016	2.07	.92	2.99
32' span, 16" O.C.	.024	Ea.	.021	3.14	1.26	4.40
24" O.C.	.016	Ea.	.014	2.10	.84	2.94
36' span, 16" O.C.	.022	Ea.	.021	3.61	1.27	4.88
24" O.C.	.015	Ea.	.015	2.46	.86	3.32
Fascia board, #2 or better, 1" x 6"	.100	L.F.	.004	.05	.17	.22
1" x 8"	.100	L.F.	.005	.06	.20	.26
1" x 10"	.100	L.F.	.005	.07	.22	.29
2" x 6"	.100	L.F.	.006	.07	.25	.32
2" x 8"	.100	L.F.	.007	.09	.31	.40
2" x 10"	.100	L.F.	.009	.13	.39	.52
Sheathing, plywood CDX, 4/12 pitch, 3/8" thick	1.170	S.F.	.012	.56	.54	1.10
1/2" thick	1.170	S.F.	.013	.61	.59	1.20
5/8" thick	1.170	S.F.	.014	.76	.63	1.39
8/12 pitch, 3/8" thick	1.330	S.F.	.014	.64	.61	1.25
1/2" thick	1.330	S.F.	.015	.69	.67	1.36
5/8" thick	1.330	S.F.	.016	.86	.72	1.58
Boards, 4/12 pitch, 1" x 6"	1.170	S.F.	.026	1.83	1.12	2.95
1" x 8"	1.170	S.F.	.021	1.65	.94	2.59
8/12 pitch, 1" x 6"	1.330	S.F.	.029	2.07	1.28	3.35
1" x 8"	1.330	S.F.	.024	1.88	1.06	2.94
Furring, 1" x 3", 12" O.C.	1.200	L.F.	.027	.30	1.20	1.50
16" O.C.	1.000	L.F.	.023	.25	1	1.25
24" O.C.	.800	L.F.	.018	.20	.80	1

3 | FRAMING — 20 | Hip Roof Framing Systems

Diagram labels: Ceiling Joists, Hip Rafter, Jack Rafters, Sheathing, Fascia Board

System Description	QUAN.	UNIT	LABOR HOURS	COST PER S.F. MAT.	COST PER S.F. INST.	COST PER S.F. TOTAL
2" X 6", 16" O.C., 4/12 PITCH						
Hip rafters, 2" x 8", 4/12 pitch	.160	L.F.	.004	.15	.16	.31
Jack rafters, 2" x 6", 16" O.C., 4/12 pitch	1.430	L.F.	.038	.90	1.66	2.56
Ceiling joists, 2" x 6", 16" O.C.	1.000	L.F.	.013	.63	.56	1.19
Fascia board, 2" x 8"	.220	L.F.	.016	.20	.68	.88
Soffit nailer (outrigger), 2" x 4", 24" O.C.	.220	L.F.	.006	.09	.25	.34
Sheathing, 1/2" exterior plywood, CDX	1.570	S.F.	.018	.82	.79	1.61
Furring strips, 1" x 3", 16" O.C.	1.000	L.F.	.023	.25	1	1.25
TOTAL		S.F.	.118	3.04	5.10	8.14
2" X 8", 16" O.C., 4/12 PITCH						
Hip rafters, 2" x 10", 4/12 pitch	.160	L.F.	.004	.21	.20	.41
Jack rafters, 2" x 8", 16" O.C., 4/12 pitch	1.430	L.F.	.047	1.32	2.03	3.35
Ceiling joists, 2" x 6", 16" O.C.	1.000	L.F.	.013	.63	.56	1.19
Fascia board, 2" x 8"	.220	L.F.	.012	.16	.53	.69
Soffit nailer (outrigger), 2" x 4", 24" O.C.	.220	L.F.	.006	.09	.25	.34
Sheathing, 1/2" exterior plywood, CDX	1.570	S.F.	.018	.82	.79	1.61
Furring strips, 1" x 3", 16" O.C.	1.000	L.F.	.023	.25	1	1.25
TOTAL		S.F.	.123	3.48	5.36	8.84

The cost of this system is based on S.F. of plan area. Measurement is area under the hip roof only. See gable roof system for added costs.

Description	QUAN.	UNIT	LABOR HOURS	COST PER S.F. MAT.	COST PER S.F. INST.	COST PER S.F. TOTAL

Hip Roof Framing Price Sheet

Hip Roof Framing Price Sheet	QUAN.	UNIT	LABOR HOURS	COST PER S.F. MAT.	COST PER S.F. INST.	COST PER S.F. TOTAL
Hip rafters, #2 or better, 2" x 6", 4/12 pitch	.160	L.F.	.003	.10	.15	.25
8/12 pitch	.210	L.F.	.006	.13	.25	.38
2" x 8", 4/12 pitch	.160	L.F.	.004	.15	.16	.31
8/12 pitch	.210	L.F.	.006	.19	.27	.46
2" x 10", 4/12 pitch	.160	L.F.	.004	.21	.20	.41
8/12 pitch roof	.210	L.F.	.008	.28	.33	.61
Jack rafters, #2 or better, 16" O.C., 2" x 6", 4/12 pitch	1.430	L.F.	.038	.90	1.66	2.56
8/12 pitch	1.800	L.F.	.061	1.13	2.65	3.78
2" x 8", 4/12 pitch	1.430	L.F.	.047	1.32	2.03	3.35
8/12 pitch	1.800	L.F.	.075	1.66	3.26	4.92
2" x 10", 4/12 pitch	1.430	L.F.	.051	1.87	2.22	4.09
8/12 pitch	1.800	L.F.	.082	2.36	3.58	5.94
24" O.C., 2" x 6", 4/12 pitch	1.150	L.F.	.031	.72	1.33	2.05
8/12 pitch	1.440	L.F.	.048	.91	2.12	3.03
2" x 8", 4/12 pitch	1.150	L.F.	.038	1.06	1.63	2.69
8/12 pitch	1.440	L.F.	.060	1.32	2.61	3.93
2" x 10", 4/12 pitch	1.150	L.F.	.041	1.51	1.78	3.29
8/12 pitch	1.440	L.F.	.066	1.89	2.87	4.76
Ceiling joists, #2 or better, 2" x 4", 16" O.C.	1.000	L.F.	.013	.41	.56	.97
24" O.C.	.750	L.F.	.010	.31	.42	.73
2" x 6", 16" O.C.	1.000	L.F.	.013	.63	.56	1.19
24" O.C.	.750	L.F.	.010	.47	.42	.89
2" x 8", 16" O.C.	1.000	L.F.	.015	.92	.63	1.55
24" O.C.	.750	L.F.	.011	.69	.47	1.16
2" x 10", 16" O.C.	1.000	L.F.	.018	1.31	.78	2.09
24" O.C.	.750	L.F.	.013	.98	.59	1.57
Fascia board, #2 or better, 1" x 6"	.220	L.F.	.009	.11	.38	.49
1" x 8"	.220	L.F.	.010	.13	.44	.57
1" x 10"	.220	L.F.	.011	.15	.49	.64
2" x 6"	.220	L.F.	.013	.16	.55	.71
2" x 8"	.220	L.F.	.016	.20	.68	.88
2" x 10"	.220	L.F.	.020	.29	.85	1.14
Soffit nailer (outrigger), 2" x 4", 16" O.C.	.280	L.F.	.007	.11	.32	.43
24" O.C.	.220	L.F.	.006	.09	.25	.34
2" x 8", 16" O.C.	.280	L.F.	.007	.19	.29	.48
24" O.C.	.220	L.F.	.005	.16	.24	.40
Sheathing, plywood CDX, 4/12 pitch, 3/8" thick	1.570	S.F.	.016	.75	.72	1.47
1/2" thick	1.570	S.F.	.018	.82	.79	1.61
5/8" thick	1.570	S.F.	.019	1.02	.85	1.87
8/12 pitch, 3/8" thick	1.900	S.F.	.020	.91	.87	1.78
1/2" thick	1.900	S.F.	.022	.99	.95	1.94
5/8" thick	1.900	S.F.	.023	1.24	1.03	2.27
Boards, 4/12 pitch, 1" x 6" boards	1.450	S.F.	.032	2.26	1.39	3.65
1" x 8" boards	1.450	S.F.	.027	2.04	1.16	3.20
8/12 pitch, 1" x 6" boards	1.750	S.F.	.039	2.73	1.68	4.41
1" x 8" boards	1.750	S.F.	.032	2.47	1.40	3.87
Furring, 1" x 3", 12" O.C.	1.200	L.F.	.027	.30	1.20	1.50
16" O.C.	1.000	L.F.	.023	.25	1	1.25
24" O.C.	.800	L.F.	.018	.20	.80	1

3 | FRAMING — 24 | Gambrel Roof Framing Systems

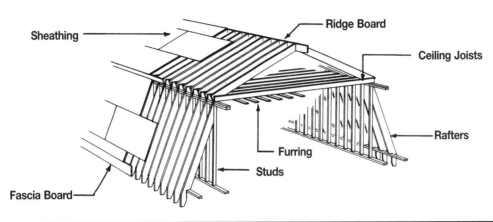

System Description	QUAN.	UNIT	LABOR HOURS	COST PER S.F. MAT.	COST PER S.F. INST.	COST PER S.F. TOTAL
2" X 6" RAFTERS, 16" O.C.						
Roof rafters, 2" x 6", 16" O.C.	1.430	L.F.	.029	.90	1.24	2.14
Ceiling joists, 2" x 6", 16" O.C.	.710	L.F.	.009	.45	.40	.85
Stud wall, 2" x 4", 16" O.C., including plates	.790	L.F.	.012	.32	.55	.87
Furring strips, 1" x 3", 16" O.C.	.710	L.F.	.016	.18	.71	.89
Ridge board, 2" x 8"	.050	L.F.	.002	.05	.08	.13
Fascia board, 2" x 6"	.100	L.F.	.006	.07	.25	.32
Sheathing, exterior grade plywood, 1/2" thick	1.450	S.F.	.017	.75	.73	1.48
TOTAL		S.F.	.091	2.72	3.96	6.68
2" X 8" RAFTERS, 16" O.C.						
Roof rafters, 2" x 8", 16" O.C.	1.430	L.F.	.031	1.32	1.33	2.65
Ceiling joists, 2" x 6", 16" O.C.	.710	L.F.	.009	.45	.40	.85
Stud wall, 2" x 4", 16" O.C., including plates	.790	L.F.	.012	.32	.55	.87
Furring strips, 1" x 3", 16" O.C.	.710	L.F.	.016	.18	.71	.89
Ridge board, 2" x 8"	.050	L.F.	.002	.05	.08	.13
Fascia board, 2" x 8"	.100	L.F.	.007	.09	.31	.40
Sheathing, exterior grade plywood, 1/2" thick	1.450	S.F.	.017	.75	.73	1.48
TOTAL		S.F.	.094	3.16	4.11	7.27

The cost of this system is based on the square foot of plan area on the first floor.

Description	QUAN.	UNIT	LABOR HOURS	COST PER S.F. MAT.	COST PER S.F. INST.	COST PER S.F. TOTAL

Gambrel Roof Framing Price Sheet	QUAN.	UNIT	LABOR HOURS	COST PER S.F.		
				MAT.	INST.	TOTAL
Roof rafters, #2 or better, 2" x 6", 16" O.C.	1.430	L.F.	.029	.90	1.24	2.14
24" O.C.	1.140	L.F.	.023	.72	.99	1.71
2" x 8", 16" O.C.	1.430	L.F.	.031	1.32	1.33	2.65
24" O.C.	1.140	L.F.	.024	1.05	1.06	2.11
2" x 10", 16" O.C.	1.430	L.F.	.046	1.87	2.02	3.89
24" O.C.	1.140	L.F.	.037	1.49	1.61	3.10
Ceiling joist, #2 or better, 2" x 4", 16" O.C.	.710	L.F.	.009	.29	.40	.69
24" O.C.	.570	L.F.	.007	.23	.32	.55
2" x 6", 16" O.C.	.710	L.F.	.009	.45	.40	.85
24" O.C.	.570	L.F.	.007	.36	.32	.68
2" x 8", 16" O.C.	.710	L.F.	.010	.65	.45	1.10
24" O.C.	.570	L.F.	.008	.52	.36	.88
Stud wall, #2 or better, 2" x 4", 16" O.C.	.790	L.F.	.012	.32	.55	.87
24" O.C.	.630	L.F.	.010	.26	.43	.69
2" x 6", 16" O.C.	.790	L.F.	.014	.50	.62	1.12
24" O.C.	.630	L.F.	.011	.40	.49	.89
Furring, 1" x 3", 16" O.C.	.710	L.F.	.016	.18	.71	.89
24" O.C.	.590	L.F.	.013	.15	.59	.74
Ridge board, #2 or better, 1" x 6"	.050	L.F.	.001	.04	.06	.10
1" x 8"	.050	L.F.	.001	.05	.06	.11
1" x 10"	.050	L.F.	.002	.07	.07	.14
2" x 6"	.050	L.F.	.002	.03	.07	.10
2" x 8"	.050	L.F.	.002	.05	.08	.13
2" x 10"	.050	L.F.	.002	.07	.09	.16
Fascia board, #2 or better, 1" x 6"	.100	L.F.	.004	.05	.17	.22
1" x 8"	.100	L.F.	.005	.06	.20	.26
1" x 10"	.100	L.F.	.005	.07	.22	.29
2" x 6"	.100	L.F.	.006	.07	.25	.32
2" x 8"	.100	L.F.	.007	.09	.31	.40
2" x 10"	.100	L.F.	.009	.13	.39	.52
Sheathing, plywood, exterior grade CDX, 3/8" thick	1.450	S.F.	.015	.70	.67	1.37
1/2" thick	1.450	S.F.	.017	.75	.73	1.48
5/8" thick	1.450	S.F.	.018	.94	.78	1.72
3/4" thick	1.450	S.F.	.019	1.13	.84	1.97
Boards, 1" x 6", laid regular	1.450	S.F.	.032	2.26	1.39	3.65
Laid diagonal	1.450	S.F.	.036	2.26	1.55	3.81
1" x 8", laid regular	1.450	S.F.	.027	2.04	1.16	3.20
Laid diagonal	1.450	S.F.	.032	2.04	1.39	3.43

3 | FRAMING — 28 | Mansard Roof Framing Systems

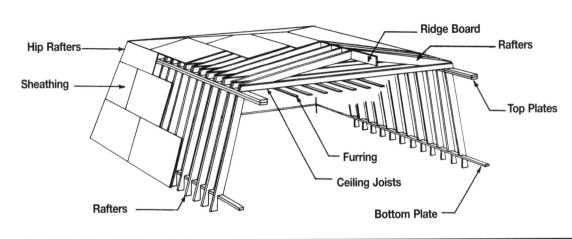

System Description	QUAN.	UNIT	LABOR HOURS	COST PER S.F. MAT.	COST PER S.F. INST.	COST PER S.F. TOTAL
2" X 6" RAFTERS, 16" O.C.						
Roof rafters, 2" x 6", 16" O.C.	1.210	L.F.	.033	.76	1.43	2.19
Rafter plates, 2" x 6", double top, single bottom	.364	L.F.	.010	.23	.43	.66
Ceiling joists, 2" x 4", 16" O.C.	.920	L.F.	.012	.38	.52	.90
Hip rafter, 2" x 6"	.070	L.F.	.002	.04	.10	.14
Jack rafter, 2" x 6", 16" O.C.	1.000	L.F.	.039	.63	1.70	2.33
Ridge board, 2" x 6"	.018	L.F.	.001	.01	.03	.04
Sheathing, exterior grade plywood, 1/2" thick	2.210	S.F.	.025	1.15	1.11	2.26
Furring strips, 1" x 3", 16" O.C.	.920	L.F.	.021	.23	.92	1.15
TOTAL		S.F.	.143	3.43	6.24	9.67
2" X 8" RAFTERS, 16" O.C.						
Roof rafters, 2" x 8", 16" O.C.	1.210	L.F.	.036	1.11	1.56	2.67
Rafter plates, 2" x 8", double top, single bottom	.364	L.F.	.011	.33	.47	.80
Ceiling joists, 2" x 6", 16" O.C.	.920	L.F.	.012	.58	.52	1.10
Hip rafter, 2" x 8"	.070	L.F.	.002	.06	.10	.16
Jack rafter, 2" x 8", 16" O.C.	1.000	L.F.	.048	.92	2.08	3
Ridge board, 2" x 8"	.018	L.F.	.001	.02	.03	.05
Sheathing, exterior grade plywood, 1/2" thick	2.210	S.F.	.025	1.15	1.11	2.26
Furring strips, 1" x 3", 16" O.C.	.920	L.F.	.021	.23	.92	1.15
TOTAL		S.F.	.156	4.40	6.79	11.19

The cost of this system is based on the square foot of plan area.

Description	QUAN.	UNIT	LABOR HOURS	COST PER S.F. MAT.	COST PER S.F. INST.	COST PER S.F. TOTAL

Mansard Roof Framing Price Sheet	QUAN.	UNIT	LABOR HOURS	COST PER S.F.		
				MAT.	INST.	TOTAL
Roof rafters, #2 or better, 2" x 6", 16" O.C.	1.210	L.F.	.033	.76	1.43	2.19
24" O.C.	.970	L.F.	.026	.61	1.14	1.75
2" x 8", 16" O.C.	1.210	L.F.	.036	1.11	1.56	2.67
24" O.C.	.970	L.F.	.029	.89	1.25	2.14
2" x 10", 16" O.C.	1.210	L.F.	.046	1.59	1.98	3.57
24" O.C.	.970	L.F.	.037	1.27	1.59	2.86
Rafter plates, #2 or better double top single bottom, 2" x 6"	.364	L.F.	.010	.23	.43	.66
2" x 8"	.364	L.F.	.011	.33	.47	.80
2" x 10"	.364	L.F.	.014	.48	.60	1.08
Ceiling joist, #2 or better, 2" x 4", 16" O.C.	.920	L.F.	.012	.38	.52	.90
24" O.C.	.740	L.F.	.009	.30	.41	.71
2" x 6", 16" O.C.	.920	L.F.	.012	.58	.52	1.10
24" O.C.	.740	L.F.	.009	.47	.41	.88
2" x 8", 16" O.C.	.920	L.F.	.013	.85	.58	1.43
24" O.C.	.740	L.F.	.011	.68	.47	1.15
Hip rafter, #2 or better, 2" x 6"	.070	L.F.	.002	.04	.10	.14
2" x 8"	.070	L.F.	.002	.06	.10	.16
2" x 10"	.070	L.F.	.003	.09	.13	.22
Jack rafter, #2 or better, 2" x 6", 16" O.C.	1.000	L.F.	.039	.63	1.70	2.33
24" O.C.	.800	L.F.	.031	.50	1.36	1.86
2" x 8", 16" O.C.	1.000	L.F.	.048	.92	2.08	3
24" O.C.	.800	L.F.	.038	.74	1.66	2.40
Ridge board, #2 or better, 1" x 6"	.018	L.F.	.001	.01	.02	.03
1" x 8"	.018	L.F.	.001	.02	.02	.04
1" x 10"	.018	L.F.	.001	.03	.03	.06
2" x 6"	.018	L.F.	.001	.01	.03	.04
2" x 8"	.018	L.F.	.001	.02	.03	.05
2" x 10"	.018	L.F.	.001	.02	.03	.05
Sheathing, plywood exterior grade CDX, 3/8" thick	2.210	S.F.	.023	1.06	1.02	2.08
1/2" thick	2.210	S.F.	.025	1.15	1.11	2.26
5/8" thick	2.210	S.F.	.027	1.44	1.19	2.63
3/4" thick	2.210	S.F.	.029	1.72	1.28	3
Boards, 1" x 6", laid regular	2.210	S.F.	.049	3.45	2.12	5.57
Laid diagonal	2.210	S.F.	.054	3.45	2.36	5.81
1" x 8", laid regular	2.210	S.F.	.040	3.12	1.77	4.89
Laid diagonal	2.210	S.F.	.049	3.12	2.12	5.24
Furring, 1" x 3", 12" O.C.	1.150	L.F.	.026	.29	1.15	1.44
24" O.C.	.740	L.F.	.017	.19	.74	.93

3 | FRAMING — 32 | Shed/Flat Roof Framing Systems

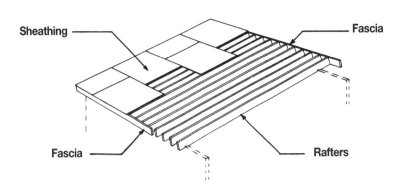

System Description	QUAN.	UNIT	LABOR HOURS	COST PER S.F. MAT.	COST PER S.F. INST.	COST PER S.F. TOTAL
2" X 6", 16" O.C., 4/12 PITCH						
Rafters, 2" x 6", 16" O.C., 4/12 pitch	1.170	L.F.	.019	.74	.82	1.56
Fascia, 2" x 6"	.100	L.F.	.006	.07	.25	.32
Bridging, 1" x 3", 6' O.C.	.080	Pr.	.005	.03	.21	.24
Sheathing, exterior grade plywood, 1/2" thick	1.230	S.F.	.014	.64	.62	1.26
TOTAL		S.F.	.044	1.48	1.90	3.38
2" X 6", 24" O.C., 4/12 PITCH						
Rafters, 2" x 6", 24" O.C., 4/12 pitch	.940	L.F.	.015	.59	.66	1.25
Fascia, 2" x 6"	.100	L.F.	.006	.07	.25	.32
Bridging, 1" x 3", 6' O.C.	.060	Pr.	.004	.02	.16	.18
Sheathing, exterior grade plywood, 1/2" thick	1.230	S.F.	.014	.64	.62	1.26
TOTAL		S.F.	.039	1.32	1.69	3.01
2" X 8", 16" O.C., 4/12 PITCH						
Rafters, 2" x 8", 16" O.C., 4/12 pitch	1.170	L.F.	.020	1.08	.85	1.93
Fascia, 2" x 8"	.100	L.F.	.007	.09	.31	.40
Bridging, 1" x 3", 6' O.C.	.080	Pr.	.005	.03	.21	.24
Sheathing, exterior grade plywood, 1/2" thick	1.230	S.F.	.014	.64	.62	1.26
TOTAL		S.F.	.046	1.84	1.99	3.83
2" X 8", 24" O.C., 4/12 PITCH						
Rafters, 2" x 8", 24" O.C., 4/12 pitch	.940	L.F.	.016	.86	.69	1.55
Fascia, 2" x 8"	.100	L.F.	.007	.09	.31	.40
Bridging, 1" x 3", 6' O.C.	.060	Pr.	.004	.02	.16	.18
Sheathing, exterior grade plywood, 1/2" thick	1.230	S.F.	.014	.64	.62	1.26
TOTAL		S.F.	.041	1.61	1.78	3.39

The cost of this system is based on the square foot of plan area.
A 1' overhang is assumed. No ceiling joists or furring are included.

Description	QUAN.	UNIT	LABOR HOURS	COST PER S.F. MAT.	COST PER S.F. INST.	COST PER S.F. TOTAL

Shed/Flat Roof Framing Price Sheet	QUAN.	UNIT	LABOR HOURS	COST PER S.F. MAT.	COST PER S.F. INST.	COST PER S.F. TOTAL
Rafters, #2 or better, 16" O.C., 2" x 4", 0 - 4/12 pitch	1.170	L.F.	.014	.55	.61	1.16
5/12 - 8/12 pitch	1.330	L.F.	.020	.63	.87	1.50
2" x 6", 0 - 4/12 pitch	1.170	L.F.	.019	.74	.82	1.56
5/12 - 8/12 pitch	1.330	L.F.	.027	.84	1.16	2
2" x 8", 0 - 4/12 pitch	1.170	L.F.	.020	1.08	.85	1.93
5/12 - 8/12 pitch	1.330	L.F.	.028	1.22	1.24	2.46
2" x 10", 0 - 4/12 pitch	1.170	L.F.	.030	1.53	1.30	2.83
5/12 - 8/12 pitch	1.330	L.F.	.043	1.74	1.88	3.62
24" O.C., 2" x 4", 0 - 4/12 pitch	.940	L.F.	.011	.45	.50	.95
5/12 - 8/12 pitch	1.060	L.F.	.021	.67	.92	1.59
2" x 6", 0 - 4/12 pitch	.940	L.F.	.015	.59	.66	1.25
5/12 - 8/12 pitch	1.060	L.F.	.021	.67	.92	1.59
2" x 8", 0 - 4/12 pitch	.940	L.F.	.016	.86	.69	1.55
5/12 - 8/12 pitch	1.060	L.F.	.023	.98	.99	1.97
2" x 10", 0 - 4/12 pitch	.940	L.F.	.024	1.23	1.04	2.27
5/12 - 8/12 pitch	1.060	L.F.	.034	1.39	1.49	2.88
Fascia, #2 or better,, 1" x 4"	.100	L.F.	.003	.04	.12	.16
1" x 6"	.100	L.F.	.004	.05	.17	.22
1" x 8"	.100	L.F.	.005	.06	.20	.26
1" x 10"	.100	L.F.	.005	.07	.22	.29
2" x 4"	.100	L.F.	.005	.06	.21	.27
2" x 6"	.100	L.F.	.006	.07	.25	.32
2" x 8"	.100	L.F.	.007	.09	.31	.40
2" x 10"	.100	L.F.	.009	.13	.39	.52
Bridging, wood 6' O.C., 1" x 3", rafters, 16" O.C.	.080	Pr.	.005	.03	.21	.24
24" O.C.	.060	Pr.	.004	.02	.16	.18
Metal, galvanized, rafters, 16" O.C.	.080	Pr.	.005	.11	.21	.32
24" O.C.	.060	Pr.	.003	.10	.15	.25
Compression type, rafters, 16" O.C.	.080	Pr.	.003	.13	.14	.27
24" O.C.	.060	Pr.	.002	.10	.10	.20
Sheathing, plywood, exterior grade, 3/8" thick, flat 0 - 4/12 pitch	1.230	S.F.	.013	.59	.57	1.16
5/12 - 8/12 pitch	1.330	S.F.	.014	.64	.61	1.25
1/2" thick, flat 0 - 4/12 pitch	1.230	S.F.	.014	.64	.62	1.26
5/12 - 8/12 pitch	1.330	S.F.	.015	.69	.67	1.36
5/8" thick, flat 0 - 4/12 pitch	1.230	S.F.	.015	.80	.66	1.46
5/12 - 8/12 pitch	1.330	S.F.	.016	.86	.72	1.58
3/4" thick, flat 0 - 4/12 pitch	1.230	S.F.	.016	.96	.71	1.67
5/12 - 8/12 pitch	1.330	S.F.	.018	1.04	.77	1.81
Boards, 1" x 6", laid regular, flat 0 - 4/12 pitch	1.230	S.F.	.027	1.92	1.18	3.10
5/12 - 8/12 pitch	1.330	S.F.	.041	2.07	1.78	3.85
Laid diagonal, flat 0 - 4/12 pitch	1.230	S.F.	.030	1.92	1.32	3.24
5/12 - 8/12 pitch	1.330	S.F.	.044	2.07	1.93	4
1" x 8", laid regular, flat 0 - 4/12 pitch	1.230	S.F.	.022	1.73	.98	2.71
5/12 - 8/12 pitch	1.330	S.F.	.034	1.88	1.46	3.34
Laid diagonal, flat 0 - 4/12 pitch	1.230	S.F.	.027	1.73	1.18	2.91
5/12 - 8/12 pitch	1.330	S.F.	.044	2.07	1.93	4

3 | FRAMING — 40 | Gable Dormer Framing Systems

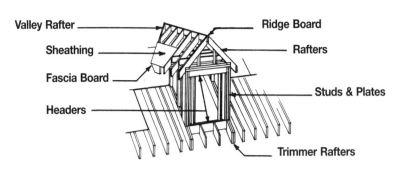

System Description	QUAN.	UNIT	LABOR HOURS	COST PER S.F. MAT.	COST PER S.F. INST.	COST PER S.F. TOTAL
2" X 6", 16" O.C.						
Dormer rafter, 2" x 6", 16" O.C.	1.330	L.F.	.036	.84	1.57	2.41
Ridge board, 2" x 6"	.280	L.F.	.009	.18	.39	.57
Trimmer rafters, 2" x 6"	.880	L.F.	.014	.55	.62	1.17
Wall studs & plates, 2" x 4", 16" O.C.	3.160	L.F.	.056	1.30	2.43	3.73
Fascia, 2" x 6"	.220	L.F.	.012	.16	.53	.69
Valley rafter, 2" x 6", 16" O.C.	.280	L.F.	.009	.18	.38	.56
Cripple rafter, 2" x 6", 16" O.C.	.560	L.F.	.022	.35	.95	1.30
Headers, 2" x 6", doubled	.670	L.F.	.030	.42	1.30	1.72
Ceiling joist, 2" x 4", 16" O.C.	1.000	L.F.	.013	.41	.56	.97
Sheathing, exterior grade plywood, 1/2" thick	3.610	S.F.	.041	1.88	1.81	3.69
TOTAL		S.F.	.242	6.27	10.54	16.81
2" X 8", 16" O.C.						
Dormer rafter, 2" x 8", 16" O.C.	1.330	L.F.	.039	1.22	1.72	2.94
Ridge board, 2" x 8"	.280	L.F.	.010	.26	.43	.69
Trimmer rafter, 2" x 8"	.880	L.F.	.015	.81	.64	1.45
Wall studs & plates, 2" x 4", 16" O.C.	3.160	L.F.	.056	1.30	2.43	3.73
Fascia, 2" x 8"	.220	L.F.	.016	.20	.68	.88
Valley rafter, 2" x 8", 16" O.C.	.280	L.F.	.010	.26	.41	.67
Cripple rafter, 2" x 8", 16" O.C.	.560	L.F.	.027	.52	1.16	1.68
Headers, 2" x 8", doubled	.670	L.F.	.032	.62	1.37	1.99
Ceiling joist, 2" x 4", 16" O.C.	1.000	L.F.	.013	.41	.56	.97
Sheathing,, exterior grade plywood, 1/2" thick	3.610	S.F.	.041	1.88	1.81	3.69
TOTAL		S.F.	.259	7.48	11.21	18.69

The cost in this system is based on the square foot of plan area.
The measurement is the plan area of the dormer only.

Description	QUAN.	UNIT	LABOR HOURS	COST PER S.F. MAT.	COST PER S.F. INST.	COST PER S.F. TOTAL

Gable Dormer Framing Price Sheet

	QUAN.	UNIT	LABOR HOURS	COST PER S.F. MAT.	COST PER S.F. INST.	COST PER S.F. TOTAL
Dormer rafters, #2 or better, 2" x 4", 16" O.C.	1.330	L.F.	.029	.67	1.26	1.93
24" O.C.	1.060	L.F.	.023	.54	1	1.54
2" x 6", 16" O.C.	1.330	L.F.	.036	.84	1.57	2.41
24" O.C.	1.060	L.F.	.029	.67	1.25	1.92
2" x 8", 16" O.C.	1.330	L.F.	.039	1.22	1.72	2.94
24" O.C.	1.060	L.F.	.031	.98	1.37	2.35
Ridge board, #2 or better, 1" x 4"	.280	L.F.	.006	.17	.26	.43
1" x 6"	.280	L.F.	.007	.22	.32	.54
1" x 8"	.280	L.F.	.008	.27	.36	.63
2" x 4"	.280	L.F.	.007	.14	.31	.45
2" x 6"	.280	L.F.	.009	.18	.39	.57
2" x 8"	.280	L.F.	.010	.26	.43	.69
Trimmer rafters, #2 or better, 2" x 4"	.880	L.F.	.011	.44	.49	.93
2" x 6"	.880	L.F.	.014	.55	.62	1.17
2" x 8"	.880	L.F.	.015	.81	.64	1.45
2" x 10"	.880	L.F.	.022	1.15	.98	2.13
Wall studs & plates, #2 or better, 2" x 4" studs, 16" O.C.	3.160	L.F.	.056	1.30	2.43	3.73
24" O.C.	2.800	L.F.	.050	1.15	2.16	3.31
2" x 6" studs, 16" O.C.	3.160	L.F.	.063	1.99	2.75	4.74
24" O.C.	2.800	L.F.	.056	1./6	2.44	4.20
Fascia, #2 or better, 1" x 4"	.220	L.F.	.006	.08	.28	.36
1" x 6"	.220	L.F.	.008	.10	.34	.44
1" x 8"	.220	L.F.	.009	.12	.40	.52
2" x 4"	.220	L.F.	.011	.14	.47	.61
2" x 6"	.220	L.F.	.014	.17	.59	.76
2" x 8"	.220	L.F.	.016	.20	.68	.88
Valley rafter, #2 or better, 2" x 4"	.280	L.F.	.007	.14	.31	.45
2" x 6"	.280	L.F.	.009	.18	.38	.56
2" x 8"	.280	L.F.	.010	.26	.41	.67
2" x 10"	.280	L.F.	.012	.37	.52	.89
Cripple rafter, #2 or better, 2" x 4", 16" O.C.	.560	L.F.	.018	.28	.77	1.05
24" O.C.	.450	L.F.	.014	.23	.61	.84
2" x 6", 16" O.C.	.560	L.F.	.022	.35	.95	1.30
24" O.C.	.450	L.F.	.018	.28	.77	1.05
2" x 8", 16" O.C.	.560	L.F.	.027	.52	1.16	1.68
24" O.C.	.450	L.F.	.021	.41	.94	1.35
Headers, #2 or better double header, 2" x 4"	.670	L.F.	.024	.34	1.05	1.39
2" x 6"	.670	L.F.	.030	.42	1.30	1.72
2" x 8"	.670	L.F.	.032	.62	1.37	1.99
2" x 10"	.670	L.F.	.034	.88	1.46	2.34
Ceiling joist, #2 or better, 2" x 4", 16" O.C.	1.000	L.F.	.013	.41	.56	.97
24" O.C.	.800	L.F.	.010	.33	.45	.78
2" x 6", 16" O.C.	1.000	L.F.	.013	.63	.56	1.19
24" O.C.	.800	L.F.	.010	.50	.45	.95
Sheathing, plywood exterior grade, 3/8" thick	3.610	S.F.	.038	1.73	1.66	3.39
1/2" thick	3.610	S.F.	.041	1.88	1.81	3.69
5/8" thick	3.610	S.F.	.044	2.35	1.95	4.30
3/4" thick	3.610	S.F.	.048	2.82	2.09	4.91
Boards, 1" x 6", laid regular	3.610	S.F.	.089	5.65	3.86	9.51
Laid diagonal	3.610	S.F.	.099	5.65	4.30	9.95
1" x 8", laid regular	3.610	S.F.	.076	5.10	3.29	8.39
Laid diagonal	3.610	S.F.	.089	5.10	3.86	8.96

3 | FRAMING — 44 | Shed Dormer Framing Systems

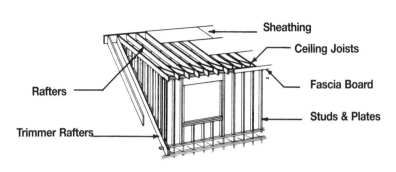

System Description	QUAN.	UNIT	LABOR HOURS	COST PER S.F. MAT.	COST PER S.F. INST.	COST PER S.F. TOTAL
2″ X 6″ RAFTERS, 16″ O.C.						
Dormer rafter, 2″ x 6″, 16″ O.C.	1.080	L.F.	.029	.68	1.27	1.95
Trimmer rafter, 2″ x 6″	.400	L.F.	.006	.25	.28	.53
Studs & plates, 2″ x 4″, 16″ O.C.	2.750	L.F.	.049	1.13	2.12	3.25
Fascia, 2″ x 6″	.250	L.F.	.014	.17	.59	.76
Ceiling joist, 2″ x 4″, 16″ O.C.	1.000	L.F.	.013	.41	.56	.97
Sheathing, exterior grade plywood, CDX, 1/2″ thick	2.940	S.F.	.034	1.53	1.47	3
TOTAL		S.F.	.145	4.17	6.29	10.46
2″ X 8″ RAFTERS, 16″ O.C.						
Dormer rafter, 2″ x 8″, 16″ O.C.	1.080	L.F.	.032	.99	1.39	2.38
Trimmer rafter, 2″ x 8″	.400	L.F.	.007	.37	.29	.66
Studs & plates, 2″ x 4″, 16″ O.C.	2.750	L.F.	.049	1.13	2.12	3.25
Fascia, 2″ x 8″	.250	L.F.	.018	.23	.78	1.01
Ceiling joist, 2″ x 6″, 16″ O.C.	1.000	L.F.	.013	.63	.56	1.19
Sheathing, exterior grade plywood, CDX, 1/2″ thick	2.940	S.F.	.034	1.53	1.47	3
TOTAL		S.F.	.153	4.88	6.61	11.49
2″ X 10″ RAFTERS, 16″ O.C.						
Dormer rafter, 2″ x 10″, 16″ O.C.	1.080	L.F.	.041	1.41	1.77	3.18
Trimmer rafter, 2″ x 10″	.400	L.F.	.010	.52	.44	.96
Studs & plates, 2″ x 4″, 16″ O.C.	2.750	L.F.	.049	1.13	2.12	3.25
Fascia, 2″ x 10″	.250	L.F.	.022	.33	.97	1.30
Ceiling joist, 2″ x 6″, 16″ O.C.	1.000	L.F.	.013	.63	.56	1.19
Sheathing, exterior grade plywood, CDX, 1/2″ thick	2.940	S.F.	.034	1.53	1.47	3
TOTAL		S.F.	.169	5.55	7.33	12.88

The cost in this system is based on the square foot of plan area.
The measurement is the plan area of the dormer only.

Description	QUAN.	UNIT	LABOR HOURS	COST PER S.F. MAT.	COST PER S.F. INST.	COST PER S.F. TOTAL

Shed Dormer Framing Price Sheet	QUAN.	UNIT	LABOR HOURS	COST PER S.F. MAT.	COST PER S.F. INST.	COST PER S.F. TOTAL
Dormer rafters, #2 or better, 2" x 4", 16" O.C.	1.080	L.F.	.023	.54	1.02	1.56
24" O.C.	.860	L.F.	.019	.43	.81	1.24
2" x 6", 16" O.C.	1.080	L.F.	.029	.68	1.27	1.95
24" O.C.	.860	L.F.	.023	.54	1.01	1.55
2" x 8", 16" O.C.	1.080	L.F.	.032	.99	1.39	2.38
24" O.C.	.860	L.F.	.025	.79	1.11	1.90
2" x 10", 16" O.C.	1.080	L.F.	.041	1.41	1.77	3.18
24" O.C.	.860	L.F.	.032	1.13	1.41	2.54
Trimmer rafter, #2 or better, 2" x 4"	.400	L.F.	.005	.20	.22	.42
2" x 6"	.400	L.F.	.006	.25	.28	.53
2" x 8"	.400	L.F.	.007	.37	.29	.66
2" x 10"	.400	L.F.	.010	.52	.44	.96
Studs & plates, #2 or better, 2" x 4", 16" O.C.	2.750	L.F.	.049	1.13	2.12	3.25
24" O.C.	2.200	L.F.	.039	.90	1.69	2.59
2" x 6", 16" O.C.	2.750	L.F.	.055	1.73	2.39	4.12
24" O.C.	2.200	L.F.	.044	1.39	1.91	3.30
Fascia, #2 or better, 1" x 4"	.250	L.F.	.006	.08	.28	.36
1" x 6"	.250	L.F.	.008	.10	.34	.44
1" x 8"	.250	L.F.	.009	.12	.40	.52
2" x 4"	.250	L.F.	.011	.14	.47	.61
2" x 6"	.250	L.F.	.014	.17	.59	.76
2" x 8"	.250	L.F.	.018	.23	.78	1.01
Ceiling joist, #2 or better, 2" x 4", 16" O.C.	1.000	L.F.	.013	.41	.56	.97
24" O.C.	.800	L.F.	.010	.33	.45	.78
2" x 6", 16" O.C.	1.000	L.F.	.013	.63	.56	1.19
24" O.C.	.800	L.F.	.010	.50	.45	.95
2" x 8", 16" O.C.	1.000	L.F.	.015	.92	.63	1.55
24" O.C.	.800	L.F.	.012	.74	.50	1.24
Sheathing, plywood exterior grade, 3/8" thick	2.940	S.F.	.031	1.41	1.35	2.76
1/2" thick	2.940	S.F.	.034	1.53	1.47	3
5/8" thick	2.940	S.F.	.036	1.91	1.59	3.50
3/4" thick	2.940	S.F.	.039	2.29	1.71	4
Boards, 1" x 6", laid regular	2.940	S.F.	.072	4.59	3.15	7.74
Laid diagonal	2.940	S.F.	.080	4.59	3.50	8.09
1" x 8", laid regular	2.940	S.F.	.062	4.15	2.68	6.83
Laid diagonal	2.940	S.F.	.072	4.15	3.15	7.30

Window Openings	QUAN.	UNIT	LABOR HOURS	COST EACH MAT.	COST EACH INST.	COST EACH TOTAL
The following are to be added to the total cost of the dormers for window openings. Do not subtract window area from the stud wall quantities.						
Headers, 2" x 6" doubled, 2' long	4.000	L.F.	.178	2.52	7.75	10.27
3' long	6.000	L.F.	.267	3.78	11.65	15.43
4' long	8.000	L.F.	.356	5.05	15.50	20.55
5' long	10.000	L.F.	.444	6.30	19.40	25.70
2" x 8" doubled, 4' long	8.000	L.F.	.376	7.35	16.40	23.75
5' long	10.000	L.F.	.471	9.20	20.50	29.70
6' long	12.000	L.F.	.565	11.05	24.50	35.55
8' long	16.000	L.F.	.753	14.70	33	47.70
2" x 10" doubled, 4' long	8.000	L.F.	.400	10.50	17.45	27.95
6' long	12.000	L.F.	.600	15.70	26	41.70
8' long	16.000	L.F.	.800	21	35	56
10' long	20.000	L.F.	1.000	26	43.50	69.50

3 | FRAMING — 48 | Partition Framing Systems

Bracing, Top Plates, Studs, Bottom Plate

System Description	QUAN.	UNIT	LABOR HOURS	COST PER S.F. MAT.	COST PER S.F. INST.	COST PER S.F. TOTAL
2" X 4", 16" O.C.						
2" x 4" studs, #2 or better, 16" O.C.	1.000	L.F.	.015	.41	.63	1.04
Plates, double top, single bottom	.375	L.F.	.005	.15	.24	.39
Cross bracing, let-in, 1" x 6"	.080	L.F.	.004	.06	.19	.25
TOTAL		S.F.	.024	.62	1.06	1.68
2" X 4", 24" O.C.						
2" x 4" studs, #2 or better, 24" O.C.	.800	L.F.	.012	.33	.50	.83
Plates, double top, single bottom	.375	L.F.	.005	.15	.24	.39
Cross bracing, let-in, 1" x 6"	.080	L.F.	.003	.06	.12	.18
TOTAL		S.F.	.020	.54	.86	1.40
2" X 6", 16" O.C.						
2" x 6" studs, #2 or better, 16" O.C.	1.000	L.F.	.016	.63	.70	1.33
Plates, double top, single bottom	.375	L.F.	.006	.24	.26	.50
Cross bracing, let-in, 1" x 6"	.080	L.F.	.004	.06	.19	.25
TOTAL		S.F.	.026	.93	1.15	2.08
2" X 6", 24" O.C.						
2" x 6" studs, #2 or better, 24" O.C.	.800	L.F.	.013	.50	.56	1.06
Plates, double top, single bottom	.375	L.F.	.006	.24	.26	.50
Cross bracing, let-in, 1" x 6"	.080	L.F.	.003	.06	.12	.18
TOTAL		S.F.	.022	.80	.94	1.74

The costs in this system are based on a square foot of wall area. Do not subtract for door or window openings.

Description	QUAN.	UNIT	LABOR HOURS	COST PER S.F. MAT.	COST PER S.F. INST.	COST PER S.F. TOTAL

Partition Framing Price Sheet	QUAN.	UNIT	LABOR HOURS	COST PER S.F.		
				MAT.	INST.	TOTAL
Wood studs, #2 or better, 2" x 4", 12" O.C.	1.250	L.F.	.018	.51	.79	1.30
16" O.C.	1.000	L.F.	.015	.41	.63	1.04
24" O.C.	.800	L.F.	.012	.33	.50	.83
32" O.C.	.650	L.F.	.009	.27	.41	.68
2" x 6", 12" O.C.	1.250	L.F.	.020	.79	.88	1.67
16" O.C.	1.000	L.F.	.016	.63	.70	1.33
24" O.C.	.800	L.F.	.013	.50	.56	1.06
32" O.C.	.650	L.F.	.010	.41	.46	.87
Plates, #2 or better double top single bottom, 2" x 4"	.375	L.F.	.005	.15	.24	.39
2" x 6"	.375	L.F.	.006	.24	.26	.50
2" x 8"	.375	L.F.	.005	.35	.24	.59
Cross bracing, let-in, 1" x 6" boards studs, 12" O.C.	.080	L.F.	.005	.07	.23	.30
16" O.C.	.080	L.F.	.004	.06	.19	.25
24" O.C.	.080	L.F.	.003	.06	.12	.18
32" O.C.	.080	L.F.	.002	.05	.10	.15
Let-in steel (T shaped) studs, 12" O.C.	.080	L.F.	.001	.06	.06	.12
16" O.C.	.080	L.F.	.001	.05	.05	.10
24" O.C.	.080	L.F.	.001	.05	.05	.10
32" O.C.	.080	L.F.	.001	.04	.04	.08
Steel straps studs, 12" O.C.	.080	L.F.	.001	.07	.05	.12
16" O.C.	.080	L.F.	.001	.07	.05	.12
24" O.C.	.080	L.F.	.001	.07	.04	.11
32" O.C.	.080	L.F.	.001	.07	.04	.11
Metal studs, load bearing 24" O.C., 20 ga. galv., 2-1/2" wide	1.000	S.F.	.015	.69	.65	1.34
3-5/8" wide	1.000	S.F.	.015	.83	.67	1.50
4" wide	1.000	S.F.	.016	.87	.68	1.55
6" wide	1.000	S.F.	.016	1.11	.69	1.80
16 ga., 2-1/2" wide	1.000	S.F.	.017	.80	.74	1.54
3-5/8" wide	1.000	S.F.	.017	.96	.76	1.72
4" wide	1.000	S.F.	.018	1.01	.78	1.79
6" wide	1.000	S.F.	.018	1.28	.80	2.08
Non-load bearing 24" O.C., 25 ga. galv., 1-5/8" wide	1.000	S.F.	.011	.23	.46	.69
2-1/2" wide	1.000	S.F.	.011	.28	.47	.75
3-5/8" wide	1.000	S.F.	.011	.32	.47	.79
4" wide	1.000	S.F.	.011	.38	.47	.85
6" wide	1.000	S.F.	.011	.47	.48	.95
20 ga., 2-1/2" wide	1.000	S.F.	.013	.45	.58	1.03
3-5/8" wide	1.000	S.F.	.014	.50	.59	1.09
4" wide	1.000	S.F.	.014	.60	.59	1.19
6" wide	1.000	S.F.	.014	.70	.60	1.30
Window & Door Openings	QUAN.	UNIT	LABOR HOURS	COST EACH		
				MAT.	INST.	TOTAL
The following costs are to be added to the total costs of the walls.						
Do not subtract openings from total wall area.						
Headers, 2" x 6" double, 2' long	4.000	L.F.	.178	2.52	7.75	10.27
3' long	6.000	L.F.	.267	3.78	11.65	15.43
4' long	8.000	L.F.	.356	5.05	15.50	20.55
5' long	10.000	L.F.	.444	6.30	19.40	25.70
2" x 8" double, 4' long	8.000	L.F.	.376	7.35	16.40	23.75
5' long	10.000	L.F.	.471	9.20	20.50	29.70
6' long	12.000	L.F.	.565	11.05	24.50	35.55
8' long	16.000	L.F.	.753	14.70	33	47.70
2" x 10" double, 4' long	8.000	L.F.	.400	10.50	17.45	27.95
6' long	12.000	L.F.	.600	15.70	26	41.70
8' long	16.000	L.F.	.800	21	35	56
10' long	20.000	L.F.	1.000	26	43.50	69.50
2" x 12" double, 8' long	16.000	L.F.	.853	25	37.50	62.50
12' long	24.000	L.F.	1.280	37.50	56	93.50

Division 4
Exterior Walls

4 | EXTERIOR WALLS 02 | Block Masonry Systems

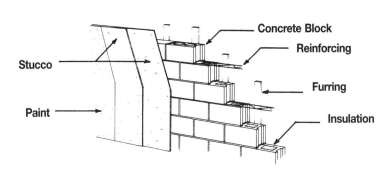

System Description	QUAN.	UNIT	LABOR HOURS	COST PER S.F. MAT.	COST PER S.F. INST.	COST PER S.F. TOTAL
6" THICK CONCRETE BLOCK WALL						
6" thick concrete block, 6" x 8" x 16"	1.000	S.F.	.100	1.96	3.99	5.95
Masonry reinforcing, truss strips every other course	.625	L.F.	.002	.14	.08	.22
Furring, 1" x 3", 16" O.C.	1.000	L.F.	.016	.25	.70	.95
Masonry insulation, poured vermiculite	1.000	S.F.	.013	.62	.58	1.20
Stucco, 2 coats	1.000	S.F.	.069	.20	2.73	2.93
Masonry paint, 2 coats	1.000	S.F.	.016	.20	.61	.81
TOTAL		S.F.	.216	3.37	8.69	12.06
8" THICK CONCRETE BLOCK WALL						
8" thick concrete block, 8" x 8" x 16"	1.000	S.F.	.107	2.14	4.26	6.40
Masonry reinforcing, truss strips every other course	.625	L.F.	.002	.14	.08	.22
Furring, 1" x 3", 16" O.C.	1.000	L.F.	.016	.25	.70	.95
Masonry insulation, poured vermiculite	1.000	S.F.	.018	.82	.77	1.59
Stucco, 2 coats	1.000	S.F.	.069	.20	2.73	2.93
Masonry paint, 2 coats	1.000	S.F.	.016	.20	.61	.81
TOTAL		S.F.	.228	3.75	9.15	12.90
12" THICK CONCRETE BLOCK WALL						
12" thick concrete block, 12" x 8" x 16"	1.000	S.F.	.141	3.18	5.50	8.68
Masonry reinforcing, truss strips every other course	.625	L.F.	.003	.16	.11	.27
Furring, 1" x 3", 16" O.C.	1.000	L.F.	.016	.25	.70	.95
Masonry insulation, poured vermiculite	1.000	S.F.	.026	1.22	1.13	2.35
Stucco, 2 coats	1.000	S.F.	.069	.20	2.73	2.93
Masonry paint, 2 coats	1.000	S.F.	.016	.20	.61	.81
TOTAL		S.F.	.271	5.21	10.78	15.99

Costs for this system are based on a square foot of wall area. Do not subtract for window openings.

Description	QUAN.	UNIT	LABOR HOURS	COST PER S.F. MAT.	COST PER S.F. INST.	COST PER S.F. TOTAL

Masonry Block Price Sheet

Masonry Block Price Sheet	QUAN.	UNIT	LABOR HOURS	COST PER S.F. MAT.	COST PER S.F. INST.	COST PER S.F. TOTAL
Block concrete, 8" x 16" regular, 4" thick	1.000	S.F.	.093	1.34	3.71	5.05
6" thick	1.000	S.F.	.100	1.96	3.99	5.95
8" thick	1.000	S.F.	.107	2.14	4.26	6.40
10" thick	1.000	S.F.	.111	2.93	4.43	7.36
12" thick	1.000	S.F.	.141	3.18	5.50	8.68
Solid block, 4" thick	1.000	S.F.	.096	1.72	3.85	5.57
6" thick	1.000	S.F.	.104	1.98	4.15	6.13
8" thick	1.000	S.F.	.111	3.16	4.43	7.59
10" thick	1.000	S.F.	.133	4.44	5.20	9.64
12" thick	1.000	S.F.	.148	4.93	5.75	10.68
Lightweight, 4" thick	1.000	S.F.	.093	1.34	3.71	5.05
6" thick	1.000	S.F.	.100	1.96	3.99	5.95
8" thick	1.000	S.F.	.107	2.14	4.26	6.40
10" thick	1.000	S.F.	.111	2.93	4.43	7.36
12" thick	1.000	S.F.	.141	3.18	5.50	8.68
Split rib profile, 4" thick	1.000	S.F.	.116	2.94	4.63	7.57
6" thick	1.000	S.F.	.123	3.41	4.91	8.32
8" thick	1.000	S.F.	.131	3.92	5.30	9.22
10" thick	1.000	S.F.	.157	4.22	6.10	10.32
12" thick	1.000	S.F.	.175	4.69	6.75	11.44
Masonry reinforcing, wire truss strips, every course, 8" block	1.375	L.F.	.004	.30	.17	.47
12" block	1.375	L.F.	.006	.34	.25	.59
Every other course, 8" block	.625	L.F.	.002	.14	.08	.22
12" block	.625	L.F.	.003	.16	.11	.27
Furring, wood, 1" x 3", 12" O.C.	1.250	L.F.	.020	.31	.88	1.19
16" O.C.	1.000	L.F.	.016	.25	.70	.95
24" O.C.	.800	L.F.	.013	.20	.56	.76
32" O.C.	.640	L.F.	.010	.16	.45	.61
Steel, 3/4" channels, 12" O.C.	1.250	L.F.	.034	.27	1.31	1.58
16" O.C.	1.000	L.F.	.030	.24	1.16	1.40
24" O.C.	.800	L.F.	.023	.16	.88	1.04
32" O.C.	.640	L.F.	.018	.13	.70	.83
Masonry insulation, vermiculite or perlite poured 4" thick	1.000	S.F.	.009	.40	.37	.77
6" thick	1.000	S.F.	.013	.62	.57	1.19
8" thick	1.000	S.F.	.018	.82	.77	1.59
10" thick	1.000	S.F.	.021	1	.93	1.93
12" thick	1.000	S.F.	.026	1.22	1.13	2.35
Block inserts polystyrene, 6" thick	1.000	S.F.		1.24		1.24
8" thick	1.000	S.F.		1.24		1.24
10" thick	1.000	S.F.		1.46		1.46
12" thick	1.000	S.F.		1.54		1.54
Stucco, 1 coat	1.000	S.F.	.057	.17	2.25	2.42
2 coats	1.000	S.F.	.069	.20	2.73	2.93
3 coats	1.000	S.F.	.081	.24	3.21	3.45
Painting, 1 coat	1.000	S.F.	.011	.13	.42	.55
2 coats	1.000	S.F.	.016	.20	.61	.81
Primer & 1 coat	1.000	S.F.	.013	.22	.50	.72
2 coats	1.000	S.F.	.018	.29	.69	.98
Lath, metal lath expanded 2.5 lb/S.Y., painted	1.000	S.F.	.010	.32	.40	.72
Galvanized	1.000	S.F.	.012	.35	.44	.79

4 | EXTERIOR WALLS — 04 | Brick/Stone Veneer Systems

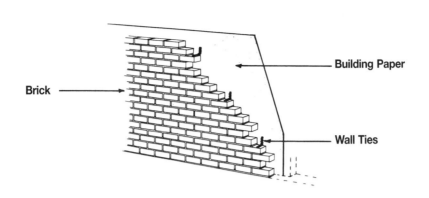

System Description	QUAN.	UNIT	LABOR HOURS	COST PER S.F. MAT.	COST PER S.F. INST.	COST PER S.F. TOTAL
SELECT COMMON BRICK						
Brick, select common, running bond	1.000	S.F.	.174	4.24	6.95	11.19
Wall ties, 7/8" x 7", 22 gauge	1.000	Ea.	.008	.07	.34	.41
Building paper, spunbonded polypropylene	1.100	S.F.	.002	.15	.10	.25
Trim, pine, painted	.125	L.F.	.004	.09	.18	.27
TOTAL		S.F.	.188	4.55	7.57	12.12
RED FACED COMMON BRICK						
Brick, common, red faced, running bond	1.000	S.F.	.182	4.24	7.25	11.49
Wall ties, 7/8" x 7", 22 gauge	1.000	Ea.	.008	.07	.34	.41
Building paper, spunbonded polypropylene	1.100	S.F.	.002	.15	.10	.25
Trim, pine, painted	.125	L.F.	.004	.09	.18	.27
TOTAL		S.F.	.196	4.55	7.87	12.42
BUFF OR GREY FACE BRICK						
Brick, buff or grey	1.000	S.F.	.182	4.48	7.25	11.73
Wall ties, 7/8" x 7", 22 gauge	1.000	Ea.	.008	.07	.34	.41
Building paper, spunbonded polypropylene	1.100	S.F.	.002	.15	.10	.25
Trim, pine, painted	.125	L.F.	.004	.09	.18	.27
TOTAL		S.F.	.196	4.79	7.87	12.66
STONE WORK, ROUGH STONE, AVERAGE						
Field stone veneer	1.000	S.F.	.223	6.20	8.91	15.11
Wall ties, 7/8" x 7", 22 gauge	1.000	Ea.	.008	.07	.34	.41
Building paper, spunbonded polypropylene	1.000	S.F.	.002	.15	.10	.25
Trim, pine, painted	.125	L.F.	.004	.09	.18	.27
TOTAL		S.F.	.237	6.51	9.53	16.04

The costs in this system are based on a square foot of wall area. Do not subtract area for window & door openings.

Description	QUAN.	UNIT	LABOR HOURS	COST PER S.F. MAT.	COST PER S.F. INST.	COST PER S.F. TOTAL

Brick/Stone Veneer Price Sheet	QUAN.	UNIT	LABOR HOURS	COST PER S.F.		
				MAT.	INST.	TOTAL
Brick						
Select common, running bond	1.000	S.F.	.174	4.24	6.95	11.19
Red faced, running bond	1.000	S.F.	.182	4.24	7.25	11.49
Buff or grey faced, running bond	1.000	S.F.	.182	4.48	7.25	11.73
Header every 6th course	1.000	S.F.	.216	4.94	8.65	13.59
English bond	1.000	S.F.	.286	6.35	11.40	17.75
Flemish bond	1.000	S.F.	.195	4.47	7.80	12.27
Common bond	1.000	S.F.	.267	5.65	10.65	16.30
Stack bond	1.000	S.F.	.182	4.48	7.25	11.73
Jumbo, running bond	1.000	S.F.	.092	4.71	3.67	8.38
Norman, running bond	1.000	S.F.	.125	5.10	4.99	10.09
Norwegian, running bond	1.000	S.F.	.107	3.78	4.26	8.04
Economy, running bond	1.000	S.F.	.129	4.50	5.15	9.65
Engineer, running bond	1.000	S.F.	.154	3.58	6.15	9.73
Roman, running bond	1.000	S.F.	.160	5.95	6.40	12.35
Utility, running bond	1.000	S.F.	.089	3.95	3.55	7.50
Glazed, running bond	1.000	S.F.	.190	10.65	7.60	18.25
Stone work, rough stone, average	1.000	S.F.	.179	6.20	8.90	15.10
Maximum	1.000	S.F.	.267	9.25	13.30	22.55
Wall ties, galvanized, corrugated 7/8" x 7", 22 gauge	1.000	Ea.	.008	.07	.34	.41
16 gauge	1.000	Ea.	.008	.25	.34	.59
Cavity wall, every 3rd course 6" long Z type, 1/4" diameter	1.330	L.F.	.010	.52	.44	.96
3/16" diameter	1.330	L.F.	.010	.26	.44	.70
8" long, Z type, 1/4" diameter	1.330	L.F.	.010	.62	.44	1.06
3/16" diameter	1.330	L.F.	.010	.29	.44	.73
Building paper, aluminum and kraft laminated foil, 1 side	1.000	S.F.	.002	.06	.09	.15
2 sides	1.000	S.F.	.002	.09	.09	.18
#15 asphalt paper	1.100	S.F.	.002	.06	.10	.16
Polyethylene, .002" thick	1.000	S.F.	.002	.01	.09	.10
.004" thick	1.000	S.F.	.002	.03	.09	.12
.006" thick	1.000	S.F.	.002	.05	.09	.14
.010" thick	1.000	S.F.	.002	.06	.09	.15
Trim, 1" x 4", cedar	.125	L.F.	.005	.24	.22	.46
Fir	.125	L.F.	.005	.11	.22	.33
Redwood	.125	L.F.	.005	.24	.22	.46
White pine	.125	L.F.	.005	.11	.22	.33

4 | EXTERIOR WALLS 08 | Wood Siding Systems

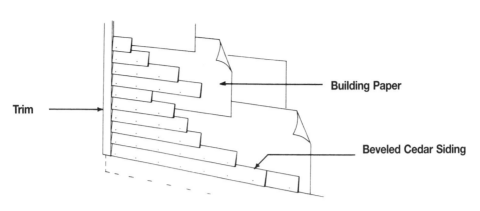

System Description	QUAN.	UNIT	LABOR HOURS	COST PER S.F.		
				MAT.	INST.	TOTAL
1/2" X 6" BEVELED CEDAR SIDING, "A" GRADE						
1/2" x 6" beveled cedar siding	1.000	S.F.	.032	3.36	1.40	4.76
Building wrap, spunbonded polypropylene	1.100	S.F.	.002	.15	.10	.25
Trim, cedar	.125	L.F.	.005	.24	.22	.46
Paint, primer & 2 coats	1.000	S.F.	.017	.19	.66	.85
TOTAL		S.F.	.056	3.94	2.38	6.32
1/2" X 8" BEVELED CEDAR SIDING, "A" GRADE						
1/2" x 8" beveled cedar siding	1.000	S.F.	.029	3.71	1.27	4.98
Building wrap, spunbonded polypropylene	1.100	S.F.	.002	.15	.10	.25
Trim, cedar	.125	L.F.	.005	.24	.22	.46
Paint, primer & 2 coats	1.000	S.F.	.017	.19	.66	.85
TOTAL		S.F.	.053	4.29	2.25	6.54
1" X 4" TONGUE & GROOVE, REDWOOD, VERTICAL GRAIN						
Redwood, clear, vertical grain, 1" x 10"	1.000	S.F.	.018	3.69	.80	4.49
Building wrap, spunbonded polypropylene	1.100	S.F.	.002	.15	.10	.25
Trim, redwood	.125	L.F.	.005	.24	.22	.46
Sealer, 1 coat, stain, 1 coat	1.000	S.F.	.013	.12	.51	.63
TOTAL		S.F.	.038	4.20	1.63	5.83
1" X 6" TONGUE & GROOVE, REDWOOD, VERTICAL GRAIN						
Redwood, clear, vertical grain, 1" x 10"	1.000	S.F.	.019	3.79	.82	4.61
Building wrap, spunbonded polypropylene	1.100	S.F.	.002	.15	.10	.25
Trim, redwood	.125	L.F.	.005	.24	.22	.46
Sealer, 1 coat, stain, 1 coat	1.000	S.F.	.013	.12	.51	.63
TOTAL		S.F.	.039	4.30	1.65	5.95

The costs in this system are based on a square foot of wall area.
Do not subtract area for door or window openings.

Description	QUAN.	UNIT	LABOR HOURS	COST PER S.F.		
				MAT.	INST.	TOTAL

Wood Siding Price Sheet	QUAN.	UNIT	LABOR HOURS	COST PER S.F. MAT.	COST PER S.F. INST.	COST PER S.F. TOTAL
Siding, beveled cedar, "A" grade, 1/2" x 6"	1.000	S.F.	.028	3.36	1.40	4.76
1/2" x 8"	1.000	S.F.	.023	3.71	1.27	4.98
"B" grade, 1/2" x 6"	1.000	S.F.	.032	3.73	1.56	5.29
1/2" x 8"	1.000	S.F.	.029	4.12	1.41	5.53
Clear grade, 1/2" x 6"	1.000	S.F.	.028	4.20	1.75	5.95
1/2" x 8"	1.000	S.F.	.023	4.64	1.59	6.23
Redwood, clear vertical grain, 1/2" x 6"	1.000	S.F.	.036	3.15	1.55	4.70
1/2" x 8"	1.000	S.F.	.032	2.54	1.40	3.94
Clear all heart vertical grain, 1/2" x 6"	1.000	S.F.	.028	3.50	1.72	5.22
1/2" x 8"	1.000	S.F.	.023	2.82	1.56	4.38
Siding board & batten, cedar, "B" grade, 1" x 10"	1.000	S.F.	.031	1.18	1.34	2.52
1" x 12"	1.000	S.F.	.031	1.18	1.34	2.52
Redwood, clear vertical grain, 1" x 6"	1.000	S.F.	.043	3.03	2.11	5.14
1" x 8"	1.000	S.F.	.018	2.86	1.86	4.72
White pine, #2 & better, 1" x 10"	1.000	S.F.	.029	1.18	1.27	2.45
1" x 12"	1.000	S.F.	.029	1.18	1.27	2.45
Siding vertical, tongue & groove, cedar "B" grade, 1" x 4"	1.000	S.F.	.033	2	.80	2.80
1" x 6"	1.000	S.F.	.024	2.06	.82	2.88
1" x 8"	1.000	S.F.	.024	2.12	.84	2.96
1" x 10"	1.000	S.F.	.021	2.18	.87	3.05
"A" grade, 1" x 4"	1.000	S.F.	.033	1.83	.73	2.56
1" x 6"	1.000	S.F.	.024	1.88	.75	2.63
1" x 8"	1.000	S.F.	.024	1.93	.77	2.70
1" x 10"	1.000	S.F.	.021	1.98	.79	2.77
Clear vertical grain, 1" x 4"	1.000	S.F.	.033	1.69	.67	2.36
1" x 6"	1.000	S.F.	.024	1.73	.69	2.42
1" x 8"	1.000	S.F.	.024	1.77	.71	2.48
1" x 10"	1.000	S.F.	.021	1.82	.73	2.55
Redwood, clear vertical grain, 1" x 4"	1.000	S.F.	.033	3.69	.80	4.49
1" x 6"	1.000	S.F.	.024	3.79	.82	4.61
1" x 8"	1.000	S.F.	.024	3.90	.84	4.74
1" x 10"	1.000	S.F.	.021	4.02	.87	4.89
Clear all heart vertical grain, 1" x 4"	1.000	S.F.	.033	3.38	.73	4.11
1" x 6"	1.000	S.F.	.024	3.47	.75	4.22
1" x 8"	1.000	S.F.	.024	3.56	.77	4.33
1" x 10"	1.000	S.F.	.021	3.65	.79	4.44
White pine, 1" x 10"	1.000	S.F.	.024	.87	.87	1.74
Siding plywood, texture 1-11 cedar, 3/8" thick	1.000	S.F.	.024	1.31	1.03	2.34
5/8" thick	1.000	S.F.	.024	2.65	1.03	3.68
Redwood, 3/8" thick	1.000	S.F.	.024	1.31	1.03	2.34
5/8" thick	1.000	S.F.	.024	2.09	1.03	3.12
Fir, 3/8" thick	1.000	S.F.	.024	.83	1.03	1.86
5/8" thick	1.000	S.F.	.024	1.22	1.03	2.25
Southern yellow pine, 3/8" thick	1.000	S.F.	.024	.83	1.03	1.86
5/8" thick	1.000	S.F.	.024	1.31	1.03	2.34
Hard board, 7/16" thick primed, plain finish	1.000	S.F.	.025	1.24	1.07	2.31
Board finish	1.000	S.F.	.023	1.01	1	2.01
Polyvinyl coated, 3/8" thick	1.000	S.F.	.021	1.21	.93	2.14
5/8" thick	1.000	S.F.	.024	1.31	1.03	2.34
Paper, #15 asphalt felt	1.100	S.F.	.002	.06	.10	.16
Trim, cedar	.125	L.F.	.005	.24	.22	.46
Fir	.125	L.F.	.005	.11	.22	.33
Redwood	.125	L.F.	.005	.24	.22	.46
White pine	.125	L.F.	.005	.11	.22	.33
Painting, primer, & 1 coat	1.000	S.F.	.013	.12	.51	.63
2 coats	1.000	S.F.	.017	.19	.66	.85
Stain, sealer, & 1 coat	1.000	S.F.	.017	.12	.66	.78
2 coats	1.000	S.F.	.019	.18	.72	.90

4 | EXTERIOR WALLS 12 | Shingle Siding Systems

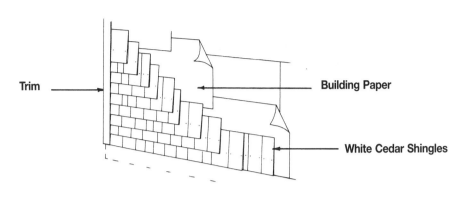

System Description	QUAN.	UNIT	LABOR HOURS	COST PER S.F.		
				MAT.	INST.	TOTAL
WHITE CEDAR SHINGLES, 5" EXPOSURE						
White cedar shingles, 16" long, grade "A", 5" exposure	1.000	S.F.	.033	2.03	1.45	3.48
Building wrap, spunbonded polypropylene	1.100	S.F.	.002	.15	.10	.25
Trim, cedar	.125	S.F.	.005	.24	.22	.46
Paint, primer & 1 coat	1.000	S.F.	.017	.12	.66	.78
TOTAL		S.F.	.057	2.54	2.43	4.97
NO. 1 PERFECTIONS, 5-1/2" EXPOSURE						
No. 1 perfections, red cedar, 5-1/2" exposure	1.000	S.F.	.029	1.84	1.27	3.11
Building wrap, spunbonded polypropylene	1.100	S.F.	.002	.15	.10	.25
Trim, cedar	.125	S.F.	.005	.24	.22	.46
Stain, sealer & 1 coat	1.000	S.F.	.017	.12	.66	.78
TOTAL		S.F.	.053	2.35	2.25	4.60
RESQUARED & REBUTTED PERFECTIONS, 5-1/2" EXPOSURE						
Resquared & rebutted perfections, 5-1/2" exposure	1.000	S.F.	.027	2.64	1.16	3.80
Building wrap, spunbonded polypropylene	1.100	S.F.	.002	.15	.10	.25
Trim, cedar	.125	S.F.	.005	.24	.22	.46
Stain, sealer & 1 coat	1.000	S.F.	.017	.12	.66	.78
TOTAL		S.F.	.051	3.15	2.14	5.29
HAND-SPLIT SHAKES, 8-1/2" EXPOSURE						
Hand-split red cedar shakes, 18" long, 8-1/2" exposure	1.000	S.F.	.040	2.10	1.74	3.84
Building wrap, spunbonded polypropylene	1.100	S.F.	.002	.15	.10	.25
Trim, cedar	.125	S.F.	.005	.24	.22	.46
Stain, sealer & 1 coat	1.000	S.F.	.017	.12	.66	.78
TOTAL		S.F.	.064	2.61	2.72	5.33

The costs in this system are based on a square foot of wall area.
Do not subtract area for door or window openings.

Description	QUAN.	UNIT	LABOR HOURS	COST PER S.F.		
				MAT.	INST.	TOTAL

Shingle Siding Price Sheet	QUAN.	UNIT	LABOR HOURS	COST PER S.F.		
				MAT.	INST.	TOTAL
Shingles wood, white cedar 16" long, "A" grade, 5" exposure	1.000	S.F.	.033	2.03	1.45	3.48
7" exposure	1.000	S.F.	.030	1.83	1.31	3.14
8-1/2" exposure	1.000	S.F.	.032	1.16	1.39	2.55
10" exposure	1.000	S.F.	.028	1.02	1.22	2.24
"B" grade, 5" exposure	1.000	S.F.	.040	1.86	1.74	3.60
7" exposure	1.000	S.F.	.028	1.30	1.22	2.52
8-1/2" exposure	1.000	S.F.	.024	1.12	1.04	2.16
10" exposure	1.000	S.F.	.020	.93	.87	1.80
Fire retardant, "A" grade, 5" exposure	1.000	S.F.	.033	2.49	1.45	3.94
7" exposure	1.000	S.F.	.028	1.62	1.22	2.84
8-1/2" exposure	1.000	S.F.	.032	1.22	1.40	2.62
10" exposure	1.000	S.F.	.025	.95	1.09	2.04
Fire retardant, 5" exposure	1.000	S.F.	.029	2.30	1.27	3.57
7" exposure	1.000	S.F.	.036	1.81	1.55	3.36
8-1/2" exposure	1.000	S.F.	.032	1.63	1.40	3.03
10" exposure	1.000	S.F.	.025	1.27	1.09	2.36
Resquared & rebutted, 5-1/2" exposure	1.000	S.F.	.027	2.64	1.16	3.80
7" exposure	1.000	S.F.	.024	2.38	1.04	3.42
8-1/2" exposure	1.000	S.F.	.021	2.11	.93	3.04
10" exposure	1.000	S.F.	.019	1.85	.81	2.66
Fire retardant, 5" exposure	1.000	S.F.	.027	3.10	1.16	4.26
7" exposure	1.000	S.F.	.024	2.79	1.04	3.83
8-1/2" exposure	1.000	S.F.	.021	2.48	.93	3.41
10" exposure	1.000	S.F.	.023	1.68	.99	2.67
Hand-split, red cedar, 24" long, 7" exposure	1.000	S.F.	.045	2.53	1.96	4.49
8-1/2" exposure	1.000	S.F.	.038	2.17	1.68	3.85
10" exposure	1.000	S.F.	.032	1.81	1.40	3.21
12" exposure	1.000	S.F.	.026	1.45	1.12	2.57
Fire retardant, 7" exposure	1.000	S.F.	.045	3.17	1.96	5.13
8-1/2" exposure	1.000	S.F.	.038	2.72	1.68	4.40
10" exposure	1.000	S.F.	.032	2.27	1.40	3.67
12" exposure	1.000	S.F.	.026	1.82	1.12	2.94
18" long, 5" exposure	1.000	S.F.	.068	3.57	2.96	6.53
7" exposure	1.000	S.F.	.048	2.52	2.09	4.61
8-1/2" exposure	1.000	S.F.	.040	2.10	1.74	3.84
10" exposure	1.000	S.F.	.036	1.89	1.57	3.46
Fire retardant, 5" exposure	1.000	S.F.	.068	4.35	2.96	7.31
7" exposure	1.000	S.F.	.048	3.07	2.09	5.16
8-1/2" exposure	1.000	S.F.	.040	2.56	1.74	4.30
10" exposure	1.000	S.F.	.036	2.30	1.57	3.87
Paper, #15 asphalt felt	1.100	S.F.	.002	.05	.09	.14
Trim, cedar	.125	S.F.	.005	.24	.22	.46
Fir	.125	S.F.	.005	.11	.22	.33
Redwood	.125	S.F.	.005	.24	.22	.46
White pine	.125	S.F.	.005	.11	.22	.33
Painting, primer, & 1 coat	1.000	S.F.	.013	.12	.51	.63
2 coats	1.000	S.F.	.017	.19	.66	.85
Staining, sealer, & 1 coat	1.000	S.F.	.017	.12	.66	.78
2 coats	1.000	S.F.	.019	.18	.72	.90

4 | EXTERIOR WALLS — 16 | Metal & Plastic Siding Systems

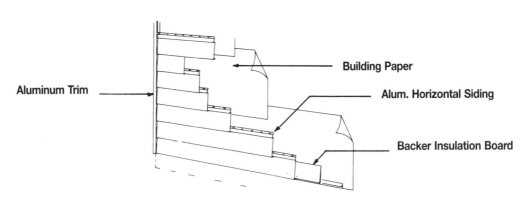

System Description	QUAN.	UNIT	LABOR HOURS	COST PER S.F. MAT.	COST PER S.F. INST.	COST PER S.F. TOTAL
ALUMINUM CLAPBOARD SIDING, 8" WIDE, WHITE						
Aluminum horizontal siding, 8" clapboard	1.000	S.F.	.031	1.80	1.35	3.15
Backer, insulation board	1.000	S.F.	.008	.87	.35	1.22
Trim, aluminum	.600	L.F.	.016	.85	.68	1.53
Building wrap, spunbonded polypropylene	1.100	S.F.	.002	.15	.10	.25
TOTAL		S.F.	.057	3.67	2.48	6.15
ALUMINUM VERTICAL BOARD & BATTEN, WHITE						
Aluminum vertical board & batten	1.000	S.F.	.027	1.98	1.18	3.16
Backer insulation board	1.000	S.F.	.008	.87	.35	1.22
Trim, aluminum	.600	L.F.	.016	.85	.68	1.53
Building wrap, spunbonded polypropylene	1.100	S.F.	.002	.15	.10	.25
TOTAL		S.F.	.053	3.85	2.31	6.16
VINYL CLAPBOARD SIDING, 8" WIDE, WHITE						
PVC vinyl horizontal siding, 8" clapboard	1.000	S.F.	.032	.98	1.41	2.39
Backer, insulation board	1.000	S.F.	.008	.87	.35	1.22
Trim, vinyl	.600	L.F.	.014	.59	.61	1.20
Building wrap, spunbonded polypropylene	1.100	S.F.	.002	.15	.10	.25
TOTAL		S.F.	.056	2.59	2.47	5.06
VINYL VERTICAL BOARD & BATTEN, WHITE						
PVC vinyl vertical board & batten	1.000	S.F.	.029	1.87	1.27	3.14
Backer, insulation board	1.000	S.F.	.008	.87	.35	1.22
Trim, vinyl	.600	L.F.	.014	.59	.61	1.20
Building wrap, spunbonded polypropylene	1.100	S.F.	.002	.15	.10	.25
TOTAL		S.F.	.053	3.48	2.33	5.81

The costs in this system are on a square foot of wall basis.
Do not subtract openings from wall area.

Description	QUAN.	UNIT	LABOR HOURS	COST PER S.F. MAT.	COST PER S.F. INST.	COST PER S.F. TOTAL

Metal & Plastic Siding Price Sheet	QUAN.	UNIT	LABOR HOURS	COST PER S.F. MAT.	COST PER S.F. INST.	COST PER S.F. TOTAL
Siding, aluminum, .024" thick, smooth, 8" wide, white	1.000	S.F.	.031	1.80	1.35	3.15
Color	1.000	S.F.	.031	1.91	1.35	3.26
Double 4" pattern, 8" wide, white	1.000	S.F.	.031	1.29	1.35	2.64
Color	1.000	S.F.	.031	1.40	1.35	2.75
Double 5" pattern, 10" wide, white	1.000	S.F.	.029	1.32	1.27	2.59
Color	1.000	S.F.	.029	1.43	1.27	2.70
Embossed, single, 8" wide, white	1.000	S.F.	.031	1.80	1.35	3.15
Color	1.000	S.F.	.031	1.91	1.35	3.26
Double 4" pattern, 8" wide, white	1.000	S.F.	.031	1.90	1.35	3.25
Color	1.000	S.F.	.031	2.01	1.35	3.36
Double 5" pattern, 10" wide, white	1.000	S.F.	.029	1.90	1.27	3.17
Color	1.000	S.F.	.029	2.01	1.27	3.28
Alum siding with insulation board, smooth, 8" wide, white	1.000	S.F.	.031	1.68	1.35	3.03
Color	1.000	S.F.	.031	1.79	1.35	3.14
Double 4" pattern, 8" wide, white	1.000	S.F.	.031	1.66	1.35	3.01
Color	1.000	S.F.	.031	1.77	1.35	3.12
Double 5" pattern, 10" wide, white	1.000	S.F.	.029	1.66	1.27	2.93
Color	1.000	S.F.	.029	1.77	1.27	3.04
Embossed, single, 8" wide, white	1.000	S.F.	.031	1.95	1.35	3.30
Color	1.000	S.F.	.031	2.06	1.35	3.41
Double 4" pattern, 8" wide, white	1.000	S.F.	.031	1.98	1.35	3.33
Color	1.000	S.F.	.031	2.09	1.35	3.44
Double 5" pattern, 10" wide, white	1.000	S.F.	.029	1.98	1.27	3.25
Color	1.000	S.F.	.029	2.09	1.27	3.36
Aluminum, shake finish, 10" wide, white	1.000	S.F.	.029	2.09	1.27	3.36
Color	1.000	S.F.	.029	2.20	1.27	3.47
Aluminum, vertical, 12" wide, white	1.000	S.F.	.027	1.98	1.18	3.16
Color	1.000	S.F.	.027	2.09	1.18	3.27
Vinyl siding, 8" wide, smooth, white	1.000	S.F.	.032	.98	1.41	2.39
Color	1.000	S.F.	.032	1.08	1.41	2.49
10" wide, Dutch lap, smooth, white	1.000	S.F.	.029	.92	1.27	2.19
Color	1.000	S.F.	.029	1.02	1.27	2.29
Double 4" pattern, 8" wide, white	1.000	S.F.	.032	.92	1.41	2.33
Color	1.000	S.F.	.032	1.02	1.41	2.43
Double 5" pattern, 10" wide, white	1.000	S.F.	.029	.80	1.27	2.07
Color	1.000	S.F.	.029	.90	1.27	2.17
Embossed, single, 8" wide, white	1.000	S.F.	.032	.97	1.41	2.38
Color	1.000	S.F.	.032	1.07	1.41	2.48
10" wide, white	1.000	S.F.	.029	.78	1.27	2.05
Color	1.000	S.F.	.029	.88	1.27	2.15
Double 4" pattern, 8" wide, white	1.000	S.F.	.032	.76	1.41	2.17
Color	1.000	S.F.	.032	.86	1.41	2.27
Double 5" pattern, 10" wide, white	1.000	S.F.	.029	.76	1.27	2.03
Color	1.000	S.F.	.029	.86	1.27	2.13
Vinyl, shake finish, 10" wide, white	1.000	S.F.	.029	2.21	1.27	3.48
Color	1.000	S.F.	.029	2.31	1.27	3.58
Vinyl, vertical, double 5" pattern, 10" wide, white	1.000	S.F.	.029	1.87	1.27	3.14
Color	1.000	S.F.	.029	1.97	1.27	3.24
Backer board, installed in siding panels 8" or 10" wide	1.000	S.F.	.008	.87	.35	1.22
4' x 8' sheets, polystyrene, 3/4" thick	1.000	S.F.	.010	.52	.44	.96
4' x 8' fiberboard, plain	1.000	S.F.	.008	.87	.35	1.22
Trim, aluminum, white	.600	L.F.	.016	.85	.68	1.53
Color	.600	L.F.	.016	.91	.68	1.59
Vinyl, white	.600	L.F.	.014	.59	.61	1.20
Color	.600	L.F.	.014	.52	.62	1.14
Paper, #15 asphalt felt	1.100	S.F.	.002	.06	.10	.16
Kraft paper, plain	1.100	S.F.	.002	.07	.10	.17
Foil backed	1.100	S.F.	.002	.10	.10	.20

4 | EXTERIOR WALLS — 20 | Insulation Systems

Description	QUAN.	UNIT	LABOR HOURS	COST PER S.F. MAT.	COST PER S.F. INST.	COST PER S.F. TOTAL
Poured insulation, cellulose fiber, R3.8 per inch (1" thick)	1.000	S.F.	.003	.06	.14	.20
Fiberglass, R4.0 per inch (1" thick)	1.000	S.F.	.003	.05	.14	.19
Mineral wool, R3.0 per inch (1" thick)	1.000	S.F.	.003	.04	.14	.18
Polystyrene, R4.0 per inch (1" thick)	1.000	S.F.	.003	.28	.14	.42
Vermiculite, R2.7 per inch (1" thick)	1.000	S.F.	.003	.16	.14	.30
Perlite, R2.7 per inch (1" thick)	1.000	S.F.	.003	.16	.14	.30
Reflective insulation, aluminum foil reinforced with scrim	1.000	S.F.	.004	.15	.18	.33
Reinforced with woven polyolefin	1.000	S.F.	.004	.19	.18	.37
With single bubble air space, R8.8	1.000	S.F.	.005	.28	.23	.51
With double bubble air space, R9.8	1.000	S.F.	.005	.29	.23	.52
Rigid insulation, fiberglass, unfaced,						
1-1/2" thick, R6.2	1.000	S.F.	.008	.87	.35	1.22
2" thick, R8.3	1.000	S.F.	.008	.75	.35	1.10
2-1/2" thick, R10.3	1.000	S.F.	.010	.87	.44	1.31
3" thick, R12.4	1.000	S.F.	.010	.87	.44	1.31
Foil faced, 1" thick, R4.3	1.000	S.F.	.008	1	.35	1.35
1-1/2" thick, R6.2	1.000	S.F.	.008	1.49	.35	1.84
2" thick, R8.7	1.000	S.F.	.009	1.86	.39	2.25
2-1/2" thick, R10.9	1.000	S.F.	.010	2.20	.44	2.64
3" thick, R13.0	1.000	S.F.	.010	2.39	.44	2.83
Foam glass, 1-1/2" thick R2.64	1.000	S.F.	.010	1.44	.44	1.88
2" thick R5.26	1.000	S.F.	.011	3.48	.48	3.96
Perlite, 1" thick R2.77	1.000	S.F.	.010	.33	.44	.77
2" thick R5.55	1.000	S.F.	.011	.65	.48	1.13
Polystyrene, extruded, blue, 2.2#/C.F., 3/4" thick R4	1.000	S.F.	.010	.52	.44	.96
1-1/2" thick R8.1	1.000	S.F.	.011	1.09	.48	1.57
2" thick R10.8	1.000	S.F.	.011	1.47	.48	1.95
Molded bead board, white, 1" thick R3.85	1.000	S.F.	.010	.24	.44	.68
1-1/2" thick, R5.6	1.000	S.F.	.011	.65	.48	1.13
2" thick, R7.7	1.000	S.F.	.011	.83	.48	1.31
Non-rigid insulation, batts						
Fiberglass, kraft faced, 3-1/2" thick, R11, 11" wide	1.000	S.F.	.005	.41	.22	.63
15" wide	1.000	S.F.	.005	.41	.22	.63
23" wide	1.000	S.F.	.005	.41	.22	.63
6" thick, R19, 11" wide	1.000	S.F.	.006	.47	.26	.73
15" wide	1.000	S.F.	.006	.47	.26	.73
23" wide	1.000	S.F.	.006	.47	.26	.73
9" thick, R30, 15" wide	1.000	S.F.	.006	1.07	.26	1.33
23" wide	1.000	S.F.	.006	1.07	.26	1.33
12" thick, R38, 15" wide	1.000	S.F.	.006	.99	.26	1.25
23" wide	1.000	S.F.	.006	.99	.26	1.25
Fiberglass, foil faced, 3-1/2" thick, R11, 15" wide	1.000	S.F.	.005	.59	.22	.81
23" wide	1.000	S.F.	.005	.59	.22	.81
6" thick, R19, 15" thick	1.000	S.F.	.005	.63	.22	.85
23" wide	1.000	S.F.	.005	.63	.22	.85
9" thick, R30, 15" wide	1.000	S.F.	.006	.91	.26	1.17
23" wide	1.000	S.F.	.006	.91	.26	1.17

Insulation Systems	QUAN.	UNIT	LABOR HOURS	COST PER S.F.		
				MAT.	INST.	TOTAL
Non-rigid insulation batts						
Fiberglass unfaced, 3-1/2" thick, R11, 15" wide	1.000	S.F.	.005	.41	.22	.63
23" wide	1.000	S.F.	.005	.41	.22	.63
6" thick, R19, 15" wide	1.000	S.F.	.006	.51	.26	.77
23" wide	1.000	S.F.	.006	.51	.26	.77
9" thick, R19, 15" wide	1.000	S.F.	.007	.91	.30	1.21
23" wide	1.000	S.F.	.007	.91	.30	1.21
12" thick, R38, 15" wide	1.000	S.F.	.007	.98	.30	1.28
23" wide	1.000	S.F.	.007	.98	.30	1.28
Mineral fiber batts, 3" thick, R11	1.000	S.F.	.005	.42	.22	.64
3-1/2" thick, R13	1.000	S.F.	.005	.42	.22	.64
6" thick, R19	1.000	S.F.	.005	.55	.22	.77
6-1/2" thick, R22	1.000	S.F.	.005	.55	.22	.77
10" thick, R30	1.000	S.F.	.006	.81	.26	1.07

4 | EXTERIOR WALLS — 28 | Double Hung Window Systems

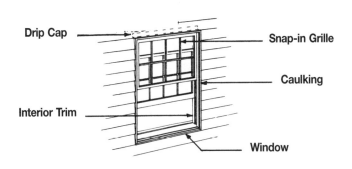

System Description	QUAN.	UNIT	LABOR HOURS	COST EACH MAT.	COST EACH INST.	COST EACH TOTAL
BUILDER'S QUALITY WOOD WINDOW 2' X 3', DOUBLE HUNG						
Window, primed, builder's quality, 2' x 3', insulating glass	1.000	Ea.	.800	209	35	244
Trim, interior casing	11.000	L.F.	.367	16.39	15.95	32.34
Paint, interior & exterior, primer & 2 coats	2.000	Face	1.778	2.28	68	70.28
Caulking	10.000	L.F.	.323	2.20	14.30	16.50
Snap-in grille	1.000	Set	.333	51	14.55	65.55
Drip cap, metal	2.000	L.F.	.040	.92	1.74	2.66
TOTAL		Ea.	3.641	281.79	149.54	431.33
PLASTIC CLAD WOOD WINDOW 3' X 4', DOUBLE HUNG						
Window, plastic clad, premium, 3' x 4', insulating glass	1.000	Ea.	.889	400	39	439
Trim, interior casing	15.000	L.F.	.500	22.35	21.75	44.10
Paint, interior, primer & 2 coats	1.000	Face	.889	1.14	34	35.14
Caulking	14.000	L.F.	.452	3.08	20.02	23.10
Snap-in grille	1.000	Set	.333	51	14.55	65.55
TOTAL		Ea.	3.063	477.57	129.32	606.89
METAL CLAD WOOD WINDOW, 3' X 5', DOUBLE HUNG						
Window, metal clad, deluxe, 3' x 5', insulating glass	1.000	Ea.	1.000	360	43.50	403.50
Trim, interior casing	17.000	L.F.	.567	25.33	24.65	49.98
Paint, interior, primer & 2 coats	1.000	Face	.889	1.14	34	35.14
Caulking	16.000	L.F.	.516	3.52	22.88	26.40
Snap-in grille	1.000	Set	.235	136	10.25	146.25
Drip cap, metal	3.000	L.F.	.060	1.38	2.61	3.99
TOTAL		Ea.	3.267	527.37	137.89	665.26

The cost of this system is on a cost per each window basis.

Description	QUAN.	UNIT	LABOR HOURS	COST EACH MAT.	COST EACH INST.	COST EACH TOTAL

Double Hung Window Price Sheet

	QUAN.	UNIT	LABOR HOURS	COST EACH		
				MAT.	INST.	TOTAL
Windows, double-hung, builder's quality, 2' x 3', single glass	1.000	Ea.	.800	199	35	234
Insulating glass	1.000	Ea.	.800	209	35	244
3' x 4', single glass	1.000	Ea.	.889	268	39	307
Insulating glass	1.000	Ea.	.889	282	39	321
4' x 4'-6", single glass	1.000	Ea.	1.000	340	43.50	383.50
Insulating glass	1.000	Ea.	1.000	365	43.50	408.50
Plastic clad premium insulating glass, 2'-6" x 3'	1.000	Ea.	.800	240	35	275
3' x 3'-6"	1.000	Ea.	.800	282	35	317
3' x 4'	1.000	Ea.	.889	400	39	439
3' x 4'-6"	1.000	Ea.	.889	375	39	414
3' x 5'	1.000	Ea.	1.000	340	43.50	383.50
3'-6" x 6'	1.000	Ea.	1.000	400	43.50	443.50
Metal clad deluxe insulating glass, 2'-6" x 3'	1.000	Ea.	.800	262	35	297
3' x 3'-6"	1.000	Ea.	.800	299	35	334
3' x 4'	1.000	Ea.	.889	315	39	354
3' x 4'-6"	1.000	Ea.	.889	330	39	369
3' x 5'	1.000	Ea.	1.000	360	43.50	403.50
3'-6" x 6'	1.000	Ea.	1.000	435	43.50	478.50
Trim, interior casing, window 2' x 3'	11.000	L.F.	.367	16.40	15.95	32.35
2'-6" x 3'	12.000	L.F.	.400	17.90	17.40	35.30
3' x 3'-6"	14.000	L.F.	.467	21	20.50	41.50
3' x 4'	15.000	L.F.	.500	22.50	22	44.50
3' x 4'-6"	16.000	L.F.	.533	24	23	47
3' x 5'	17.000	L.F.	.567	25.50	24.50	50
3'-6" x 6'	20.000	L.F.	.667	30	29	59
4' x 4'-6"	18.000	L.F.	.600	27	26	53
Paint or stain, interior or exterior, 2' x 3' window, 1 coat	1.000	Face	.444	.40	16.95	17.35
2 coats	1.000	Face	.727	.80	28	28.80
Primer & 1 coat	1.000	Face	.727	.75	28	28.75
Primer & 2 coats	1.000	Face	.889	1.14	34	35.14
3' x 4' window, 1 coat	1.000	Face	.667	.89	25.50	26.39
2 coats	1.000	Face	.667	1.04	25.50	26.54
Primer & 1 coat	1.000	Face	.727	1.16	28	29.16
Primer & 2 coats	1.000	Face	.889	1.14	34	35.14
4' x 4'-6" window, 1 coat	1.000	Face	.667	.89	25.50	26.39
2 coats	1.000	Face	.667	1.04	25.50	26.54
Primer & 1 coat	1.000	Face	.727	1.16	28	29.16
Primer & 2 coats	1.000	Face	.889	1.14	34	35.14
Caulking, window, 2' x 3'	10.000	L.F.	.323	2.20	14.30	16.50
2'-6" x 3'	11.000	L.F.	.355	2.42	15.75	18.17
3' x 3'-6"	13.000	L.F.	.419	2.86	18.60	21.46
3' x 4'	14.000	L.F.	.452	3.08	20	23.08
3' x 4'-6"	15.000	L.F.	.484	3.30	21.50	24.80
3' x 5'	16.000	L.F.	.516	3.52	23	26.52
3'-6" x 6'	19.000	L.F.	.613	4.18	27	31.18
4' x 4'-6"	17.000	L.F.	.548	3.74	24.50	28.24
Grilles, glass size to, 16" x 24" per sash	1.000	Set	.333	51	14.55	65.55
32" x 32" per sash	1.000	Set	.235	136	10.25	146.25
Drip cap, aluminum, 2' long	2.000	L.F.	.040	.92	1.74	2.66
3' long	3.000	L.F.	.060	1.38	2.61	3.99
4' long	4.000	L.F.	.080	1.84	3.48	5.32
Wood, 2' long	2.000	L.F.	.067	2.98	2.90	5.88
3' long	3.000	L.F.	.100	4.47	4.35	8.82
4' long	4.000	L.F.	.133	5.95	5.80	11.75

4 | EXTERIOR WALLS — 32 | Casement Window Systems

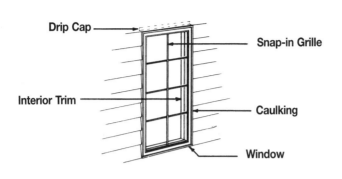

System Description	QUAN.	UNIT	LABOR HOURS	COST EACH MAT.	COST EACH INST.	COST EACH TOTAL
BUILDER'S QUALITY WINDOW, WOOD, 2' BY 3', CASEMENT						
Window, primed, builder's quality, 2' x 3', insulating glass	1.000	Ea.	.800	310	35	345
Trim, interior casing	11.000	L.F.	.367	16.39	15.95	32.34
Paint, interior & exterior, primer & 2 coats	2.000	Face	1.778	2.28	68	70.28
Caulking	10.000	L.F.	.323	2.20	14.30	16.50
Snap-in grille	1.000	Ea.	.267	29	11.65	40.65
Drip cap, metal	2.000	L.F.	.040	.92	1.74	2.66
TOTAL		Ea.	3.575	360.79	146.64	507.43
PLASTIC CLAD WOOD WINDOW, 2' X 4', CASEMENT						
Window, plastic clad, premium, 2' x 4', insulating glass	1.000	Ea.	.889	340	39	379
Trim, interior casing	13.000	L.F.	.433	19.37	18.85	38.22
Paint, interior, primer & 2 coats	1.000	Ea.	.889	1.14	34	35.14
Caulking	12.000	L.F.	.387	2.64	17.16	19.80
Snap-in grille	1.000	Ea.	.267	29	11.65	40.65
TOTAL		Ea.	2.865	392.15	120.66	512.81
METAL CLAD WOOD WINDOW, 2' X 5', CASEMENT						
Window, metal clad, deluxe, 2' x 5', insulating glass	1.000	Ea.	1.000	310	43.50	353.50
Trim, interior casing	15.000	L.F.	.500	22.35	21.75	44.10
Paint, interior, primer & 2 coats	1.000	Ea.	.889	1.14	34	35.14
Caulking	14.000	L.F.	.452	3.08	20.02	23.10
Snap-in grille	1.000	Ea.	.250	42	10.90	52.90
Drip cap, metal	12.000	L.F.	.040	.92	1.74	2.66
TOTAL		Ea.	3.131	379.49	131.91	511.40

The cost of this system is on a cost per each window basis.

Description	QUAN.	UNIT	LABOR HOURS	COST EACH MAT.	COST EACH INST.	COST EACH TOTAL

Casement Window Price Sheet	QUAN.	UNIT	LABOR HOURS	COST EACH		
				MAT.	INST.	TOTAL
Window, casement, builders quality, 2' x 3', single glass	1.000	Ea.	.800	289	35	324
Insulating glass	1.000	Ea.	.800	310	35	345
2' x 4'-6", single glass	1.000	Ea.	.727	855	31.50	886.50
Insulating glass	1.000	Ea.	.727	720	31.50	751.50
2' x 6', single glass	1.000	Ea.	.889	440	43.50	483.50
Insulating glass	1.000	Ea.	.889	1,100	39	1,139
Plastic clad premium insulating glass, 2' x 3'	1.000	Ea.	.800	256	35	291
2' x 4'	1.000	Ea.	.889	410	39	449
2' x 5'	1.000	Ea.	1.000	428	43.50	471.50
2' x 6'	1.000	Ea.	1.000	440	43.50	483.50
Metal clad deluxe insulating glass, 2' x 3'	1.000	Ea.	.800	227	35	262
2' x 4'	1.000	Ea.	.889	273	39	312
2' x 5'	1.000	Ea.	1.000	310	43.50	353.50
2' x 6'	1.000	Ea.	1.000	355	43.50	398.50
Trim, interior casing, window 2' x 3'	11.000	L.F.	.367	16.40	15.95	32.35
2' x 4'	13.000	L.F.	.433	19.35	18.85	38.20
2' x 4'-6"	14.000	L.F.	.467	21	20.50	41.50
2' x 5'	15.000	L.F.	.500	22.50	22	44.50
2' x 6'	17.000	L.F.	.567	25.50	24.50	50
Paint or stain, interior or exterior, 2' x 3' window, 1 coat	1.000	Face	.444	.40	16.95	17.35
2 coats	1.000	Face	.727	.80	28	28.80
Primer & 1 coat	1.000	Face	.727	.75	28	28.75
Primer & 2 coats	1.000	Face	.889	1.14	34	35.14
2' x 4' window, 1 coat	1.000	Face	.444	.40	16.95	17.35
2 coats	1.000	Face	.727	.80	28	28.80
Primer & 1 coat	1.000	Face	.727	.75	28	28.75
Primer & 2 coats	1.000	Face	.889	1.14	34	35.14
2' x 6' window, 1 coat	1.000	Face	.667	.89	25.50	26.39
2 coats	1.000	Face	.667	1.04	25.50	26.54
Primer & 1 coat	1.000	Face	.727	1.16	28	29.16
Primer & 2 coats	1.000	Face	.889	1.14	34	35.14
Caulking, window, 2' x 3'	10.000	L.F.	.323	2.20	14.30	16.50
2' x 4'	12.000	L.F.	.387	2.64	17.15	19.79
2' x 4'-6"	13.000	L.F.	.419	2.86	18.60	21.46
2' x 5'	14.000	L.F.	.452	3.08	20	23.08
2' x 6'	16.000	L.F.	.516	3.52	23	26.52
Grilles, glass size, to 20" x 36"	1.000	Ea.	.267	29	11.65	40.65
To 20" x 56"	1.000	Ea.	.250	42	10.90	52.90
Drip cap, metal, 2' long	2.000	L.F.	.040	.92	1.74	2.66
Wood, 2' long	2.000	L.F.	.067	2.98	2.90	5.88

4 | EXTERIOR WALLS — 36 | Awning Window Systems

System Description	QUAN.	UNIT	LABOR HOURS	COST EACH MAT.	COST EACH INST.	COST EACH TOTAL
BUILDER'S QUALITY WINDOW, WOOD, 34" X 22", AWNING						
Window, builder quality, 34" x 22", insulating glass	1.000	Ea.	.800	271	35	306
Trim, interior casing	10.500	L.F.	.350	15.65	15.23	30.88
Paint, interior & exterior, primer & 2 coats	2.000	Face	1.778	2.28	68	70.28
Caulking	9.500	L.F.	.306	2.09	13.59	15.68
Snap-in grille	1.000	Ea.	.267	23.50	11.65	35.15
Drip cap, metal	3.000	L.F.	.060	1.38	2.61	3.99
TOTAL		Ea.	3.561	315.90	146.08	461.98
PLASTIC CLAD WOOD WINDOW, 40" X 28", AWNING						
Window, plastic clad, premium, 40" x 28", insulating glass	1.000	Ea.	.889	355	39	394
Trim interior casing	13.500	L.F.	.450	20.12	19.58	39.70
Paint, interior, primer & 2 coats	1.000	Face	.889	1.14	34	35.14
Caulking	12.500	L.F.	.403	2.75	17.88	20.63
Snap-in grille	1.000	Ea.	.267	23.50	11.65	35.15
TOTAL		Ea.	2.898	402.51	122.11	524.62
METAL CLAD WOOD WINDOW, 48" X 36", AWNING						
Window, metal clad, deluxe, 48" x 36", insulating glass	1.000	Ea.	1.000	380	43.50	423.50
Trim, interior casing	15.000	L.F.	.500	22.35	21.75	44.10
Paint, interior, primer & 2 coats	1.000	Face	.889	1.14	34	35.14
Caulking	14.000	L.F.	.452	3.08	20.02	23.10
Snap-in grille	1.000	Ea.	.250	34.50	10.90	45.40
Drip cap, metal	4.000	L.F.	.080	1.84	3.48	5.32
TOTAL		Ea.	3.171	442.91	133.65	576.56

The cost of this system is on a cost per each window basis.

Description	QUAN.	UNIT	LABOR HOURS	COST EACH MAT.	COST EACH INST.	COST EACH TOTAL

Awning Window Price Sheet	QUAN.	UNIT	LABOR HOURS	COST EACH		
				MAT.	INST.	TOTAL
Windows, awning, builder's quality, 34" x 22", insulated glass	1.000	Ea.	.800	257	35	292
Low E glass	1.000	Ea.	.800	271	35	306
40" x 28", insulated glass	1.000	Ea.	.889	325	39	364
Low E glass	1.000	Ea.	.889	345	39	384
48" x 36", insulated glass	1.000	Ea.	1.000	475	43.50	518.50
Low E glass	1.000	Ea.	1.000	500	43.50	543.50
Plastic clad premium insulating glass, 34" x 22"	1.000	Ea.	.800	273	35	308
40" x 22"	1.000	Ea.	.800	298	35	333
36" x 28"	1.000	Ea.	.889	320	39	359
36" x 36"	1.000	Ea.	.889	355	39	394
48" x 28"	1.000	Ea.	1.000	380	43.50	423.50
60" x 36"	1.000	Ea.	1.000	550	43.50	593.50
Metal clad deluxe insulating glass, 34" x 22"	1.000	Ea.	.800	255	35	290
40" x 22"	1.000	Ea.	.800	300	35	335
36" x 25"	1.000	Ea.	.889	278	39	317
40" x 30"	1.000	Ea.	.889	345	39	384
48" x 28"	1.000	Ea.	1.000	355	43.50	398.50
60" x 36"	1.000	Ea.	1.000	380	43.50	423.50
Trim, interior casing window, 34" x 22"	10.500	L.F.	.350	15.65	15.25	30.90
40" x 22"	11.500	L.F.	.383	17.15	16.70	33.85
36" x 28"	12.500	L.F.	.417	18.65	18.15	36.80
40" x 28"	13.500	L.F.	.450	20	19.60	39.60
48" x 28"	14.500	L.F.	.483	21.50	21	42.50
48" x 36"	15.000	L.F.	.500	22.50	22	44.50
Paint or stain, interior or exterior, 34" x 22", 1 coat	1.000	Face	.444	.40	16.95	17.35
2 coats	1.000	Face	.727	.80	28	28.80
Primer & 1 coat	1.000	Face	.727	.75	28	28.75
Primer & 2 coats	1.000	Face	.889	1.14	34	35.14
36" x 28", 1 coat	1.000	Face	.444	.40	16.95	17.35
2 coats	1.000	Face	.727	.80	28	28.80
Primer & 1 coat	1.000	Face	.727	.75	28	28.75
Primer & 2 coats	1.000	Face	.889	1.14	34	35.14
48" x 36", 1 coat	1.000	Face	.667	.89	25.50	26.39
2 coats	1.000	Face	.667	1.04	25.50	26.54
Primer & 1 coat	1.000	Face	.727	1.16	28	29.16
Primer & 2 coats	1.000	Face	.889	1.14	34	35.14
Caulking, window, 34" x 22"	9.500	L.F.	.306	2.09	13.60	15.69
40" x 22"	10.500	L.F.	.339	2.31	15	17.31
36" x 28"	11.500	L.F.	.371	2.53	16.45	18.98
40" x 28"	12.500	L.F.	.403	2.75	17.90	20.65
48" x 28"	13.500	L.F.	.436	2.97	19.30	22.27
48" x 36"	14.000	L.F.	.452	3.08	20	23.08
Grilles, glass size, to 28" by 16"	1.000	Ea.	.267	23.50	11.65	35.15
To 44" by 24"	1.000	Ea.	.250	34.50	10.90	45.40
Drip cap, aluminum, 3' long	3.000	L.F.	.060	1.38	2.61	3.99
3'-6" long	3.500	L.F.	.070	1.61	3.05	4.66
4' long	4.000	L.F.	.080	1.84	3.48	5.32
Wood, 3' long	3.000	L.F.	.100	4.47	4.35	8.82
3'-6" long	3.500	L.F.	.117	5.20	5.10	10.30
4' long	4.000	L.F.	.133	5.95	5.80	11.75

4 | EXTERIOR WALLS — 40 | Sliding Window Systems

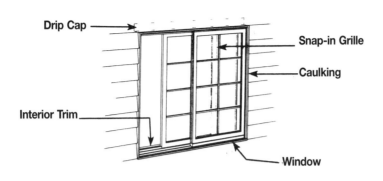

System Description	QUAN.	UNIT	LABOR HOURS	COST EACH MAT.	COST EACH INST.	COST EACH TOTAL
BUILDER'S QUALITY WOOD WINDOW, 3' X 2', SLIDING						
Window, primed, builder's quality, 3' x 2', insul. glass	1.000	Ea.	.800	280	35	315
Trim, interior casing	11.000	L.F.	.367	16.39	15.95	32.34
Paint, interior & exterior, primer & 2 coats	2.000	Face	1.778	2.28	68	70.28
Caulking	10.000	L.F.	.323	2.20	14.30	16.50
Snap-in grille	1.000	Set	.333	27.50	14.55	42.05
Drip cap, metal	3.000	L.F.	.060	1.38	2.61	3.99
TOTAL		Ea.	3.661	329.75	150.41	480.16
PLASTIC CLAD WOOD WINDOW, 4' X 3'-6", SLIDING						
Window, plastic clad, premium, 4' x 3'-6", insulating glass	1.000	Ea.	.889	690	39	729
Trim, interior casing	16.000	L.F.	.533	23.84	23.20	47.04
Paint, interior, primer & 2 coats	1.000	Face	.889	1.14	34	35.14
Caulking	17.000	L.F.	.548	3.74	24.31	28.05
Snap-in grille	1.000	Set	.333	27.50	14.55	42.05
TOTAL		Ea.	3.192	746.22	135.06	881.28
METAL CLAD WOOD WINDOW, 6' X 5', SLIDING						
Window, metal clad, deluxe, 6' x 5', insulating glass	1.000	Ea.	1.000	740	43.50	783.50
Trim, interior casing	23.000	L.F.	.767	34.27	33.35	67.62
Paint, interior, primer & 2 coats	1.000	Face	.889	1.14	34	35.14
Caulking	22.000	L.F.	.710	4.84	31.46	36.30
Snap-in grille	1.000	Set	.364	42	15.85	57.85
Drip cap, metal	6.000	L.F.	.120	2.76	5.22	7.98
TOTAL		Ea.	3.850	825.01	163.38	988.39

The cost of this system is on a cost per each window basis.

Description	QUAN.	UNIT	LABOR HOURS	COST EACH MAT.	COST EACH INST.	COST EACH TOTAL

Sliding Window Price Sheet	QUAN.	UNIT	LABOR HOURS	COST EACH		
				MAT.	INST.	TOTAL
Windows, sliding, builder's quality, 3' x 3', single glass	1.000	Ea.	.800	256	35	291
Insulating glass	1.000	Ea.	.800	280	35	315
4' x 3'-6", single glass	1.000	Ea.	.889	273	39	312
Insulating glass	1.000	Ea.	.889	310	39	349
6' x 5', single glass	1.000	Ea.	1.000	440	43.50	483.50
Insulating glass	1.000	Ea.	1.000	485	43.50	528.50
Plastic clad premium insulating glass, 3' x 3'	1.000	Ea.	.800	560	35	595
4' x 3'-6"	1.000	Ea.	.889	690	39	729
5' x 4'	1.000	Ea.	.889	835	39	874
6' x 5'	1.000	Ea.	1.000	1,050	43.50	1,093.50
Metal clad deluxe insulating glass, 3' x 3'	1.000	Ea.	.800	340	35	375
4' x 3'-6"	1.000	Ea.	.889	415	39	454
5' x 4'	1.000	Ea.	.889	500	39	539
6' x 5'	1.000	Ea.	1.000	740	43.50	783.50
Trim, interior casing, window 3' x 2'	11.000	L.F.	.367	16.40	15.95	32.35
3' x 3'	13.000	L.F.	.433	19.35	18.85	38.20
4' x 3'-6"	16.000	L.F.	.533	24	23	47
5' x 4'	19.000	L.F.	.633	28.50	27.50	56
6' x 5'	23.000	L.F.	.767	34.50	33.50	68
Paint or stain, interior or exterior, 3' x 2' window, 1 coat	1.000	Face	.444	.40	16.95	17.35
2 coats	1.000	Face	.727	.80	28	28.80
Primer & 1 coat	1.000	Face	.727	.75	28	28.75
Primer & 2 coats	1.000	Face	.889	1.14	34	35.14
4' x 3'-6" window, 1 coat	1.000	Face	.667	.89	25.50	26.39
2 coats	1.000	Face	.667	1.04	25.50	26.54
Primer & 1 coat	1.000	Face	.727	1.16	28	29.16
Primer & 2 coats	1.000	Face	.889	1.14	34	35.14
6' x 5' window, 1 coat	1.000	Face	.889	2.28	34	36.28
2 coats	1.000	Face	1.333	4.16	51	55.16
Primer & 1 coat	1.000	Face	1.333	3.99	51	54.99
Primer & 2 coats	1.000	Face	1.600	6.05	61	67.05
Caulking, window, 3' x 2'	10.000	L.F.	.323	2.20	14.30	16.50
3' x 3'	12.000	L.F.	.387	2.64	17.15	19.79
4' x 3'-6"	15.000	L.F.	.484	3.30	21.50	24.80
5' x 4'	18.000	L.F.	.581	3.96	25.50	29.46
6' x 5'	22.000	L.F.	.710	4.84	31.50	36.34
Grilles, glass size, to 14" x 36"	1.000	Set	.333	27.50	14.55	42.05
To 36" x 36"	1.000	Set	.364	42	15.85	57.85
Drip cap, aluminum, 3' long	3.000	L.F.	.060	1.38	2.61	3.99
4' long	4.000	L.F.	.080	1.84	3.48	5.32
5' long	5.000	L.F.	.100	2.30	4.35	6.65
6' long	6.000	L.F.	.120	2.76	5.20	7.96
Wood, 3' long	3.000	L.F.	.100	4.47	4.35	8.82
4' long	4.000	L.F.	.133	5.95	5.80	11.75
5' long	5.000	L.F.	.167	7.45	7.25	14.70
6' long	6.000	L.F.	.200	8.95	8.70	17.65

4 | EXTERIOR WALLS   44 | Bow/Bay Window Systems

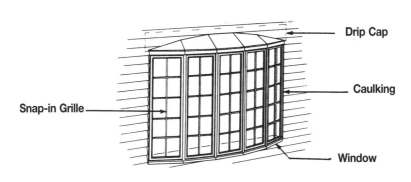

System Description	QUAN.	UNIT	LABOR HOURS	COST EACH		
				MAT.	INST.	TOTAL
AWNING TYPE BOW WINDOW, BUILDER'S QUALITY, 8' X 5'						
Window, primed, builder's quality, 8' x 5', insulating glass	1.000	Ea.	1.600	1,400	70	1,470
Trim, interior casing	27.000	L.F.	.900	40.23	39.15	79.38
Paint, interior & exterior, primer & 1 coat	2.000	Face	3.200	12.10	122	134.10
Drip cap, vinyl	1.000	Ea.	.533	82.50	23.50	106
Caulking	26.000	L.F.	.839	5.72	37.18	42.90
Snap-in grilles	1.000	Set	1.067	116	46.60	162.60
TOTAL		Ea.	8.139	1,656.55	338.43	1,994.98
CASEMENT TYPE BOW WINDOW, PLASTIC CLAD, 10' X 6'						
Window, plastic clad, premium, 10' x 6', insulating glass	1.000	Ea.	2.286	2,075	99.50	2,174.50
Trim, interior casing	33.000	L.F.	1.100	49.17	47.85	97.02
Paint, interior, primer & 1 coat	1.000	Face	1.778	2.28	68	70.28
Drip cap, vinyl	1.000	Ea.	.615	90	27	117
Caulking	32.000	L.F.	1.032	7.04	45.76	52.80
Snap-in grilles	1.000	Set	1.333	145	58.25	203.25
TOTAL		Ea.	8.144	2,368.49	346.36	2,714.85
DOUBLE HUNG TYPE, METAL CLAD, 9' X 5'						
Window, metal clad, deluxe, 9' x 5', insulating glass	1.000	Ea.	2.667	1,400	116	1,516
Trim, interior casing	29.000	L.F.	.967	43.21	42.05	85.26
Paint, interior, primer & 1 coat	1.000	Face	1.778	2.28	68	70.28
Drip cap, vinyl	1.000	Set	.615	90	27	117
Caulking	28.000	L.F.	.903	6.16	40.04	46.20
Snap-in grilles	1.000	Set	1.067	116	46.60	162.60
TOTAL		Ea.	7.997	1,657.65	339.69	1,997.34

The cost of this system is on a cost per each window basis.

Description	QUAN.	UNIT	LABOR HOURS	COST EACH		
				MAT.	INST.	TOTAL

Bow/Bay Window Price Sheet	QUAN.	UNIT	LABOR HOURS	COST EACH		
				MAT.	INST.	TOTAL
Windows, bow awning type, builder's quality, 8' x 5', insulating glass	1.000	Ea.	1.600	1,325	70	1,395
Low E glass	1.000	Ea.	1.600	1,400	70	1,470
12' x 6', insulating glass	1.000	Ea.	2.667	1,475	116	1,591
Low E glass	1.000	Ea.	2.667	1,550	116	1,666
Plastic clad premium insulating glass, 6' x 4'	1.000	Ea.	1.600	1,350	70	1,420
9' x 4'	1.000	Ea.	2.000	1,600	87	1,687
10' x 5'	1.000	Ea.	2.286	2,600	99.50	2,699.50
12' x 6'	1.000	Ea.	2.667	3,300	116	3,416
Metal clad deluxe insulating glass, 6' x 4'	1.000	Ea.	1.600	940	70	1,010
9' x 4'	1.000	Ea.	2.000	1,325	87	1,412
10' x 5'	1.000	Ea.	2.286	1,825	99.50	1,924.50
12' x 6'	1.000	Ea.	2.667	2,525	116	2,641
Bow casement type, builder's quality, 8' x 5', single glass	1.000	Ea.	1.600	1,825	70	1,895
Insulating glass	1.000	Ea.	1.600	2,225	70	2,295
12' x 6', single glass	1.000	Ea.	2.667	2,300	116	2,416
Insulating glass	1.000	Ea.	2.667	2,350	116	2,466
Plastic clad premium insulating glass, 8' x 5'	1.000	Ea.	1.600	1,400	70	1,470
10' x 5'	1.000	Ea.	2.000	2,000	87	2,087
10' x 6'	1.000	Ea.	2.286	2,075	99.50	2,174.50
12' x 6'	1.000	Ea.	2.667	2,475	116	2,591
Metal clad deluxe insulating glass, 8' x 5'	1.000	Ea.	1.600	1,625	70	1,695
10' x 5'	1.000	Ea.	2.000	1,750	87	1,837
10' x 6'	1.000	Ea.	2.286	2,050	99.50	2,149.50
12' x 6'	1.000	Ea.	2.667	2,850	116	2,966
Bow, double hung type, builder's quality, 8' x 4', single glass	1.000	Ea.	1.600	1,275	70	1,345
Insulating glass	1.000	Ea.	1.600	1,375	70	1,445
9' x 5', single glass	1.000	Ea.	2.667	1,400	116	1,516
Insulating glass	1.000	Ea.	2.667	1,475	116	1,591
Plastic clad premium insulating glass, 7' x 4'	1.000	Ea.	1.600	1,325	70	1,395
8' x 4'	1.000	Ea.	2.000	1,375	87	1,462
8' x 5'	1.000	Ea.	2.286	1,425	99.50	1,524.50
9' x 5'	1.000	Ea.	2.667	1,475	116	1,591
Metal clad deluxe insulating glass, 7' x 4'	1.000	Ea.	1.600	1,225	70	1,295
8' x 4'	1.000	Ea.	2.000	1,275	87	1,362
8' x 5'	1.000	Ea.	2.286	1,325	99.50	1,424.50
9' x 5'	1.000	Ea.	2.667	1,400	116	1,516
Trim, interior casing, window 7' x 4'	1.000	Ea.	.767	34.50	33.50	68
8' x 5'	1.000	Ea.	.900	40	39	79
10' x 6'	1.000	Ea.	1.100	49	48	97
12' x 6'	1.000	Ea.	1.233	55	53.50	108.50
Paint or stain, interior, or exterior, 7' x 4' window, 1 coat	1.000	Face	.889	2.28	34	36.28
Primer & 1 coat	1.000	Face	1.333	3.99	51	54.99
8' x 5' window, 1 coat	1.000	Face	.889	2.28	34	36.28
Primer & 1 coat	1.000	Face	1.333	3.99	51	54.99
10' x 6' window, 1 coat	1.000	Face	1.333	1.78	51	52.78
Primer & 1 coat	1.000	Face	1.778	2.28	68	70.28
12' x 6' window, 1 coat	1.000	Face	1.778	4.56	68	72.56
Primer & 1 coat	1.000	Face	2.667	8	102	110
Drip cap, vinyl moulded window, 7' long	1.000	Ea.	.533	72	20.50	92.50
8' long	1.000	Ea.	.533	82.50	23.50	106
10' long	1.000	Ea.	.615	90	26	116
12' long	1.000	Ea.	.615	90	26	116
Caulking, window, 7' x 4'	1.000	Ea.	.710	4.84	31.50	36.34
8' x 5'	1.000	Ea.	.839	5.70	37	42.70
10' x 6'	1.000	Ea.	1.032	7.05	46	53.05
12' x 6'	1.000	Ea.	1.161	7.90	51.50	59.40
Grilles, window, 7' x 4'	1.000	Set	.800	87	35	122
8' x 5'	1.000	Set	1.067	116	46.50	162.50
10' x 6'	1.000	Set	1.333	145	58.50	203.50
12' x 6'	1.000	Set	1.600	174	70	244

4 | EXTERIOR WALLS — 48 | Fixed Window Systems

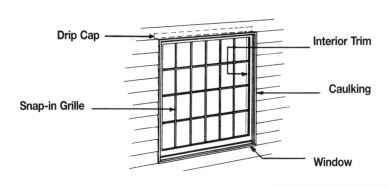

System Description	QUAN.	UNIT	LABOR HOURS	COST EACH MAT.	COST EACH INST.	COST EACH TOTAL
BUILDER'S QUALITY PICTURE WINDOW, 4' X 4'						
Window, primed, builder's quality, 4' x 4', insulating glass	1.000	Ea.	1.333	425	58	483
Trim, interior casing	17.000	L.F.	.567	25.33	24.65	49.98
Paint, interior & exterior, primer & 2 coats	2.000	Face	1.778	2.28	68	70.28
Caulking	16.000	L.F.	.516	3.52	22.88	26.40
Snap-in grille	1.000	Ea.	.267	155	11.65	166.65
Drip cap, metal	4.000	L.F.	.080	1.84	3.48	5.32
TOTAL		Ea.	4.541	612.97	188.66	801.63
PLASTIC CLAD WOOD WINDOW, 4'-6" X 6'-6"						
Window, plastic clad, prem., 4'-6" x 6'-6", insul. glass	1.000	Ea.	1.455	965	63.50	1,028.50
Trim, interior casing	23.000	L.F.	.767	34.27	33.35	67.62
Paint, interior, primer & 2 coats	1.000	Face	.889	1.14	34	35.14
Caulking	22.000	L.F.	.710	4.84	31.46	36.30
Snap-in grille	1.000	Ea.	.267	155	11.65	166.65
TOTAL		Ea.	4.088	1,160.25	173.96	1,334.21
METAL CLAD WOOD WINDOW, 6'-6" X 6'-6"						
Window, metal clad, deluxe, 6'-6" x 6'-6", insulating glass	1.000	Ea.	1.600	675	70	745
Trim interior casing	27.000	L.F.	.900	40.23	39.15	79.38
Paint, interior, primer & 2 coats	1.000	Face	1.600	6.05	61	67.05
Caulking	26.000	L.F.	.839	5.72	37.18	42.90
Snap-in grille	1.000	Ea.	.267	155	11.65	166.65
Drip cap, metal	6.500	L.F.	.130	2.99	5.66	8.65
TOTAL		Ea.	5.336	884.99	224.64	1,109.63

The cost of this system is on a cost per each window basis.

Description	QUAN.	UNIT	LABOR HOURS	COST EACH MAT.	COST EACH INST.	COST EACH TOTAL

Fixed Window Price Sheet

Fixed Window Price Sheet	QUAN.	UNIT	LABOR HOURS	COST EACH MAT.	COST EACH INST.	COST EACH TOTAL
Window-picture, builder's quality, 4' x 4', single glass	1.000	Ea.	1.333	385	58	443
Insulating glass	1.000	Ea.	1.333	425	58	483
4' x 4'-6", single glass	1.000	Ea.	1.455	475	63.50	538.50
Insulating glass	1.000	Ea.	1.455	495	63.50	558.50
5' x 4', single glass	1.000	Ea.	1.455	555	63.50	618.50
Insulating glass	1.000	Ea.	1.455	575	63.50	638.50
6' x 4'-6", single glass	1.000	Ea.	1.600	600	70	670
Insulating glass	1.000	Ea.	1.600	620	70	690
Plastic clad premium insulating glass, 4' x 4'	1.000	Ea.	1.333	515	58	573
4'-6" x 6'-6"	1.000	Ea.	1.455	965	63.50	1,028.50
5'-6" x 6'-6"	1.000	Ea.	1.600	1,075	70	1,145
6'-6" x 6'-6"	1.000	Ea.	1.600	1,075	70	1,145
Metal clad deluxe insulating glass, 4' x 4'	1.000	Ea.	1.333	360	58	418
4'-6" x 6'-6"	1.000	Ea.	1.455	535	63.50	598.50
5'-6" x 6'-6"	1.000	Ea.	1.600	585	70	655
6'-6" x 6'-6"	1.000	Ea.	1.600	675	70	745
Trim, interior casing, window 4' x 4'	17.000	L.F.	.567	25.50	24.50	50
4'-6" x 4'-6"	19.000	L.F.	.633	28.50	27.50	56
5'-0" x 4'-0"	19.000	L.F.	.633	28.50	27.50	56
4'-6" x 6'-6"	23.000	L.F.	.767	34.50	33.50	68
5'-6" x 6'-6"	25.000	L.F.	.833	37.50	36.50	74
6'-6" x 6'-6"	27.000	L.F.	.900	40	39	79
Paint or stain, interior or exterior, 4' x 4' window, 1 coat	1.000	Face	.667	.89	25.50	26.39
2 coats	1.000	Face	.667	1.04	25.50	26.54
Primer & 1 coat	1.000	Face	.727	1.16	28	29.16
Primer & 2 coats	1.000	Face	.889	1.14	34	35.14
4'-6" x 6'-6" window, 1 coat	1.000	Face	.667	.89	25.50	26.39
2 coats	1.000	Face	.667	1.04	25.50	26.54
Primer & 1 coat	1.000	Face	.727	1.16	28	29.16
Primer & 2 coats	1.000	Face	.889	1.14	34	35.14
6'-6" x 6'-6" window, 1 coat	1.000	Face	.889	2.28	34	36.28
2 coats	1.000	Face	1.333	4.16	51	55.16
Primer & 1 coat	1.000	Face	1.333	3.99	51	54.99
Primer & 2 coats	1.000	Face	1.600	6.05	61	67.05
Caulking, window, 4' x 4'	1.000	Ea.	.516	3.52	23	26.52
4'-6" x 4'-6"	1.000	Ea.	.581	3.96	25.50	29.46
5'-0" x 4'-0"	1.000	Ea.	.581	3.96	25.50	29.46
4'-6" x 6'-6"	1.000	Ea.	.710	4.84	31.50	36.34
5'-6" x 6'-6"	1.000	Ea.	.774	5.30	34.50	39.80
6'-6" x 6'-6"	1.000	Ea.	.839	5.70	37	42.70
Grilles, glass size, to 48" x 48"	1.000	Ea.	.267	155	11.65	166.65
To 60" x 68"	1.000	Ea.	.286	119	12.45	131.45
Drip cap, aluminum, 4' long	4.000	L.F.	.080	1.84	3.48	5.32
4'-6" long	4.500	L.F.	.090	2.07	3.92	5.99
5' long	5.000	L.F.	.100	2.30	4.35	6.65
6' long	6.000	L.F.	.120	2.76	5.20	7.96
Wood, 4' long	4.000	L.F.	.133	5.95	5.80	11.75
4'-6" long	4.500	L.F.	.150	6.70	6.55	13.25
5' long	5.000	L.F.	.167	7.45	7.25	14.70
6' long	6.000	L.F.	.200	8.95	8.70	17.65

4 | EXTERIOR WALLS — 52 | Entrance Door Systems

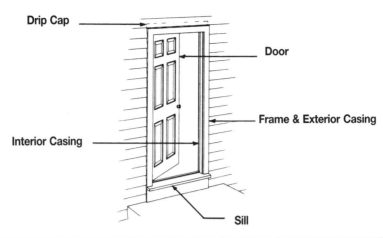

System Description	QUAN.	UNIT	LABOR HOURS	MAT.	INST.	TOTAL
COLONIAL, 6 PANEL, 3' X 6'-8", WOOD						
Door, 3' x 6'-8" x 1-3/4" thick, pine, 6 panel colonial	1.000	Ea.	1.067	440	46.50	486.50
Frame, 5-13/16" deep, incl. exterior casing & drip cap	17.000	L.F.	.725	204	31.62	235.62
Interior casing, 2-1/2" wide	18.000	L.F.	.600	26.82	26.10	52.92
Sill, 8/4 x 8" deep	3.000	L.F.	.480	60	21	81
Butt hinges, brass, 4-1/2" x 4-1/2"	1.500	Pr.		21.90		21.90
Lockset	1.000	Ea.	.571	37	25	62
Weatherstripping, metal, spring type, bronze	1.000	Set	1.053	19.70	46	65.70
Paint, interior & exterior, primer & 2 coats	2.000	Face	1.778	12	68	80
TOTAL		Ea.	6.274	821.42	264.22	1,085.64
SOLID CORE BIRCH, FLUSH, 3' X 6'-8"						
Door, 3'-0" x 6'-8", 1-3/4" thick, birch, flush solid core	1.000	Ea.	1.067	111	46.50	157.50
Frame, 5-13/16" deep, incl. exterior casing & drip cap	17.000	L.F.	.725	204	31.62	235.62
Interior casing, 2-1/2" wide	18.000	L.F.	.600	26.82	26.10	52.92
Sill, 8/4 x 8" deep	3.000	L.F.	.480	60	21	81
Butt hinges, brass, 4-1/2" x 4-1/2"	1.500	Pr.		21.90		21.90
Lockset	1.000	Ea.	.571	37	25	62
Weatherstripping, metal, spring type, bronze	1.000	Set	1.053	19.70	46	65.70
Paint, Interior & exterior, primer & 2 coats	2.000	Face	1.778	11.20	68	79.20
TOTAL		Ea.	6.274	491.62	264.22	755.84

These systems are on a cost per each door basis.

Description	QUAN.	UNIT	LABOR HOURS	MAT.	INST.	TOTAL

Entrance Door Price Sheet	QUAN.	UNIT	LABOR HOURS	COST EACH		
				MAT.	INST.	TOTAL
Door exterior wood 1-3/4" thick, pine, dutch door, 2'-8" x 6'-8" minimum	1.000	Ea.	1.333	740	58	798
Maximum	1.000	Ea.	1.600	785	70	855
3'-0" x 6'-8", minimum	1.000	Ea.	1.333	755	58	813
Maximum	1.000	Ea.	1.600	825	70	895
Colonial, 6 panel, 2'-8" x 6'-8"	1.000	Ea.	1.000	415	43.50	458.50
3'-0" x 6'-8"	1.000	Ea.	1.067	440	46.50	486.50
8 panel, 2'-6" x 6'-8"	1.000	Ea.	1.000	615	43.50	658.50
3'-0" x 6'-8"	1.000	Ea.	1.067	590	46.50	636.50
Flush, birch, solid core, 2'-8" x 6'-8"	1.000	Ea.	1.000	107	43.50	150.50
3'-0" x 6'-8"	1.000	Ea.	1.067	111	46.50	157.50
Porch door, 2'-8" x 6'-8"	1.000	Ea.	1.000	560	43.50	603.50
3'-0" x 6'-8"	1.000	Ea.	1.067	560	46.50	606.50
Hand carved mahogany, 2'-8" x 6'-8"	1.000	Ea.	1.067	520	46.50	566.50
3'-0" x 6'-8"	1.000	Ea.	1.067	555	46.50	601.50
Rosewood, 2'-8" x 6'-8"	1.000	Ea.	1.067	810	46.50	856.50
3'-0" x 6-8"	1.000	Ea.	1.067	840	46.50	886.50
Door, metal clad wood 1-3/8" thick raised panel, 2'-8" x 6'-8"	1.000	Ea.	1.067	293	41	334
3'-0" x 6'-8"	1.000	Ea.	1.067	263	46.50	309.50
Deluxe metal door, 3'-0" x 6'-8"	1.000	Ea.	1.231	263	46.50	309.50
3'-0" x 6'-8"	1.000	Ea.	1.231	263	46.50	309.50
Frame, pine, including exterior trim & drip cap, 5/4, x 4-9/16" deep	17.000	L.F.	.725	99.50	31.50	131
5-13/16" deep	17.000	L.F.	.725	204	31.50	235.50
6-9/16" deep	17.000	L.F.	.725	155	31.50	186.50
Safety glass lites, add	1.000	Ea.		71.50		71.50
Interior casing, 2'-8" x 6'-8" door	18.000	L.F.	.600	27	26	53
3'-0" x 6'-8" door	19.000	L.F.	.633	28.50	27.50	56
Sill, oak, 8/4 x 8" deep	3.000	L.F.	.480	60	21	81
8/4 x 10" deep	3.000	L.F.	.533	69	23.50	92.50
Butt hinges, steel plated, 4-1/2" x 4-1/2", plain	1.500	Pr.		22		22
Ball bearing	1.500	Pr.		49		49
Bronze, 4-1/2" x 4-1/2", plain	1.500	Pr.		25		25
Ball bearing	1.500	Pr.		52		52
Lockset, minimum	1.000	Ea.	.571	37	25	62
Maximum	1.000	Ea.	1.000	156	43.50	199.50
Weatherstripping, metal, interlocking, zinc	1.000	Set	2.667	15.35	116	131.35
Bronze	1.000	Set	2.667	24	116	140
Spring type, bronze	1.000	Set	1.053	19.70	46	65.70
Rubber, minimum	1.000	Set	1.053	5.10	46	51.10
Maximum	1.000	Set	1.143	5.85	50	55.85
Felt minimum	1.000	Set	.571	2.37	25	27.37
Maximum	1.000	Set	.615	2.55	27	29.55
Paint or stain, flush door, interior or exterior, 1 coat	2.000	Face	.941	4.26	36	40.26
2 coats	2.000	Face	1.455	8.50	56	64.50
Primer & 1 coat	2.000	Face	1.455	7.20	56	63.20
Primer & 2 coats	2.000	Face	1.778	11.20	68	79.20
Paneled door, interior & exterior, 1 coat	2.000	Face	1.143	4.56	44	48.56
2 coats	2.000	Face	2.000	9.10	76	85.10
Primer & 1 coat	2.000	Face	1.455	7.70	56	63.70
Primer & 2 coats	2.000	Face	1.778	12	68	80

4 | EXTERIOR WALLS — 53 | Sliding Door Systems

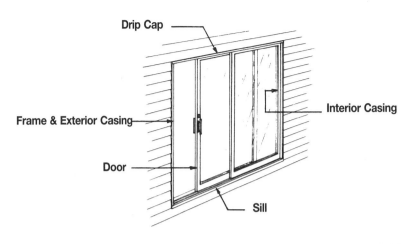

System Description	QUAN.	UNIT	LABOR HOURS	COST EACH MAT.	COST EACH INST.	COST EACH TOTAL
WOOD SLIDING DOOR, 8' WIDE, PREMIUM						
Wood, tempered insul. glass, 8' wide, premium	1.000	Ea.	5.333	1,475	233	1,708
Interior casing	22.000	L.F.	.733	32.78	31.90	64.68
Exterior casing	22.000	L.F.	.733	32.78	31.90	64.68
Sill, oak, 8/4 x 8" deep	8.000	L.F.	1.280	160	56	216
Drip cap	8.000	L.F.	.160	3.68	6.96	10.64
Paint, interior & exterior, primer & 2 coats	2.000	Face	2.816	16.72	107.36	124.08
TOTAL		Ea.	11.055	1,720.96	467.12	2,188.08
ALUMINUM SLIDING DOOR, 8' WIDE, PREMIUM						
Aluminum, tempered insul. glass, 8' wide, premium	1.000	Ea.	5.333	1,675	233	1,908
Interior casing	22.000	L.F.	.733	32.78	31.90	64.68
Exterior casing	22.000	L.F.	.733	32.78	31.90	64.68
Sill, oak, 8/4 x 8" deep	8.000	L.F.	1.280	160	56	216
Drip cap	8.000	L.F.	.160	3.68	6.96	10.64
Paint, interior & exterior, primer & 2 coats	2.000	Face	2.816	16.72	107.36	124.08
TOTAL		Ea.	11.055	1,920.96	467.12	2,388.08

The cost of this system is on a cost per each door basis.

Description	QUAN.	UNIT	LABOR HOURS	COST EACH MAT.	COST EACH INST.	COST EACH TOTAL

Sliding Door Price Sheet	QUAN.	UNIT	LABOR HOURS	MAT.	INST.	TOTAL
Sliding door, wood, 5/8" thick, tempered insul. glass, 6' wide, premium	1.000	Ea.	4.000	1,275	174	1,449
Economy	1.000	Ea.	4.000	870	174	1,044
8' wide, wood premium	1.000	Ea.	5.333	1,475	233	1,708
Economy	1.000	Ea.	5.333	1,000	233	1,233
12' wide, wood premium	1.000	Ea.	6.400	3,175	279	3,454
Economy	1.000	Ea.	6.400	2,300	279	2,579
Aluminum, 5/8" thick, tempered insul. glass, 6' wide, premium	1.000	Ea.	4.000	1,550	174	1,724
Economy	1.000	Ea.	4.000	835	174	1,009
8' wide, premium	1.000	Ea.	5.333	1,675	233	1,908
Economy	1.000	Ea.	5.333	1,425	233	1,658
12' wide, premium	1.000	Ea.	6.400	2,700	279	2,979
Economy	1.000	Ea.	6.400	1,650	279	1,929
Interior casing, 6' wide door	20.000	L.F.	.667	30	29	59
8' wide door	22.000	L.F.	.733	33	32	65
12' wide door	26.000	L.F.	.867	38.50	37.50	76
Exterior casing, 6' wide door	20.000	L.F.	.667	30	29	59
8' wide door	22.000	L.F.	.733	33	32	65
12' wide door	26.000	L.F.	.867	38.50	37.50	76
Sill, oak, 8/4 x 8" deep, 6' wide door	6.000	L.F.	.960	120	42	162
8' wide door	8.000	L.F.	1.280	160	56	216
12' wide door	12.000	L.F.	1.920	240	84	324
8/4 x 10" deep, 6' wide door	6.000	L.F.	1.067	138	46.50	184.50
8' wide door	8.000	L.F.	1.422	184	62	246
12' wide door	12.000	L.F.	2.133	276	93	369
Drip cap, 6' wide door	6.000	L.F.	.120	2.76	5.20	7.96
8' wide door	8.000	L.F.	.160	3.68	6.95	10.63
12' wide door	12.000	L.F.	.240	5.50	10.45	15.95
Paint or stain, interior & exterior, 6' wide door, 1 coat	2.000	Face	1.600	5.60	61	66.60
2 coats	2.000	Face	1.600	5.60	61	66.60
Primer & 1 coat	2.000	Face	1.778	10.40	68	78.40
Primer & 2 coats	2.000	Face	2.560	15.20	97.50	112.70
8' wide door, 1 coat	2.000	Face	1.760	6.15	67	73.15
2 coats	2.000	Face	1.760	6.15	67	73.15
Primer & 1 coat	2.000	Face	1.955	11.45	75	86.45
Primer & 2 coats	2.000	Face	2.816	16.70	107	123.70
12' wide door, 1 coat	2.000	Face	2.080	7.30	79	86.30
2 coats	2.000	Face	2.080	7.30	79	86.30
Primer & 1 coat	2.000	Face	2.311	13.50	88.50	102
Primer & 2 coats	2.000	Face	3.328	19.75	127	146.75
Aluminum door, trim only, interior & exterior, 6' door, 1 coat	2.000	Face	.800	2.80	30.50	33.30
2 coats	2.000	Face	.800	2.80	30.50	33.30
Primer & 1 coat	2.000	Face	.889	5.20	34	39.20
Primer & 2 coats	2.000	Face	1.280	7.60	49	56.60
8' wide door, 1 coat	2.000	Face	.880	3.08	33.50	36.58
2 coats	2.000	Face	.880	3.08	33.50	36.58
Primer & 1 coat	2.000	Face	.978	5.70	37.50	43.20
Primer & 2 coats	2.000	Face	1.408	8.35	53.50	61.85
12' wide door, 1 coat	2.000	Face	1.040	3.64	39.50	43.14
2 coats	2.000	Face	1.040	3.64	39.50	43.14
Primer & 1 coat	2.000	Face	1.155	6.75	44	50.75
Primer & 2 coats	2.000	Face	1.664	9.90	63.50	73.40

4 | EXTERIOR WALLS — 56 | Residential Garage Door Systems

System Description	QUAN.	UNIT	LABOR HOURS	COST EACH MAT.	COST EACH INST.	COST EACH TOTAL
OVERHEAD, SECTIONAL GARAGE DOOR, 9' X 7'						
Wood, overhead sectional door, std., incl. hardware, 9' x 7'	1.000	Ea.	2.000	575	87	662
Jamb & header blocking, 2" x 6"	25.000	L.F.	.901	15.75	39.25	55
Exterior trim	25.000	L.F.	.833	37.25	36.25	73.50
Paint, interior & exterior, primer & 2 coats	2.000	Face	3.556	24	136	160
Weatherstripping, molding type	1.000	Set	.767	34.27	33.35	67.62
Drip cap	9.000	L.F.	.180	4.14	7.83	11.97
TOTAL		Ea.	8.237	690.41	339.68	1,030.09
OVERHEAD, SECTIONAL GARAGE DOOR, 16' X 7'						
Wood, overhead sectional, std., incl. hardware, 16' x 7'	1.000	Ea.	2.667	1,150	116	1,266
Jamb & header blocking, 2" x 6"	30.000	L.F.	1.081	18.90	47.10	66
Exterior trim	30.000	L.F.	1.000	44.70	43.50	88.20
Paint, interior & exterior, primer & 2 coats	2.000	Face	5.333	36	204	240
Weatherstripping, molding type	1.000	Set	1.000	44.70	43.50	88.20
Drip cap	16.000	L.F.	.320	7.36	13.92	21.28
TOTAL		Ea.	11.401	1,301.66	468.02	1,769.68
OVERHEAD, SWING-UP TYPE, GARAGE DOOR, 16' X 7'						
Wood, overhead, swing-up, std., incl. hardware, 16' x 7'	1.000	Ea.	2.667	760	116	876
Jamb & header blocking, 2" x 6"	30.000	L.F.	1.081	18.90	47.10	66
Exterior trim	30.000	L.F.	1.000	44.70	43.50	88.20
Paint, interior & exterior, primer & 2 coats	2.000	Face	5.333	36	204	240
Weatherstripping, molding type	1.000	Set	1.000	44.70	43.50	88.20
Drip cap	16.000	L.F.	.320	7.36	13.92	21.28
TOTAL		Ea.	11.401	911.66	468.02	1,379.68

This system is on a cost per each door basis.

Description	QUAN.	UNIT	LABOR HOURS	COST EACH MAT.	COST EACH INST.	COST EACH TOTAL

Resi Garage Door Price Sheet	QUAN.	UNIT	LABOR HOURS	COST EACH MAT.	COST EACH INST.	COST EACH TOTAL
Overhead, sectional, including hardware, fiberglass, 9' x 7', standard	1.000	Ea.	3.030	690	132	822
Deluxe	1.000	Ea.	3.030	880	132	1,012
16' x 7', standard	1.000	Ea.	2.667	1,250	116	1,366
Deluxe	1.000	Ea.	2.667	1,575	116	1,691
Hardboard, 9' x 7', standard	1.000	Ea.	2.000	465	87	552
Deluxe	1.000	Ea.	2.000	625	87	712
16' x 7', standard	1.000	Ea.	2.667	910	116	1,026
Deluxe	1.000	Ea.	2.667	1,075	116	1,191
Metal, 9' x 7', standard	1.000	Ea.	3.030	545	132	677
Deluxe	1.000	Ea.	2.000	735	87	822
16' x 7', standard	1.000	Ea.	5.333	695	233	928
Deluxe	1.000	Ea.	2.667	1,125	116	1,241
Wood, 9' x 7', standard	1.000	Ea.	2.000	575	87	662
Deluxe	1.000	Ea.	2.000	1,650	87	1,737
16' x 7', standard	1.000	Ea.	2.667	1,150	116	1,266
Deluxe	1.000	Ea.	2.667	2,425	116	2,541
Overhead swing-up type including hardware, fiberglass, 9' x 7', standard	1.000	Ea.	2.000	795	87	882
Deluxe	1.000	Ea.	2.000	835	87	922
16' x 7', standard	1.000	Ea.	2.667	1,000	116	1,116
Deluxe	1.000	Ea.	2.667	1,025	116	1,141
Hardboard, 9' x 7', standard	1.000	Ea.	2.000	365	87	452
Deluxe	1.000	Ea.	2.000	480	87	567
16' x 7', standard	1.000	Ea.	2.667	510	116	626
Deluxe	1.000	Ea.	2.667	755	116	871
Metal, 9' x 7', standard	1.000	Ea.	2.000	400	87	487
Deluxe	1.000	Ea.	2.000	725	87	812
16' x 7', standard	1.000	Ea.	2.667	625	116	741
Deluxe	1.000	Ea.	2.667	1,000	116	1,116
Wood, 9' x 7', standard	1.000	Ea.	2.000	435	87	522
Deluxe	1.000	Ea.	2.000	785	87	872
16' x 7', standard	1.000	Ea.	2.667	760	116	876
Deluxe	1.000	Ea.	2.667	1,075	116	1,191
Jamb & header blocking, 2" x 6", 9' x 7' door	25.000	L.F.	.901	15.75	39.50	55.25
16' x 7' door	30.000	L.F.	1.081	18.90	47	65.90
2" x 8", 9' x 7' door	25.000	L.F.	1.000	23	43.50	66.50
16' x 7' door	30.000	L.F.	1.200	27.50	52	79.50
Exterior trim, 9' x 7' door	25.000	L.F.	.833	37.50	36.50	74
16' x 7' door	30.000	L.F.	1.000	44.50	43.50	88
Paint or stain, interior & exterior, 9' x 7' door, 1 coat	1.000	Face	2.286	9.10	88	97.10
2 coats	1.000	Face	4.000	18.25	152	170.25
Primer & 1 coat	1.000	Face	2.909	15.45	112	127.45
Primer & 2 coats	1.000	Face	3.556	24	136	160
16' x 7' door, 1 coat	1.000	Face	3.429	13.70	132	145.70
2 coats	1.000	Face	6.000	27.50	228	255.50
Primer & 1 coat	1.000	Face	4.364	23	168	191
Primer & 2 coats	1.000	Face	5.333	36	204	240
Weatherstripping, molding type, 9' x 7' door	1.000	Set	.767	34.50	33.50	68
16' x 7' door	1.000	Set	1.000	44.50	43.50	88
Drip cap, 9' door	9.000	L.F.	.180	4.14	7.85	11.99
16' door	16.000	L.F.	.320	7.35	13.90	21.25
Garage door opener, economy	1.000	Ea.	1.000	330	43.50	373.50
Deluxe, including remote control	1.000	Ea.	1.000	485	43.50	528.50

4 | EXTERIOR WALLS 58 | Aluminum Window Systems

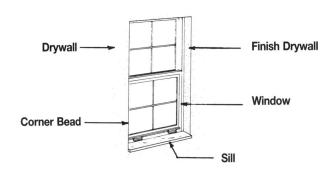

System Description	QUAN.	UNIT	LABOR HOURS	COST EACH MAT.	COST EACH INST.	COST EACH TOTAL
SINGLE HUNG, 2' X 3' OPENING						
Window, 2' x 3' opening, insulating glass	1.000	Ea.	1.600	240	86.50	326.50
Blocking, 1" x 3" furring strip nailers	10.000	L.F.	.146	2.50	6.30	8.80
Drywall, 1/2" thick, standard	5.000	S.F.	.040	1.75	1.75	3.50
Corner bead, 1" x 1", galvanized steel	8.000	L.F.	.160	1.20	6.96	8.16
Finish drywall, tape and finish corners inside and outside	16.000	L.F.	.269	1.60	11.68	13.28
Sill, slate	2.000	L.F.	.400	23.10	15.50	38.60
TOTAL		Ea.	2.615	270.15	128.69	398.84
SLIDING, 3' X 2' OPENING						
Window, 3' x 2' opening, enameled, insulating glass	1.000	Ea.	1.600	224	86.50	310.50
Blocking, 1" x 3" furring strip nailers	10.000	L.F.	.146	2.50	6.30	8.80
Drywall, 1/2" thick, standard	5.000	S.F.	.040	1.75	1.75	3.50
Corner bead, 1" x 1", galvanized steel	7.000	L.F.	.140	1.05	6.09	7.14
Finish drywall, tape and finish corners inside and outside	14.000	L.F.	.236	1.40	10.22	11.62
Sill, slate	3.000	L.F.	.600	34.65	23.25	57.90
TOTAL		Ea.	2.762	265.35	134.11	399.46
AWNING, 3'-1" X 3'-2"						
Window, 3'-1" x 3'-2" opening, enameled, insul. glass	1.000	Ea.	1.600	257	86.50	343.50
Blocking, 1" x 3" furring strip, nailers	12.500	L.F.	.182	3.13	7.88	11.01
Drywall, 1/2" thick, standard	4.500	S.F.	.036	1.58	1.58	3.16
Corner bead, 1" x 1", galvanized steel	9.250	L.F.	.185	1.39	8.05	9.44
Finish drywall, tape and finish corners, inside and outside	18.500	L.F.	.312	1.85	13.51	15.36
Sill, slate	3.250	L.F.	.650	37.54	25.19	62.73
TOTAL		Ea.	2.965	302.49	142.71	445.20

Description	QUAN.	UNIT	LABOR HOURS	COST PER S.F. MAT.	COST PER S.F. INST.	COST PER S.F. TOTAL

Aluminum Window Price Sheet	QUAN.	UNIT	LABOR HOURS	COST EACH MAT.	COST EACH INST.	COST EACH TOTAL
Window, aluminum, awning, 3'-1" x 3'-2", standard glass	1.000	Ea.	1.600	330	86.50	416.50
Insulating glass	1.000	Ea.	1.600	330	86.50	416.50
4'-5" x 5'-3", standard glass	1.000	Ea.	2.000	375	108	483
Insulating glass	1.000	Ea.	2.000	390	108	498
Casement, 3'-1" x 3'-2", standard glass	1.000	Ea.	1.600	350	86.50	436.50
Insulating glass	1.000	Ea.	1.600	445	86.50	531.50
Single hung, 2' x 3', standard glass	1.000	Ea.	1.600	198	86.50	284.50
Insulating glass	1.000	Ea.	1.600	240	86.50	326.50
2'-8" x 6'-8", standard glass	1.000	Ea.	2.000	350	108	458
Insulating glass	1.000	Ea.	2.000	450	108	558
3'-4" x 5'-0", standard glass	1.000	Ea.	1.778	286	96	382
Insulating glass	1.000	Ea.	1.778	315	96	411
Sliding, 3' x 2', standard glass	1.000	Ea.	1.600	209	86.50	295.50
Insulating glass	1.000	Ea.	1.600	224	86.50	310.50
5' x 3', standard glass	1.000	Ea.	1.778	320	96	416
Insulating glass	1.000	Ea.	1.778	370	96	466
8' x 4', standard glass	1.000	Ea.	2.667	335	144	479
Insulating glass	1.000	Ea.	2.667	540	144	684
Blocking, 1" x 3" furring, opening 3' x 2'	10.000	L.F.	.146	2.50	6.30	8.80
3' x 3'	12.500	L.F.	.182	3.13	7.90	11.03
3' x 5'	16.000	L.F.	.233	4	10.10	14.10
4' x 4'	16.000	L.F.	.233	4	10.10	14.10
4' x 5'	18.000	L.F.	.262	4.50	11.35	15.85
4' x 6'	20.000	L.F.	.291	5	12.60	17.60
4' x 8'	24.000	L.F.	.349	6	15.10	21.10
6'-8" x 2'-8"	19.000	L.F.	.276	4.75	11.95	16.70
Drywall, 1/2" thick, standard, opening 3' x 2'	5.000	S.F.	.040	1.75	1.75	3.50
3' x 3'	6.000	S.F.	.048	2.10	2.10	4.20
3' x 5'	8.000	S.F.	.064	2.80	2.80	5.60
4' x 4'	8.000	S.F.	.064	2.80	2.80	5.60
4' x 5'	9.000	S.F.	.072	3.15	3.15	6.30
4' x 6'	10.000	S.F.	.080	3.50	3.50	7
4' x 8'	12.000	S.F.	.096	4.20	4.20	8.40
6'-8" x 2'	9.500	S.F.	.076	3.33	3.33	6.66
Corner bead, 1" x 1", galvanized steel, opening 3' x 2'	7.000	L.F.	.140	1.05	6.10	7.15
3' x 3'	9.000	L.F.	.180	1.35	7.85	9.20
3' x 5'	11.000	L.F.	.220	1.65	9.55	11.20
4' x 4'	12.000	L.F.	.240	1.80	10.45	12.25
4' x 5'	13.000	L.F.	.260	1.95	11.30	13.25
4' x 6'	14.000	L.F.	.280	2.10	12.20	14.30
4' x 8'	16.000	L.F.	.320	2.40	13.90	16.30
6'-8" x 2'	15.000	L.F.	.300	2.25	13.05	15.30
Tape and finish corners, inside and outside, opening 3' x 2'	14.000	L.F.	.204	1.40	10.20	11.60
3' x 3'	18.000	L.F.	.262	1.80	13.15	14.95
3' x 5'	22.000	L.F.	.320	2.20	16.05	18.25
4' x 4'	24.000	L.F.	.349	2.40	17.50	19.90
4' x 5'	26.000	L.F.	.378	2.60	19	21.60
4' x 6'	28.000	L.F.	.407	2.80	20.50	23.30
4' x 8'	32.000	L.F.	.466	3.20	23.50	26.70
6'-8" x 2'	30.000	L.F.	.437	3	22	25
Sill, slate, 2' long	2.000	L.F.	.400	23	15.50	38.50
3' long	3.000	L.F.	.600	34.50	23.50	58
4' long	4.000	L.F.	.800	46	31	77
Wood, 1-5/8" x 6-1/4", 2' long	2.000	L.F.	.128	14	5.60	19.60
3' long	3.000	L.F.	.192	21	8.35	29.35
4' long	4.000	L.F.	.256	28	11.15	39.15

4 | EXTERIOR WALLS | 60 | Storm Door & Window Systems

System Description	QUAN.	UNIT	LABOR HOURS	COST EACH MAT.	COST EACH INST.	COST EACH TOTAL
Storm door, aluminum, combination, storm & screen, anodized, 2'-6" x 6'-8"	1.000	Ea.	1.067	179	46.50	225.50
2'-8" x 6'-8"	1.000	Ea.	1.143	210	50	260
3'-0" x 6'-8"	1.000	Ea.	1.143	210	50	260
Mill finish, 2'-6" x 6'-8"	1.000	Ea.	1.067	237	46.50	283.50
2'-8" x 6'-8"	1.000	Ea.	1.143	237	50	287
3'-0" x 6'-8"	1.000	Ea.	1.143	257	50	307
Painted, 2'-6" x 6'-8"	1.000	Ea.	1.067	288	46.50	334.50
2'-8" x 6'-8"	1.000	Ea.	1.143	288	50	338
3'-0" x 6'-8"	1.000	Ea.	1.143	284	50	334
Wood, combination, storm & screen, crossbuck, 2'-6" x 6'-9"	1.000	Ea.	1.455	350	63.50	413.50
2'-8" x 6'-9"	1.000	Ea.	1.600	300	70	370
3'-0" x 6'-9"	1.000	Ea.	1.778	305	77.50	382.50
Full lite, 2'-6" x 6'-9"	1.000	Ea.	1.455	320	63.50	383.50
2'-8" x 6'-9"	1.000	Ea.	1.600	320	70	390
3'-0" x 6'-9"	1.000	Ea.	1.778	330	77.50	407.50
Windows, aluminum, combination storm & screen, basement, 1'-10" x 1'-0"	1.000	Ea.	.533	32.50	23.50	56
2'-9" x 1'-6"	1.000	Ea.	.533	35.50	23.50	59
3'-4" x 2'-0"	1.000	Ea.	.533	42.50	23.50	66
Double hung, anodized, 2'-0" x 3'-5"	1.000	Ea.	.533	84	23.50	107.50
2'-6" x 5'-0"	1.000	Ea.	.571	113	25	138
4'-0" x 6'-0"	1.000	Ea.	.640	238	28	266
Painted, 2'-0" x 3'-5"	1.000	Ea.	.533	100	23.50	123.50
2'-6" x 5'-0"	1.000	Ea.	.571	161	25	186
4'-0" x 6'-0"	1.000	Ea.	.640	288	28	316
Fixed window, anodized, 4'-6" x 4'-6"	1.000	Ea.	.640	128	28	156
5'-8" x 4'-6"	1.000	Ea.	.800	146	35	181
Painted, 4'-6" x 4'-6"	1.000	Ea.	.640	128	28	156
5'-8" x 4'-6"	1.000	Ea.	.800	146	35	181

4 | EXTERIOR WALLS 64 | Shutters/Blinds Systems

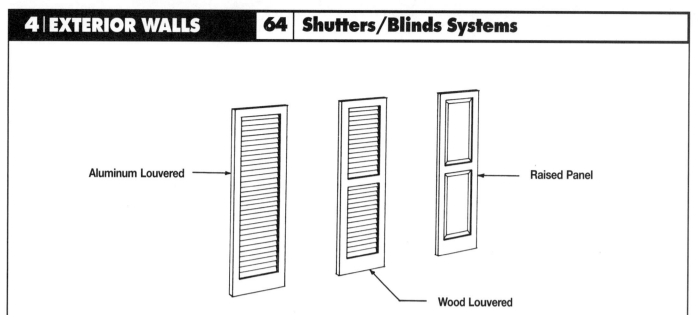

System Description	QUAN.	UNIT	LABOR HOURS	COST PER PAIR		
				MAT.	INST.	TOTAL
Shutters, exterior blinds, aluminum, louvered, 1'-4" wide, 3"-0" long	1.000	Set	.800	52.50	35	87.50
4'-0" long	1.000	Set	.800	63	35	98
5'-4" long	1.000	Set	.800	83	35	118
6'-8" long	1.000	Set	.889	106	39	145
Wood, louvered, 1'-2" wide, 3'-3" long	1.000	Set	.800	100	35	135
4'-7" long	1.000	Set	.800	135	35	170
5'-3" long	1.000	Set	.800	153	35	188
1'-6" wide, 3'-3" long	1.000	Set	.800	106	35	141
4'-7" long	1.000	Set	.800	149	35	184
Polystyrene, louvered, 1'-2" wide, 3'-3" long	1.000	Set	.800	30	35	65
4'-7" long	1.000	Set	.800	37	35	72
5'-3" long	1.000	Set	.800	42.50	35	77.50
6'-8" long	1.000	Set	.889	50.50	39	89.50
Vinyl, louvered, 1'-2" wide, 4'-7" long	1.000	Set	.720	33.50	31.50	65
1'-4" x 6'-8" long	1.000	Set	.889	52	39	91

Division 5
Roofing

5 | ROOFING — 04 | Gable End Roofing Systems

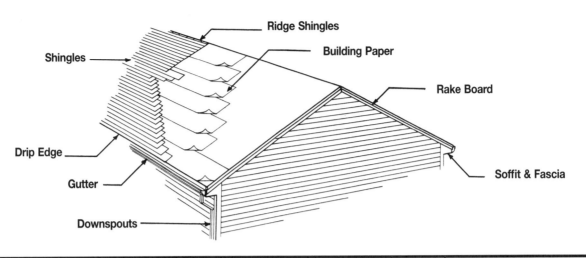

System Description	QUAN.	UNIT	LABOR HOURS	COST PER S.F. MAT.	COST PER S.F. INST.	COST PER S.F. TOTAL
ASPHALT, ROOF SHINGLES, CLASS A						
Shingles, inorganic class A, 210-235 lb./sq., 4/12 pitch	1.160	S.F.	.017	.55	.70	1.25
Drip edge, metal, 5" wide	.150	L.F.	.003	.07	.13	.20
Building paper, #15 felt	1.300	S.F.	.002	.06	.07	.13
Ridge shingles, asphalt	.042	L.F.	.001	.07	.04	.11
Soffit & fascia, white painted aluminum, 1' overhang	.083	L.F.	.012	.26	.53	.79
Rake trim, 1" x 6"	.040	L.F.	.002	.05	.07	.12
Rake trim, prime and paint	.040	L.F.	.002	.01	.07	.08
Gutter, seamless, aluminum painted	.083	L.F.	.006	.17	.27	.44
Downspouts, aluminum painted	.035	L.F.	.002	.06	.08	.14
TOTAL		S.F.	.047	1.30	1.96	3.26
WOOD, CEDAR SHINGLES NO. 1 PERFECTIONS, 18" LONG						
Shingles, wood, cedar, No. 1 perfections, 4/12 pitch	1.160	S.F.	.035	2.21	1.52	3.73
Drip edge, metal, 5" wide	.150	L.F.	.003	.07	.13	.20
Building paper, #15 felt	1.300	S.F.	.002	.06	.07	.13
Ridge shingles, cedar	.042	L.F.	.001	.14	.05	.19
Soffit & fascia, white painted aluminum, 1' overhang	.083	L.F.	.012	.26	.53	.79
Rake trim, 1" x 6"	.040	L.F.	.002	.05	.07	.12
Rake trim, prime and paint	.040	L.F.	.002	.01	.07	.08
Gutter, seamless, aluminum, painted	.083	L.F.	.006	.17	.27	.44
Downspouts, aluminum, painted	.035	L.F.	.002	.06	.08	.14
TOTAL		S.F.	.065	3.03	2.79	5.82

The prices in these systems are based on a square foot of plan area.
All quantities have been adjusted accordingly.

Description	QUAN.	UNIT	LABOR HOURS	COST PER S.F. MAT.	COST PER S.F. INST.	COST PER S.F. TOTAL

Gable End Roofing Price Sheet	QUAN.	UNIT	LABOR HOURS	COST PER S.F. MAT.	COST PER S.F. INST.	COST PER S.F. TOTAL
Shingles, asphalt, inorganic, class A, 210-235 lb./sq., 4/12 pitch	1.160	S.F.	.017	.55	.70	1.25
8/12 pitch	1.330	S.F.	.019	.60	.76	1.36
Laminated, multi-layered, 240-260 lb./sq., 4/12 pitch	1.160	S.F.	.021	.69	.86	1.55
8/12 pitch	1.330	S.F.	.023	.75	.93	1.68
Premium laminated, multi-layered, 260-300 lb./sq., 4/12 pitch	1.160	S.F.	.027	.89	1.10	1.99
8/12 pitch	1.330	S.F.	.030	.96	1.20	2.16
Clay tile, Spanish tile, red, 4/12 pitch	1.160	S.F.	.053	3.90	2.16	6.06
8/12 pitch	1.330	S.F.	.058	4.23	2.34	6.57
Mission tile, red, 4/12 pitch	1.160	S.F.	.083	9.25	3.38	12.63
8/12 pitch	1.330	S.F.	.090	10	3.67	13.67
French tile, red, 4/12 pitch	1.160	S.F.	.071	8.40	2.88	11.28
8/12 pitch	1.330	S.F.	.077	9.10	3.12	12.22
Slate, Buckingham, Virginia, black, 4/12 pitch	1.160	S.F.	.055	5.20	2.22	7.42
8/12 pitch	1.330	S.F.	.059	5.65	2.41	8.06
Vermont, black or grey, 4/12 pitch	1.160	S.F.	.055	5.90	2.22	8.12
8/12 pitch	1.330	S.F.	.059	6.35	2.41	8.76
Wood, No. 1 red cedar, 5X, 16" long, 5" exposure, 4/12 pitch	1.160	S.F.	.038	2.50	1.68	4.18
8/12 pitch	1.330	S.F.	.042	2.70	1.82	4.52
Fire retardant, 4/12 pitch	1.160	S.F.	.038	3.05	1.68	4.73
8/12 pitch	1.330	S.F.	.042	3.30	1.82	5.12
18" long, No.1 perfections, 5" exposure, 4/12 pitch	1.160	S.F.	.035	2.21	1.52	3.73
8/12 pitch	1.330	S.F.	.038	2.39	1.65	4.04
Fire retardant, 4/12 pitch	1.160	S.F.	.035	2.76	1.52	4.28
8/12 pitch	1.330	S.F.	.038	2.99	1.65	4.64
Resquared & rebutted, 18" long, 6" exposure, 4/12 pitch	1.160	S.F.	.032	3.17	1.39	4.56
8/12 pitch	1.330	S.F.	.035	3.43	1.51	4.94
Fire retardant, 4/12 pitch	1.160	S.F.	.032	3.72	1.39	5.11
8/12 pitch	1.330	S.F.	.035	4.03	1.51	5.54
Wood shakes hand split, 24" long, 10" exposure, 4/12 pitch	1.160	S.F.	.038	2.17	1.68	3.85
8/12 pitch	1.330	S.F.	.042	2.35	1.82	4.17
Fire retardant, 4/12 pitch	1.160	S.F.	.038	2.72	1.68	4.40
8/12 pitch	1.330	S.F.	.042	2.95	1.82	4.77
18" long, 8" exposure, 4/12 pitch	1.160	S.F.	.048	2.52	2.09	4.61
8/12 pitch	1.330	S.F.	.052	2.73	2.26	4.99
Fire retardant, 4/12 pitch	1.160	S.F.	.048	3.07	2.09	5.16
8/12 pitch	1.330	S.F.	.052	3.33	2.26	5.59
Drip edge, metal, 5" wide	.150	L.F.	.003	.07	.13	.20
8" wide	.150	L.F.	.003	.09	.13	.22
Building paper, #15 asphalt felt	1.300	S.F.	.002	.06	.07	.13
Ridge shingles, asphalt	.042	L.F.	.001	.07	.04	.11
Clay	.042	L.F.	.002	.47	.07	.54
Slate	.042	L.F.	.002	.42	.07	.49
Wood, shingles	.042	L.F.	.001	.14	.05	.19
Shakes	.042	L.F.	.001	.14	.05	.19
Soffit & fascia, aluminum, vented, 1' overhang	.083	L.F.	.012	.26	.53	.79
2' overhang	.083	L.F.	.013	.39	.58	.97
Vinyl, vented, 1' overhang	.083	L.F.	.011	.15	.48	.63
2' overhang	.083	L.F.	.012	.21	.53	.74
Wood, board fascia, plywood soffit, 1' overhang	.083	L.F.	.004	.02	.14	.16
2' overhang	.083	L.F.	.006	.03	.21	.24
Rake trim, painted, 1" x 6"	.040	L.F.	.004	.06	.14	.20
1" x 8"	.040	L.F.	.004	.09	.15	.24
Gutter, 5" box, aluminum, seamless, painted	.083	L.F.	.006	.17	.27	.44
Vinyl	.083	L.F.	.006	.10	.26	.36
Downspout, 2" x 3", aluminum, one story house	.035	L.F.	.001	.06	.07	.13
Two story house	.060	L.F.	.003	.10	.12	.22
Vinyl, one story house	.035	L.F.	.002	.06	.08	.14
Two story house	.060	L.F.	.003	.10	.12	.22

5 | ROOFING　　08 | Hip Roof Roofing Systems

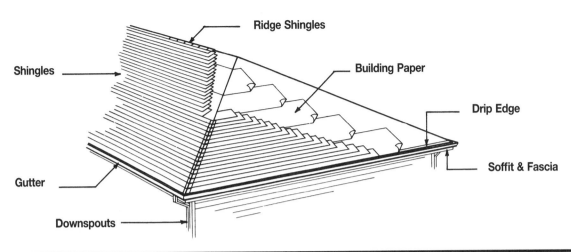

System Description	QUAN.	UNIT	LABOR HOURS	COST PER S.F. MAT.	COST PER S.F. INST.	COST PER S.F. TOTAL
ASPHALT, ROOF SHINGLES, CLASS A						
Shingles, inorganic, class A, 210-235 lb./sq. 4/12 pitch	1.570	S.F.	.023	.74	.94	1.68
Drip edge, metal, 5" wide	.122	L.F.	.002	.06	.11	.17
Building paper, #15 asphalt felt	1.800	S.F.	.002	.08	.10	.18
Ridge shingles, asphalt	.075	L.F.	.002	.12	.07	.19
Soffit & fascia, white painted aluminum, 1' overhang	.120	L.F.	.017	.37	.76	1.13
Gutter, seamless, aluminum, painted	.120	L.F.	.008	.24	.39	.63
Downspouts, aluminum, painted	.035	L.F.	.002	.06	.08	.14
TOTAL		S.F.	.056	1.67	2.45	4.12
WOOD, CEDAR SHINGLES, NO. 1 PERFECTIONS, 18" LONG						
Shingles, red cedar, No. 1 perfections, 5" exp., 4/12 pitch	1.570	S.F.	.047	2.94	2.03	4.97
Drip edge, metal, 5" wide	.122	L.F.	.002	.06	.11	.17
Building paper, #15 asphalt felt	1.800	S.F.	.002	.08	.10	.18
Ridge shingles, wood, cedar	.075	L.F.	.002	.25	.09	.34
Soffit & fascia, white painted aluminum, 1' overhang	.120	L.F.	.017	.37	.76	1.13
Gutter, seamless, aluminum, painted	.120	L.F.	.008	.24	.39	.63
Downspouts, aluminum, painted	.035	L.F.	.002	.06	.08	.14
TOTAL		S.F.	.080	4	3.56	7.56

The prices in these systems are based on a square foot of plan area.
All quantities have been adjusted accordingly.

Description	QUAN.	UNIT	LABOR HOURS	COST PER S.F. MAT.	COST PER S.F. INST.	COST PER S.F. TOTAL

Hip Roof - Roofing Price Sheet	QUAN.	UNIT	LABOR HOURS	COST PER S.F. MAT.	INST.	TOTAL
Shingles, asphalt, inorganic, class A, 210-235 lb./sq., 4/12 pitch	1.570	S.F.	.023	.74	.94	1.68
8/12 pitch	1.850	S.F.	.028	.87	1.11	1.98
Laminated, multi-layered, 240-260 lb./sq., 4/12 pitch	1.570	S.F.	.028	.92	1.14	2.06
8/12 pitch	1.850	S.F.	.034	1.09	1.36	2.45
Prem. laminated, multi-layered, 260-300 lb./sq., 4/12 pitch	1.570	S.F.	.037	1.18	1.47	2.65
8/12 pitch	1.850	S.F.	.043	1.41	1.75	3.16
Clay tile, Spanish tile, red, 4/12 pitch	1.570	S.F.	.071	5.20	2.88	8.08
8/12 pitch	1.850	S.F.	.084	6.20	3.42	9.62
Mission tile, red, 4/12 pitch	1.570	S.F.	.111	12.30	4.51	16.81
8/12 pitch	1.850	S.F.	.132	14.65	5.35	20
French tile, red, 4/12 pitch	1.570	S.F.	.095	11.20	3.84	15.04
8/12 pitch	1.850	S.F.	.113	13.30	4.56	17.86
Slate, Buckingham, Virginia, black, 4/12 pitch	1.570	S.F.	.073	6.95	2.96	9.91
8/12 pitch	1.850	S.F.	.087	8.25	3.52	11.77
Vermont, black or grey, 4/12 pitch	1.570	S.F.	.073	7.85	2.96	10.81
8/12 pitch	1.850	S.F.	.087	9.30	3.52	12.82
Wood, red cedar, No.1 5X, 16" long, 5" exposure, 4/12 pitch	1.570	S.F.	.051	3.33	2.24	5.57
8/12 pitch	1.850	S.F.	.061	3.95	2.66	6.61
Fire retardant, 4/12 pitch	1.570	S.F.	.051	4.07	2.24	6.31
8/12 pitch	1.850	S.F.	.061	4.82	2.66	7.48
18" long, No.1 perfections, 5" exposure, 4/12 pitch	1.570	S.F.	.047	2.94	2.03	4.97
8/12 pitch	1.850	S.F.	.055	3.50	2.41	5.91
Fire retardant, 4/12 pitch	1.570	S.F.	.047	3.68	2.03	5.71
8/12 pitch	1.850	S.F.	.055	4.37	2.41	6.78
Resquared & rebutted, 18" long, 6" exposure, 4/12 pitch	1.570	S.F.	.043	4.22	1.86	6.08
8/12 pitch	1.850	S.F.	.051	5	2.20	7.20
Fire retardant, 4/12 pitch	1.570	S.F.	.043	4.96	1.86	6.82
8/12 pitch	1.850	S.F.	.051	5.90	2.20	8.10
Wood shakes hand split, 24" long, 10" exposure, 4/12 pitch	1.570	S.F.	.051	2.90	2.24	5.14
8/12 pitch	1.850	S.F.	.061	3.44	2.66	6.10
Fire retardant, 4/12 pitch	1.570	S.F.	.051	3.64	2.24	5.88
8/12 pitch	1.850	S.F.	.061	4.31	2.66	6.97
18" long, 8" exposure, 4/12 pitch	1.570	S.F.	.064	3.36	2.78	6.14
8/12 pitch	1.850	S.F.	.076	3.99	3.31	7.30
Fire retardant, 4/12 pitch	1.570	S.F.	.064	4.10	2.78	6.88
8/12 pitch	1.850	S.F.	.076	4.86	3.31	8.17
Drip edge, metal, 5" wide	.122	L.F.	.002	.06	.11	.17
8" wide	.122	L.F.	.002	.07	.11	.18
Building paper, #15 asphalt felt	1.800	S.F.	.002	.08	.10	.18
Ridge shingles, asphalt	.075	L.F.	.002	.12	.07	.19
Clay	.075	L.F.	.003	.83	.12	.95
Slate	.075	L.F.	.003	.74	.12	.86
Wood, shingles	.075	L.F.	.002	.25	.09	.34
Shakes	.075	L.F.	.002	.25	.09	.34
Soffit & fascia, aluminum, vented, 1' overhang	.120	L.F.	.017	.37	.76	1.13
2' overhang	.120	L.F.	.019	.57	.84	1.41
Vinyl, vented, 1' overhang	.120	L.F.	.016	.22	.70	.92
2' overhang	.120	L.F.	.017	.31	.76	1.07
Wood, board fascia, plywood soffit, 1' overhang	.120	L.F.	.004	.02	.14	.16
2' overhang	.120	L.F.	.006	.03	.21	.24
Gutter, 5" box, aluminum, seamless, painted	.120	L.F.	.008	.24	.39	.63
Vinyl	.120	L.F.	.009	.14	.38	.52
Downspout, 2" x 3", aluminum, one story house	.035	L.F.	.002	.06	.08	.14
Two story house	.060	L.F.	.003	.10	.12	.22
Vinyl, one story house	.035	L.F.	.001	.06	.07	.13
Two story house	.060	L.F.	.003	.10	.12	.22
	QUAN.	UNIT	LABOR HOURS	MAT.	INST.	TOTAL

5 | ROOFING   12 | Gambrel Roofing Systems

Diagram labels: Shingles, Ridge Shingles, Building Paper, Rake Boards, Soffit, Drip Edge

System Description	QUAN.	UNIT	LABOR HOURS	COST PER S.F. MAT.	COST PER S.F. INST.	COST PER S.F. TOTAL
ASPHALT, ROOF SHINGLES, CLASS A						
Shingles, asphalt, inorganic, class A, 210-235 lb./sq.	1.450	S.F.	.022	.69	.88	1.57
Drip edge, metal, 5" wide	.146	L.F.	.003	.07	.13	.20
Building paper, #15 asphalt felt	1.500	S.F.	.002	.07	.08	.15
Ridge shingles, asphalt	.042	L.F.	.001	.07	.04	.11
Soffit & fascia, painted aluminum, 1' overhang	.083	L.F.	.012	.26	.53	.79
Rake trim, 1" x 6"	.063	L.F.	.003	.08	.11	.19
Rake trim, prime and paint	.063	L.F.	.003	.02	.11	.13
Gutter, seamless, alumunum, painted	.083	L.F.	.006	.17	.27	.44
Downspouts, aluminum, painted	.042	L.F.	.002	.07	.09	.16
TOTAL		S.F.	.054	1.50	2.24	3.74
WOOD, CEDAR SHINGLES, NO. 1 PERFECTIONS, 18" LONG						
Shingles, wood, red cedar, No. 1 perfections, 5" exposure	1.450	S.F.	.044	2.76	1.91	4.67
Drip edge, metal, 5" wide	.146	L.F.	.003	.07	.13	.20
Building paper, #15 asphalt felt	1.500	S.F.	.002	.07	.08	.15
Ridge shingles, wood	.042	L.F.	.001	.14	.05	.19
Soffit & fascia, white painted aluminum, 1' overhang	.083	L.F.	.012	.26	.53	.79
Rake trim, 1" x 6"	.063	L.F.	.003	.08	.11	.19
Rake trim, prime and paint	.063	L.F.	.001	.01	.05	.06
Gutter, seamless, aluminum, painted	.083	L.F.	.006	.17	.27	.44
Downspouts, aluminum, painted	.042	L.F.	.002	.07	.09	.16
TOTAL		S.F.	.074	3.63	3.22	6.85

The prices in this system are based on a square foot of plan area.
All quantities have been adjusted accordingly.

Description	QUAN.	UNIT	LABOR HOURS	COST PER S.F. MAT.	COST PER S.F. INST.	COST PER S.F. TOTAL

RSMeans Complete Residential Costs are Available in the 2007 Means CostWorks®!

MEANS CostWorks® 2007
Current industry-standard construction costs with point-and-click access!

From RSMeans, the most quoted name in construction costs!

Your 12-month subscription includes...
- FREE downloadable Quarterly Updates
- Ability to toggle between CSI MasterFormat™ 2004 and the older 16-division version
- 17,000 terms from Means Construction Dictionary

And much more!

RSMeans

This Industry-Standard Estimating Resource Just Got Better!

Residential costs are now delivered electronically in a fast & efficient CD format. *Means CostWorks* provides reliable construction cost information for the residential contractor or for anyone involved in small home construction projects.

RSMeans

Increase your profits & guarantee accurate estimates!

See how **Means CostWorks** can take your residential estimating to a new level!

Order Today!

Call **1-800-334-3509** or visit **www.rsmeans.com**

Recently added Residential Repair & Remodeling content from one of our most popular Contractor's Pricing Guides organizes information in a different way than other RSMeans cost data titles:

Costs are grouped the way you build – from frame to finish – covering every step from demolition and installation through painting and cleaning. All information is formatted by category and cost element breakouts/tasks.

You'll find unit prices for all aspects of residential repair & remodeling that contain simplified estimating methods with mark-ups. Includes crew tables, location factors, and other supporting text.

From the benchmark *RSMeans Residential Cost Data* title:

Over 9,000 current unit costs and 100 updated assemblies appear in the standard cost data format that you have relied upon for years for simplified system selection and accurate design-stage estimating.

And, like all RSMeans cost data, costs are localized to over 930 locations nationwide – so you can easily determine best-cost solutions.

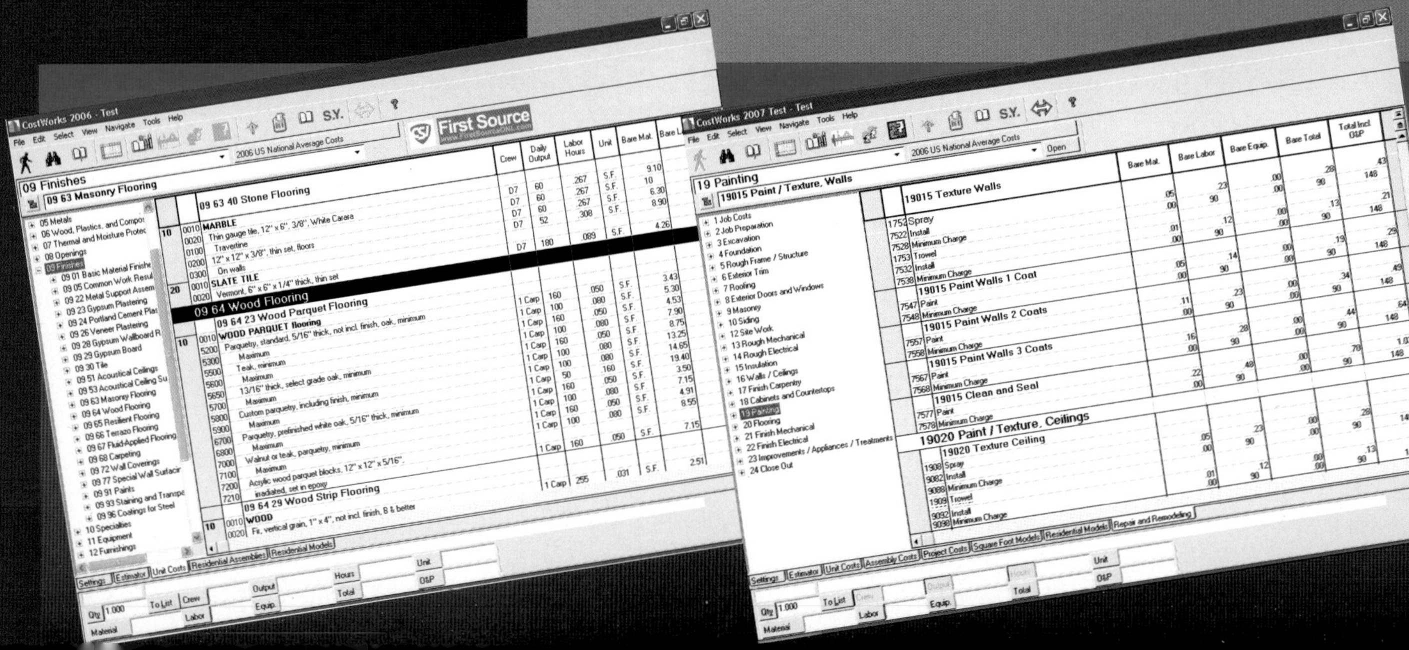

Gambrel Roofing Price Sheet	QUAN.	UNIT	LABOR HOURS	COST PER S.F.		
				MAT.	INST.	TOTAL
Shingles, asphalt, standard, inorganic, class A, 210-235 lb./sq.	1.450	S.F.	.022	.69	.88	1.57
Laminated, multi-layered, 240-260 lb./sq.	1.450	S.F.	.027	.86	1.07	1.93
Premium laminated, multi-layered, 260-300 lb./sq.	1.450	S.F.	.034	1.11	1.38	2.49
Slate, Buckingham, Virginia, black	1.450	S.F.	.069	6.55	2.78	9.33
Vermont, black or grey	1.450	S.F.	.069	7.35	2.78	10.13
Wood, red cedar, No.1 5X, 16" long, 5" exposure, plain	1.450	S.F.	.048	3.12	2.10	5.22
Fire retardant	1.450	S.F.	.048	3.81	2.10	5.91
18" long, No.1 perfections, 6" exposure, plain	1.450	S.F.	.044	2.76	1.91	4.67
Fire retardant	1.450	S.F.	.044	3.45	1.91	5.36
Resquared & rebutted, 18" long, 6" exposure, plain	1.450	S.F.	.040	3.96	1.74	5.70
Fire retardant	1.450	S.F.	.040	4.65	1.74	6.39
Shakes, hand split, 24" long, 10" exposure, plain	1.450	S.F.	.048	2.72	2.10	4.82
Fire retardant	1.450	S.F.	.048	3.41	2.10	5.51
18" long, 8" exposure, plain	1.450	S.F.	.060	3.15	2.61	5.76
Fire retardant	1.450	S.F.	.060	3.84	2.61	6.45
Drip edge, metal, 5" wide	.146	L.F.	.003	.07	.13	.20
8" wide	.146	L.F.	.003	.08	.13	.21
Building paper, #15 asphalt felt	1.500	S.F.	.002	.07	.08	.15
Ridge shingles, asphalt	.042	L.F.	.001	.07	.04	.11
Slate	.042	L.F.	.002	.42	.07	.49
Wood, shingles	.042	L.F.	.001	.14	.05	.19
Shakes	.042	L.F.	.001	.14	.05	.19
Soffit & fascia, aluminum, vented, 1' overhang	.083	L.F.	.012	.26	.53	.79
2' overhang	.083	L.F.	.013	.39	.58	.97
Vinyl vented, 1' overhang	.083	L.F.	.011	.15	.48	.63
2' overhang	.083	L.F.	.012	.21	.53	.74
Wood board fascia, plywood soffit, 1' overhang	.083	L.F.	.004	.02	.14	.16
2' overhang	.083	L.F.	.006	.03	.21	.24
Rake trim, painted, 1" x 6"	.063	L.F.	.006	.10	.22	.32
1" x 8"	.063	L.F.	.007	.12	.28	.40
Gutter, 5" box, aluminum, seamless, painted	.083	L.F.	.006	.17	.27	.44
Vinyl	.083	L.F.	.006	.10	.26	.36
Downspout 2" x 3", aluminum, one story house	.042	L.F.	.002	.07	.09	.16
Two story house	.070	L.F.	.003	.12	.14	.26
Vinyl, one story house	.042	L.F.	.002	.07	.09	.16
Two story house	.070	L.F.	.003	.12	.14	.26

5 | ROOFING   16 | Mansard Roofing Systems

System Description	QUAN.	UNIT	LABOR HOURS	COST PER S.F. MAT.	COST PER S.F. INST.	COST PER S.F. TOTAL
ASPHALT, ROOF SHINGLES, CLASS A						
Shingles, standard inorganic class A 210-235 lb./sq.	2.210	S.F.	.032	1.01	1.29	2.30
Drip edge, metal, 5" wide	.122	L.F.	.002	.06	.11	.17
Building paper, #15 asphalt felt	2.300	S.F.	.003	.11	.13	.24
Ridge shingles, asphalt	.090	L.F.	.002	.14	.09	.23
Soffit & fascia, white painted aluminum, 1' overhang	.122	L.F.	.018	.38	.77	1.15
Gutter, seamless, aluminum, painted	.122	L.F.	.008	.25	.39	.64
Downspouts, aluminum, painted	.042	L.F.	.002	.07	.09	.16
TOTAL		S.F.	.067	2.02	2.87	4.89
WOOD, CEDAR SHINGLES, NO. 1 PERFECTIONS, 18" LONG						
Shingles, wood, red cedar, No. 1 perfections, 5" exposure	2.210	S.F.	.064	4.05	2.79	6.84
Drip edge, metal, 5" wide	.122	L.F.	.002	.06	.11	.17
Building paper, #15 asphalt felt	2.300	S.F.	.003	.11	.13	.24
Ridge shingles, wood	.090	L.F.	.003	.30	.11	.41
Soffit & fascia, white painted aluminum, 1' overhang	.122	L.F.	.018	.38	.77	1.15
Gutter, seamless, aluminum, painted	.122	L.F.	.008	.25	.39	.64
Downspouts, aluminum, painted	.042	L.F.	.002	.07	.09	.16
TOTAL		S.F.	.100	5.22	4.39	9.61

The prices in these systems are based on a square foot of plan area.
All quantities have been adjusted accordingly.

Description	QUAN.	UNIT	LABOR HOURS	COST PER S.F. MAT.	COST PER S.F. INST.	COST PER S.F. TOTAL

Mansard Roofing Price Sheet	QUAN.	UNIT	LABOR HOURS	COST PER S.F.		
				MAT.	INST.	TOTAL
Shingles, asphalt, standard, inorganic, class A, 210-235 lb./sq.	2.210	S.F.	.032	1.01	1.29	2.30
Laminated, multi-layered, 240-260 lb./sq.	2.210	S.F.	.039	1.27	1.57	2.84
Premium laminated, multi-layered, 260-300 lb./sq.	2.210	S.F.	.050	1.63	2.02	3.65
Slate Buckingham, Virginia, black	2.210	S.F.	.101	9.55	4.07	13.62
Vermont, black or grey	2.210	S.F.	.101	10.80	4.07	14.87
Wood, red cedar, No.1 5X, 16" long, 5" exposure, plain	2.210	S.F.	.070	4.58	3.08	7.66
Fire retardant	2.210	S.F.	.070	5.60	3.08	8.68
18" long, No.1 perfections 6" exposure, plain	2.210	S.F.	.064	4.05	2.79	6.84
Fire retardant	2.210	S.F.	.064	5.05	2.79	7.84
Resquared & rebutted, 18" long, 6" exposure, plain	2.210	S.F.	.059	5.80	2.55	8.35
Fire retardant	2.210	S.F.	.059	6.80	2.55	9.35
Shakes, hand split, 24" long 10" exposure, plain	2.210	S.F.	.070	3.98	3.08	7.06
Fire retardant	2.210	S.F.	.070	4.99	3.08	8.07
18" long, 8" exposure, plain	2.210	S.F.	.088	4.62	3.83	8.45
Fire retardant	2.210	S.F.	.088	5.65	3.83	9.48
Drip edge, metal, 5" wide	.122	S.F.	.002	.06	.11	.17
8" wide	.122	S.F.	.002	.07	.11	.18
Building paper, #15 asphalt felt	2.300	S.F.	.003	.11	.13	.24
Ridge shingles, asphalt	.090	L.F.	.002	.14	.09	.23
Slate	.090	L.F.	.004	.89	.15	1.04
Wood, shingles	.090	L.F.	.003	.30	.11	.41
Shakes	.090	L.F.	.003	.30	.11	.41
Soffit & fascia, aluminum vented, 1' overhang	.122	L.F.	.018	.38	.77	1.15
2' overhang	.122	L.F.	.020	.58	.85	1.43
Vinyl vented, 1' overhang	.122	L.F.	.016	.22	.71	.93
2' overhang	.122	L.F.	.018	.31	.77	1.08
Wood board fascia, plywood soffit, 1' overhang	.122	L.F.	.013	.38	.53	.91
2' overhang	.122	L.F.	.019	.49	.81	1.30
Gutter, 5" box, aluminum, seamless, painted	.122	L.F.	.008	.25	.39	.64
Vinyl	.122	L.F.	.009	.15	.39	.54
Downspout 2" x 3", aluminum, one story house	.042	L.F.	.002	.07	.09	.16
Two story house	.070	L.F.	.003	.11	.14	.25
Vinyl, one story house	.042	L.F.	.002	.07	.09	.16
Two story house	.070	L.F.	.003	.11	.14	.25

5 | ROOFING — 20 | Shed Roofing Systems

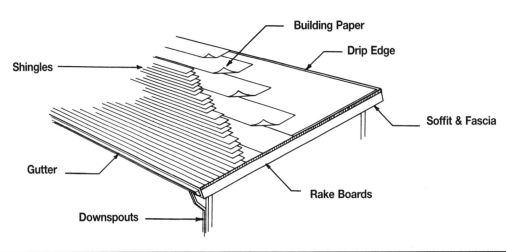

System Description	QUAN.	UNIT	LABOR HOURS	COST PER S.F. MAT.	COST PER S.F. INST.	COST PER S.F. TOTAL
ASPHALT, ROOF SHINGLES, CLASS A						
Shingles, inorganic class A 210-235 lb./sq. 4/12 pitch	1.230	S.F.	.019	.60	.76	1.36
Drip edge, metal, 5" wide	.100	L.F.	.002	.05	.09	.14
Building paper, #15 asphalt felt	1.300	S.F.	.002	.06	.07	.13
Soffit & fascia, white painted aluminum, 1' overhang	.080	L.F.	.012	.25	.51	.76
Rake trim, 1" x 6"	.043	L.F.	.002	.05	.07	.12
Rake trim, prime and paint	.043	L.F.	.002	.01	.07	.08
Gutter, seamless, aluminum, painted	.040	L.F.	.003	.08	.13	.21
Downspouts, painted aluminum	.020	L.F.	.001	.03	.04	.07
TOTAL		S.F.	.043	1.13	1.74	2.87
WOOD, CEDAR SHINGLES, NO. 1 PERFECTIONS, 18" LONG						
Shingles, red cedar, No. 1 perfections, 5" exp., 4/12 pitch	1.230	S.F.	.035	2.21	1.52	3.73
Drip edge, metal, 5" wide	.100	L.F.	.002	.05	.09	.14
Building paper, #15 asphalt felt	1.300	S.F.	.002	.06	.07	.13
Soffit & fascia, white painted aluminum, 1' overhang	.080	L.F.	.012	.25	.51	.76
Rake trim, 1" x 6"	.043	L.F.	.002	.05	.07	.12
Rake trim, prime and paint	.043	L.F.	.001	.01	.03	.04
Gutter, seamless, aluminum, painted	.040	L.F.	.003	.08	.13	.21
Downspouts, painted aluminum	.020	L.F.	.001	.03	.04	.07
TOTAL		S.F.	.058	2.74	2.46	5.20

The prices in these systems are based on a square foot of plan area.
All quantities have been adjusted accordingly.

Description	QUAN.	UNIT	LABOR HOURS	COST PER S.F. MAT.	COST PER S.F. INST.	COST PER S.F. TOTAL

Shed Roofing Price Sheet	QUAN.	UNIT	LABOR HOURS	COST PER S.F.		
				MAT.	INST.	TOTAL
Shingles, asphalt, inorganic, class A, 210-235 lb./sq., 4/12 pitch	1.230	S.F.	.017	.55	.70	1.25
8/12 pitch	1.330	S.F.	.019	.60	.76	1.36
Laminated, multi-layered, 240-260 lb./sq. 4/12 pitch	1.230	S.F.	.021	.69	.86	1.55
8/12 pitch	1.330	S.F.	.023	.75	.93	1.68
Premium laminated, multi-layered, 260-300 lb./sq. 4/12 pitch	1.230	S.F.	.027	.89	1.10	1.99
8/12 pitch	1.330	S.F.	.030	.96	1.20	2.16
Clay tile, Spanish tile, red, 4/12 pitch	1.230	S.F.	.053	3.90	2.16	6.06
8/12 pitch	1.330	S.F.	.058	4.23	2.34	6.57
Mission tile, red, 4/12 pitch	1.230	S.F.	.083	9.25	3.38	12.63
8/12 pitch	1.330	S.F.	.090	10	3.67	13.67
French tile, red, 4/12 pitch	1.230	S.F.	.071	8.40	2.88	11.28
8/12 pitch	1.330	S.F.	.077	9.10	3.12	12.22
Slate, Buckingham, Virginia, black, 4/12 pitch	1.230	S.F.	.055	5.20	2.22	7.42
8/12 pitch	1.330	S.F.	.059	5.65	2.41	8.06
Vermont, black or grey, 4/12 pitch	1.230	S.F.	.055	5.90	2.22	8.12
8/12 pitch	1.330	S.F.	.059	6.35	2.41	8.76
Wood, red cedar, No.1 5X, 16" long, 5" exposure, 4/12 pitch	1.230	S.F.	.038	2.50	1.68	4.18
8/12 pitch	1.330	S.F.	.042	2.70	1.82	4.52
Fire retardant, 4/12 pitch	1.230	S.F.	.038	3.05	1.68	4.73
8/12 pitch	1.330	S.F.	.042	3.30	1.82	5.12
18" long, 6" exposure, 4/12 pitch	1.230	S.F.	.035	2.21	1.52	3.73
8/12 pitch	1.330	S.F.	.038	2.39	1.65	4.04
Fire retardant, 4/12 pitch	1.230	S.F.	.035	2.76	1.52	4.28
8/12 pitch	1.330	S.F.	.038	2.99	1.65	4.64
Resquared & rebutted, 18" long, 6" exposure, 4/12 pitch	1.230	S.F.	.032	3.17	1.39	4.56
8/12 pitch	1.330	S.F.	.035	3.43	1.51	4.94
Fire retardant, 4/12 pitch	1.230	S.F.	.032	3.72	1.39	5.11
8/12 pitch	1.330	S.F.	.035	4.03	1.51	5.54
Wood shakes, hand split, 24" long, 10" exposure, 4/12 pitch	1.230	S.F.	.038	2.17	1.68	3.85
8/12 pitch	1.330	S.F.	.042	2.35	1.82	4.17
Fire retardant, 4/12 pitch	1.230	S.F.	.038	2.72	1.68	4.40
8/12 pitch	1.330	S.F.	.042	2.95	1.82	4.77
18" long, 8" exposure, 4/12 pitch	1.230	S.F.	.048	2.52	2.09	4.61
8/12 pitch	1.330	S.F.	.052	2.73	2.26	4.99
Fire retardant, 4/12 pitch	1.230	S.F.	.048	3.07	2.09	5.16
8/12 pitch	1.330	S.F.	.052	3.33	2.26	5.59
Drip edge, metal, 5" wide	.100	L.F.	.002	.05	.09	.14
8" wide	.100	L.F.	.002	.06	.09	.15
Building paper, #15 asphalt felt	1.300	S.F.	.002	.06	.07	.13
Soffit & fascia, aluminum vented, 1' overhang	.080	L.F.	.012	.25	.51	.76
2' overhang	.080	L.F.	.013	.38	.56	.94
Vinyl vented, 1' overhang	.080	L.F.	.011	.15	.46	.61
2' overhang	.080	L.F.	.012	.20	.51	.71
Wood board fascia, plywood soffit, 1' overhang	.080	L.F.	.010	.26	.39	.65
2' overhang	.080	L.F.	.014	.33	.59	.92
Rake, trim, painted, 1" x 6"	.043	L.F.	.004	.06	.14	.20
1" x 8"	.043	L.F.	.004	.06	.14	.20
Gutter, 5" box, aluminum, seamless, painted	.040	L.F.	.003	.08	.13	.21
Vinyl	.040	L.F.	.003	.05	.13	.18
Downspout 2" x 3", aluminum, one story house	.020	L.F.	.001	.03	.04	.07
Two story house	.020	L.F.	.001	.05	.07	.12
Vinyl, one story house	.020	L.F.	.001	.03	.04	.07
Two story house	.020	L.F.	.001	.05	.07	.12

5 | ROOFING   24 | Gable Dormer Roofing Systems

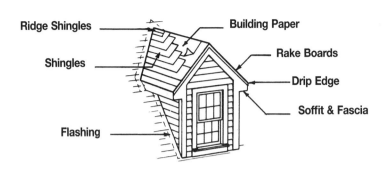

System Description	QUAN.	UNIT	LABOR HOURS	COST PER S.F. MAT.	COST PER S.F. INST.	COST PER S.F. TOTAL
ASPHALT, ROOF SHINGLES, CLASS A						
Shingles, standard inorganic class A 210-235 lb./sq	1.400	S.F.	.020	.64	.82	1.46
Drip edge, metal, 5" wide	.220	L.F.	.004	.10	.19	.29
Building paper, #15 asphalt felt	1.500	S.F.	.002	.07	.08	.15
Ridge shingles, asphalt	.280	L.F.	.007	.44	.27	.71
Soffit & fascia, aluminum, vented	.220	L.F.	.032	.68	1.40	2.08
Flashing, aluminum, mill finish, .013" thick	1.500	S.F.	.083	.71	3.33	4.04
TOTAL		S.F.	.148	2.64	6.09	8.73
WOOD, CEDAR, NO. 1 PERFECTIONS						
Shingles, red cedar, No.1 perfections, 18" long, 5" exp.	1.400	S.F.	.041	2.58	1.78	4.36
Drip edge, metal, 5" wide	.220	L.F.	.004	.10	.19	.29
Building paper, #15 asphalt felt	1.500	S.F.	.002	.07	.08	.15
Ridge shingles, wood	.280	L.F.	.008	.92	.35	1.27
Soffit & fascia, aluminum, vented	.220	L.F.	.032	.68	1.40	2.08
Flashing, aluminum, mill finish, .013" thick	1.500	S.F.	.083	.71	3.33	4.04
TOTAL		S.F.	.170	5.06	7.13	12.19
SLATE, BUCKINGHAM, BLACK						
Shingles, Buckingham, Virginia, black	1.400	S.F.	.064	6.09	2.59	8.68
Drip edge, metal, 5" wide	.220	L.F.	.004	.10	.19	.29
Building paper, #15 asphalt felt	1.500	S.F.	.002	.07	.08	.15
Ridge shingles, slate	.280	L.F.	.011	2.77	.45	3.22
Soffit & fascia, aluminum, vented	.220	L.F.	.032	.68	1.40	2.08
Flashing, copper, 16 oz.	1.500	S.F.	.104	8.33	4.20	12.53
TOTAL		S.F.	.217	18.04	8.91	26.95

The prices in these systems are based on a square foot of plan area under the dormer roof.

Description	QUAN.	UNIT	LABOR HOURS	COST PER S.F. MAT.	COST PER S.F. INST.	COST PER S.F. TOTAL

Gable Dormer Roofing Price Sheet	QUAN.	UNIT	LABOR HOURS	COST PER S.F. MAT.	COST PER S.F. INST.	COST PER S.F. TOTAL
Shingles, asphalt, standard, inorganic, class A, 210-235 lb./sq.	1.400	S.F.	.020	.64	.82	1.46
Laminated, multi-layered, 240-260 lb./sq.	1.400	S.F.	.025	.81	1	1.81
Premium laminated, multi-layered, 260-300 lb./sq.	1.400	S.F.	.032	1.04	1.29	2.33
Clay tile, Spanish tile, red	1.400	S.F.	.062	4.55	2.52	7.07
Mission tile, red	1.400	S.F.	.097	10.80	3.95	14.75
French tile, red	1.400	S.F.	.083	9.80	3.36	13.16
Slate Buckingham, Virginia, black	1.400	S.F.	.064	6.10	2.59	8.69
Vermont, black or grey	1.400	S.F.	.064	6.85	2.59	9.44
Wood, red cedar, No.1 5X, 16" long, 5" exposure	1.400	S.F.	.045	2.91	1.96	4.87
Fire retardant	1.400	S.F.	.045	3.55	1.96	5.51
18" long, No.1 perfections, 5" exposure	1.400	S.F.	.041	2.58	1.78	4.36
Fire retardant	1.400	S.F.	.041	3.22	1.78	5
Resquared & rebutted, 18" long, 5" exposure	1.400	S.F.	.037	3.70	1.62	5.32
Fire retardant	1.400	S.F.	.037	4.34	1.62	5.96
Shakes hand split, 24" long, 10" exposure	1.400	S.F.	.045	2.53	1.96	4.49
Fire retardant	1.400	S.F.	.045	3.17	1.96	5.13
18" long, 8" exposure	1.400	S.F.	.056	2.94	2.44	5.38
Fire retardant	1.400	S.F.	.056	3.58	2.44	6.02
Drip edge, metal, 5" wide	.220	L.F.	.004	.10	.19	.29
8" wide	.220	L.F.	.004	.13	.19	.32
Building paper, #15 asphalt felt	1.500	S.F.	.002	.07	.08	.15
Ridge shingles, asphalt	.280	L.F.	.007	.44	.27	.71
Clay	.280	L.F.	.011	3.11	.45	3.56
Slate	.280	L.F.	.011	2.77	.45	3.22
Wood	.280	L.F.	.008	.92	.35	1.27
Soffit & fascia, aluminum, vented	.220	L.F.	.032	.68	1.40	2.08
Vinyl, vented	.220	L.F.	.029	.40	1.28	1.68
Wood, board fascia, plywood soffit	.220	L.F.	.026	.72	1.08	1.80
Flashing, aluminum, .013" thick	1.500	S.F.	.083	.71	3.33	4.04
.032" thick	1.500	S.F.	.083	1.95	3.33	5.28
.040" thick	1.500	S.F.	.083	2.93	3.33	6.26
.050" thick	1.500	S.F.	.083	3.32	3.33	6.65
Copper, 16 oz.	1.500	S.F.	.104	8.35	4.20	12.55
20 oz.	1.500	S.F.	.109	10.75	4.40	15.15
24 oz.	1.500	S.F.	.114	13.15	4.61	17.76
32 oz.	1.500	S.F.	.120	17.50	4.83	22.33

5 | ROOFING

28 | Shed Dormer Roofing Systems

System Description	QUAN.	UNIT	LABOR HOURS	COST PER S.F.		
				MAT.	INST.	TOTAL
ASPHALT, ROOF SHINGLES, CLASS A						
Shingles, standard inorganic class A 210-235 lb./sq.	1.100	S.F.	.016	.51	.64	1.15
Drip edge, aluminum, 5" wide	.250	L.F.	.005	.12	.22	.34
Building paper, #15 asphalt felt	1.200	S.F.	.002	.06	.07	.13
Soffit & fascia, aluminum, vented, 1' overhang	.250	L.F.	.036	.78	1.59	2.37
Flashing, aluminum, mill finish, 0.013" thick	.800	L.F.	.044	.38	1.78	2.16
TOTAL		S.F.	.103	1.85	4.30	6.15
WOOD, CEDAR, NO. 1 PERFECTIONS, 18" LONG						
Shingles, wood, red cedar, #1 perfections, 5" exposure	1.100	S.F.	.032	2.02	1.40	3.42
Drip edge, aluminum, 5" wide	.250	L.F.	.005	.12	.22	.34
Building paper, #15 asphalt felt	1.200	S.F.	.002	.06	.07	.13
Soffit & fascia, aluminum, vented, 1' overhang	.250	L.F.	.036	.78	1.59	2.37
Flashing, aluminum, mill finish, 0.013" thick	.800	L.F.	.044	.38	1.78	2.16
TOTAL		S.F.	.119	3.36	5.06	8.42
SLATE, BUCKINGHAM, BLACK						
Shingles, slate, Buckingham, black	1.100	S.F.	.050	4.79	2.04	6.83
Drip edge, aluminum, 5" wide	.250	L.F.	.005	.12	.22	.34
Building paper, #15 asphalt felt	1.200	S.F.	.002	.06	.07	.13
Soffit & fascia, aluminum, vented, 1' overhang	.250	L.F.	.036	.78	1.59	2.37
Flashing, copper, 16 oz.	.800	L.F.	.056	4.44	2.24	6.68
TOTAL		S.F.	.149	10.19	6.16	16.35

The prices in this system are based on a square foot of plan area under the dormer roof.

Description	QUAN.	UNIT	LABOR HOURS	COST PER S.F.		
				MAT.	INST.	TOTAL

Shed Dormer Roofing Price Sheet	QUAN.	UNIT	LABOR HOURS	COST PER S.F.		
				MAT.	INST.	TOTAL
Shingles, asphalt, standard, inorganic, class A, 210-235 lb./sq.	1.100	S.F.	.016	.51	.64	1.15
Laminated, multi-layered, 240-260 lb./sq.	1.100	S.F.	.020	.63	.79	1.42
Premium laminated, multi-layered, 260-300 lb./sq.	1.100	S.F.	.025	.81	1.01	1.82
Clay tile, Spanish tile, red	1.100	S.F.	.049	3.58	1.98	5.56
Mission tile, red	1.100	S.F.	.077	8.45	3.10	11.55
French tile, red	1.100	S.F.	.065	7.70	2.64	10.34
Slate Buckingham, Virginia, black	1.100	S.F.	.050	4.79	2.04	6.83
Vermont, black or grey	1.100	S.F.	.050	5.40	2.04	7.44
Wood, red cedar, No. 1 5X, 16" long, 5" exposure	1.100	S.F.	.035	2.29	1.54	3.83
Fire retardant	1.100	S.F.	.035	2.80	1.54	4.34
18" long, No.1 perfections, 5" exposure	1.100	S.F.	.032	2.02	1.40	3.42
Fire retardant	1.100	S.F.	.032	2.53	1.40	3.93
Resquared & rebutted, 18" long, 5" exposure	1.100	S.F.	.029	2.90	1.28	4.18
Fire retardant	1.100	S.F.	.029	3.41	1.28	4.69
Shakes hand split, 24" long, 10" exposure	1.100	S.F.	.035	1.99	1.54	3.53
Fire retardant	1.100	S.F.	.035	2.50	1.54	4.04
18" long, 8" exposure	1.100	S.F.	.044	2.31	1.91	4.22
Fire retardant	1.100	S.F.	.044	2.82	1.91	4.73
Drip edge, metal, 5" wide	.250	L.F.	.005	.12	.22	.34
8" wide	.250	L.F.	.005	.14	.22	.36
Building paper, #15 asphalt felt	1.200	S.F.	.002	.06	.07	.13
Soffit & fascia, aluminum, vented	.250	L.F.	.036	.78	1.59	2.37
Vinyl, vented	.250	L.F.	.033	.46	1.45	1.91
Wood, board fascia, plywood soffit	.250	L.F.	.030	.81	1.24	2.05
Flashing, aluminum, .013" thick	.800	L.F.	.044	.38	1.78	2.16
.032" thick	.800	L.F.	.044	1.04	1.78	2.82
.040" thick	.800	L.F.	.044	1.56	1.78	3.34
.050" thick	.800	L.F.	.044	1.77	1.78	3.55
Copper, 16 oz.	.800	L.F.	.056	4.44	2.24	6.68
20 oz.	.800	L.F.	.058	5.70	2.34	8.04
24 oz.	.800	L.F.	.061	7	2.46	9.46
32 oz.	.800	L.F.	.064	9.30	2.58	11.88

5 | ROOFING — 32 | Skylight/Skywindow Systems

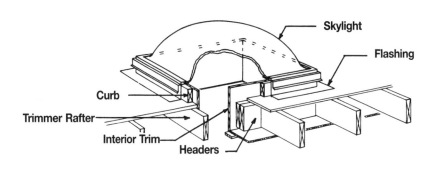

System Description	QUAN.	UNIT	LABOR HOURS	COST EACH MAT.	COST EACH INST.	COST EACH TOTAL
SKYLIGHT, FIXED, 32" X 32"						
Skylight, fixed bubble, insulating, 32" x 32"	1.000	Ea.	1.422	149.33	56.89	206.22
Trimmer rafters, 2" x 6"	28.000	L.F.	.448	17.64	19.60	37.24
Headers, 2" x 6"	6.000	L.F.	.267	3.78	11.64	15.42
Curb, 2" x 4"	12.000	L.F.	.154	4.92	6.72	11.64
Flashing, aluminum, .013" thick	13.500	S.F.	.745	6.35	29.97	36.32
Trim, stock pine, 11/16" x 2-1/2"	12.000	L.F.	.400	17.88	17.40	35.28
Trim primer coat, oil base, brushwork	12.000	L.F.	.148	.36	5.64	6
Trim paint, 1 coat, brushwork	12.000	L.F.	.148	.36	5.64	6
TOTAL		Ea.	3.732	200.62	153.50	354.12
SKYLIGHT, FIXED, 48" X 48"						
Skylight, fixed bubble, insulating, 48" x 48"	1.000	Ea.	1.296	304	51.84	355.84
Trimmer rafters, 2" x 6"	28.000	L.F.	.448	17.64	19.60	37.24
Headers, 2" x 6"	8.000	L.F.	.356	5.04	15.52	20.56
Curb, 2" x 4"	16.000	L.F.	.205	6.56	8.96	15.52
Flashing, aluminum, .013" thick	16.000	S.F.	.883	7.52	35.52	43.04
Trim, stock pine, 11/16" x 2-1/2"	16.000	L.F.	.533	23.84	23.20	47.04
Trim primer coat, oil base, brushwork	16.000	L.F.	.197	.48	7.52	8
Trim paint, 1 coat, brushwork	16.000	L.F.	.197	.48	7.52	8
TOTAL		Ea.	4.115	365.56	169.68	535.24
SKYWINDOW, OPERATING, 24" X 48"						
Skywindow, operating, thermopane glass, 24" x 48"	1.000	Ea.	3.200	600	128	728
Trimmer rafters, 2" x 6"	28.000	L.F.	.448	17.64	19.60	37.24
Headers, 2" x 6"	8.000	L.F.	.267	3.78	11.64	15.42
Curb, 2" x 4"	14.000	L.F.	.179	5.74	7.84	13.58
Flashing, aluminum, .013" thick	14.000	S.F.	.772	6.58	31.08	37.66
Trim, stock pine, 11/16" x 2-1/2"	14.000	L.F.	.467	20.86	20.30	41.16
Trim primer coat, oil base, brushwork	14.000	L.F.	.172	.42	6.58	7
Trim paint, 1 coat, brushwork	14.000	L.F.	.172	.42	6.58	7
TOTAL		Ea.	5.677	655.44	231.62	887.06

The prices in these systems are on a cost each basis.

Description	QUAN.	UNIT	LABOR HOURS	COST EACH MAT.	COST EACH INST.	COST EACH TOTAL

Skylight/Skywindow Price Sheet	QUAN.	UNIT	LABOR HOURS	COST EACH		
				MAT.	INST.	TOTAL
Skylight, fixed bubble insulating, 24" x 24"	1.000	Ea.	.800	84	32	116
32" x 32"	1.000	Ea.	1.422	149	57	206
32" x 48"	1.000	Ea.	.864	203	34.50	237.50
48" x 48"	1.000	Ea.	1.296	305	52	357
Ventilating bubble insulating, 36" x 36"	1.000	Ea.	2.667	420	107	527
52" x 52"	1.000	Ea.	2.667	625	107	732
28" x 52"	1.000	Ea.	3.200	490	128	618
36" x 52"	1.000	Ea.	3.200	530	128	658
Skywindow, operating, thermopane glass, 24" x 48"	1.000	Ea.	3.200	600	128	728
32" x 48"	1.000	Ea.	3.556	630	142	772
Trimmer rafters, 2" x 6"	28.000	L.F.	.448	17.65	19.60	37.25
2" x 8"	28.000	L.F.	.472	26	20.50	46.50
2" x 10"	28.000	L.F.	.711	36.50	31	67.50
Headers, 24" window, 2" x 6"	4.000	L.F.	.178	2.52	7.75	10.27
2" x 8"	4.000	L.F.	.188	3.68	8.20	11.88
2" x 10"	4.000	L.F.	.200	5.25	8.70	13.95
32" window, 2" x 6"	6.000	L.F.	.267	3.78	11.65	15.43
2" x 8"	6.000	L.F.	.282	5.50	12.30	17.80
2" x 10"	6.000	L.F.	.300	7.85	13.10	20.95
48" window, 2" x 6"	8.000	L.F.	.356	5.05	15.50	20.55
2" x 8"	8.000	L.F.	.376	7.35	16.40	23.75
2" x 10"	8.000	L.F.	.400	10.50	17.45	27.95
Curb, 2" x 4", skylight, 24" x 24"	8.000	L.F.	.102	3.28	4.48	7.76
32" x 32"	12.000	L.F.	.154	4.92	6.70	11.62
32" x 48"	14.000	L.F.	.179	5.75	7.85	13.60
48" x 48"	16.000	L.F.	.205	6.55	8.95	15.50
Flashing, aluminum .013" thick, skylight, 24" x 24"	9.000	S.F.	.497	4.23	20	24.23
32" x 32"	13.500	S.F.	.745	6.35	30	36.35
32" x 48"	14.000	S.F.	.772	6.60	31	37.60
48" x 48"	16.000	S.F.	.883	7.50	35.50	43
Copper 16 oz., skylight, 24" x 24"	9.000	S.F.	.626	50	25	75
32" x 32"	13.500	S.F.	.939	75	38	113
32" x 48"	14.000	S.F.	.974	77.50	39	116.50
48" x 48"	16.000	S.F.	1.113	89	45	134
Trim, interior casing painted, 24" x 24"	8.000	L.F.	.347	12.80	14.65	27.45
32" x 32"	12.000	L.F.	.520	19.20	22	41.20
32" x 48"	14.000	L.F.	.607	22.50	25.50	48
48" x 48"	16.000	L.F.	.693	25.50	29.50	55

5 | ROOFING — 34 | Built-up Roofing Systems

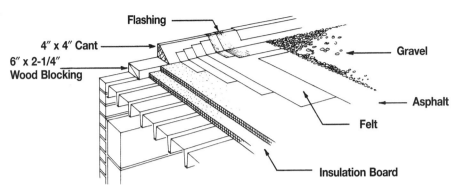

System Description	QUAN.	UNIT	LABOR HOURS	COST PER S.F.		
				MAT.	INST.	TOTAL
ASPHALT, ORGANIC, 4-PLY, INSULATED DECK						
Membrane, asphalt, 4-plies #15 felt, gravel surfacing	1.000	S.F.	.025	.91	1.13	2.04
Insulation board, 2-layers of 1-1/16" glass fiber	2.000	S.F.	.016	2.02	.64	2.66
Wood blocking, 2" x 6"	.040	L.F.	.004	.08	.19	.27
Treated 4" x 4" cant strip	.040	L.F.	.001	.07	.04	.11
Flashing, aluminum, 0.040" thick	.050	S.F.	.003	.10	.11	.21
TOTAL		S.F.	.049	3.18	2.11	5.29
ASPHALT, INORGANIC, 3-PLY, INSULATED DECK						
Membrane, asphalt, 3-plies type IV glass felt, gravel surfacing	1.000	S.F.	.028	.85	1.24	2.09
Insulation board, 2-layers of 1-1/16" glass fiber	2.000	S.F.	.016	2.02	.64	2.66
Wood blocking, 2" x 6"	.040	L.F.	.004	.08	.19	.27
Treated 4" x 4" cant strip	.040	L.F.	.001	.07	.04	.11
Flashing, aluminum, 0.040" thick	.050	S.F.	.003	.10	.11	.21
TOTAL		S.F.	.052	3.12	2.22	5.34
COAL TAR, ORGANIC, 4-PLY, INSULATED DECK						
Membrane, coal tar, 4-plies #15 felt, gravel surfacing	1.000	S.F.	.027	1.45	1.18	2.63
Insulation board, 2-layers of 1-1/16" glass fiber	2.000	S.F.	.016	2.02	.64	2.66
Wood blocking, 2" x 6"	.040	L.F.	.004	.08	.19	.27
Treated 4" x 4" cant strip	.040	L.F.	.001	.07	.04	.11
Flashing, aluminum, 0.040" thick	.050	S.F.	.003	.10	.11	.21
TOTAL		S.F.	.051	3.72	2.16	5.88
COAL TAR, INORGANIC, 3-PLY, INSULATED DECK						
Membrane, coal tar, 3-plies type IV glass felt, gravel surfacing	1.000	S.F.	.029	1.18	1.30	2.48
Insulation board, 2-layers of 1-1/16" glass fiber	2.000	S.F.	.016	2.02	.64	2.66
Wood blocking, 2" x 6"	.040	L.F.	.004	.08	.19	.27
Treated 4" x 4" cant strip	.040	L.F.	.001	.07	.04	.11
Flashing, aluminum, 0.040" thick	.050	S.F.	.003	.10	.11	.21
TOTAL		S.F.	.053	3.45	2.28	5.73

Built-Up Roofing Price Sheet

	QUAN.	UNIT	LABOR HOURS	COST PER S.F. MAT.	COST PER S.F. INST.	COST PER S.F. TOTAL
Membrane, asphalt, 4-plies #15 organic felt, gravel surfacing	1.000	S.F.	.025	.91	1.13	2.04
Asphalt base sheet & 3-plies #15 asphalt felt	1.000	S.F.	.025	.69	1.13	1.82
3-plies type IV glass fiber felt	1.000	S.F.	.028	.85	1.24	2.09
4-plies type IV glass fiber felt	1.000	S.F.	.028	1.04	1.24	2.28
Coal tar, 4-plies #15 organic felt, gravel surfacing	1.000	S.F.	.027			
4-plies tarred felt	1.000	S.F.	.027	1.45	1.18	2.63
3-plies type IV glass fiber felt	1.000	S.F.	.029	1.18	1.30	2.48
4-plies type IV glass fiber felt	1.000	S.F.	.027	1.64	1.18	2.82
Roll, asphalt, 1-ply #15 organic felt, 2-plies mineral surfaced	1.000	S.F.	.021	.57	.93	1.50
3-plies type IV glass fiber, 1-ply mineral surfaced	1.000	S.F.	.022	.88	1	1.88
Insulation boards, glass fiber, 1-1/16" thick	1.000	S.F.	.008	1.01	.32	1.33
2-1/16" thick	1.000	S.F.	.010	1.47	.40	1.87
2-7/16" thick	1.000	S.F.	.010	1.68	.40	2.08
Expanded perlite, 1" thick	1.000	S.F.	.010	.50	.40	.90
1-1/2" thick	1.000	S.F.	.010	.52	.40	.92
2" thick	1.000	S.F.	.011	.86	.46	1.32
Fiberboard, 1" thick	1.000	S.F.	.010	.46	.40	.86
1-1/2" thick	1.000	S.F.	.010	.69	.40	1.09
2" thick	1.000	S.F.	.010	.92	.40	1.32
Extruded polystyrene, 15 PSI compressive strength, 2" thick R10	1.000	S.F.	.006	.65	.26	.91
3" thick R15	1.000	S.F.	.008	1.33	.32	1.65
4" thick R20	1.000	S.F.	.008	1.72	.32	2.04
Tapered for drainage	1.000	S.F.	.005	.55	.21	.76
40 PSI compressive strength, 1" thick R5	1.000	S.F.	.005	.52	.21	.73
2" thick R10	1.000	S.F.	.006	1	.26	1.26
3" thick R15	1.000	S.F.	.008	1.47	.32	1.79
4" thick R20	1.000	S.F.	.008	1.97	.32	2.29
Fiberboard high density, 1/2" thick R1.3	1.000	S.F.	.008	.24	.32	.56
1" thick R2.5	1.000	S.F.	.010	.48	.40	.88
1 1/2" thick R3.8	1.000	S.F.	.010	.72	.40	1.12
Polyisocyanurate, 1 1/2" thick R10.87	1.000	S.F.	.006	.69	.26	.95
2" thick R14.29	1.000	S.F.	.007	.91	.29	1.20
3 1/2" thick R25	1.000	S.F.	.008	2.12	.32	2.44
Tapered for drainage	1.000	S.F.	.006	2.12	.23	2.35
Expanded polystyrene, 1" thick	1.000	S.F.	.005	.32	.21	.53
2" thick R10	1.000	S.F.	.006	.65	.26	.91
3" thick	1.000	S.F.	.006	.96	.26	1.22
Wood blocking, treated, 6" x 2" & 4" x 4" cant	.040	L.F.	.002	.11	.10	.21
6" x 4-1/2" & 4" x 4" cant	.040	L.F.	.005	.17	.23	.40
6" x 5" & 4" x 4" cant	.040	L.F.	.007	.20	.29	.49
Flashing, aluminum, 0.019" thick	.050	S.F.	.003	.05	.11	.16
0.032" thick	.050	S.F.	.003	.07	.11	.18
0.040" thick	.050	S.F.	.003	.10	.11	.21
Copper sheets, 16 oz., under 500 lbs.	.050	S.F.	.003	.28	.14	.42
Over 500 lbs.	.050	S.F.	.003	.37	.10	.47
20 oz., under 500 lbs.	.050	S.F.	.004	.36	.15	.51
Over 500 lbs.	.050	S.F.	.003	.37	.11	.48
Stainless steel, 32 gauge	.050	S.F.	.003	.17	.10	.27
28 gauge	.050	S.F.	.003	.21	.10	.31
26 gauge	.050	S.F.	.003	.26	.10	.36
24 gauge	.050	S.F.	.003	.34	.10	.44

Division 6
Interiors

6 | INTERIORS — 04 | Drywall & Thincoat Wall Systems

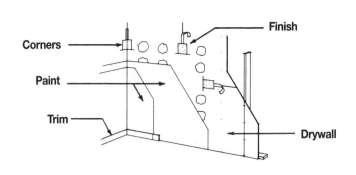

System Description	QUAN.	UNIT	LABOR HOURS	COST PER S.F. MAT.	COST PER S.F. INST.	COST PER S.F. TOTAL
1/2" DRYWALL, TAPED & FINISHED						
Gypsum wallboard, 1/2" thick, standard	1.000	S.F.	.008	.35	.35	.70
Finish, taped & finished joints	1.000	S.F.	.008	.04	.35	.39
Corners, taped & finished, 32 L.F. per 12' x 12' room	.083	L.F.	.002	.01	.07	.08
Painting, primer & 2 coats	1.000	S.F.	.011	.19	.40	.59
Paint trim, to 6" wide, primer + 1 coat enamel	.125	L.F.	.001	.01	.05	.06
Trim, baseboard	.125	L.F.	.005	.33	.22	.55
TOTAL		S.F.	.035	.93	1.44	2.37
THINCOAT, SKIM-COAT, ON 1/2" BACKER DRYWALL						
Gypsum wallboard, 1/2" thick, thincoat backer	1.000	S.F.	.008	.35	.35	.70
Thincoat plaster	1.000	S.F.	.011	.09	.44	.53
Corners, taped & finished, 32 L.F. per 12' x 12' room	.083	L.F.	.002	.01	.07	.08
Painting, primer & 2 coats	1.000	S.F.	.011	.19	.40	.59
Paint trim, to 6" wide, primer + 1 coat enamel	.125	L.F.	.001	.01	.05	.06
Trim, baseboard	.125	L.F.	.005	.33	.22	.55
TOTAL		S.F.	.038	.98	1.53	2.51
5/8" DRYWALL, TAPED & FINISHED						
Gypsum wallboard, 5/8" thick, standard	1.000	S.F.	.008	.43	.35	.78
Finish, taped & finished joints	1.000	S.F.	.008	.04	.35	.39
Corners, taped & finished, 32 L.F. per 12' x 12' room	.083	L.F.	.002	.01	.07	.08
Painting, primer & 2 coats	1.000	S.F.	.011	.19	.40	.59
Trim, baseboard	.125	L.F.	.005	.33	.22	.55
Paint trim, to 6" wide, primer + 1 coat enamel	.125	L.F.	.001	.01	.05	.06
TOTAL		S.F.	.035	1.01	1.44	2.45

The costs in this system are based on a square foot of wall.
Do not deduct for openings.

Description	QUAN.	UNIT	LABOR HOURS	COST PER S.F. MAT.	COST PER S.F. INST.	COST PER S.F. TOTAL

Drywall & Thincoat Wall Price Sheet

	QUAN.	UNIT	LABOR HOURS	COST PER S.F. MAT.	COST PER S.F. INST.	COST PER S.F. TOTAL
Gypsum wallboard, 1/2" thick, standard	1.000	S.F.	.008	.35	.35	.70
Fire resistant	1.000	S.F.	.008	.42	.35	.77
Water resistant	1.000	S.F.	.008	.43	.35	.78
5/8" thick, standard	1.000	S.F.	.008	.43	.35	.78
Fire resistant	1.000	S.F.	.008	.42	.35	.77
Water resistant	1.000	S.F.	.008	.43	.35	.78
Gypsum wallboard backer for thincoat system, 1/2" thick	1.000	S.F.	.008	.35	.35	.70
5/8" thick	1.000	S.F.	.008	.43	.35	.78
Gypsum wallboard, taped & finished	1.000	S.F.	.008	.04	.35	.39
Texture spray	1.000	S.F.	.010	.06	.38	.44
Thincoat plaster, including tape	1.000	S.F.	.011	.09	.44	.53
Gypsum wallboard corners, taped & finished, 32 L.F. per 4' x 4' room	.250	L.F.	.004	.03	.18	.21
6' x 6' room	.110	L.F.	.002	.01	.08	.09
10' x 10' room	.100	L.F.	.001	.01	.07	.08
12' x 12' room	.083	L.F.	.001	.01	.06	.07
16' x 16' room	.063	L.F.	.001	.01	.05	.06
Thincoat system, 32 L.F. per 4' x 4' room	.250	L.F.	.003	.02	.11	.13
6' x 6' room	.110	L.F.	.001	.01	.05	.06
10' x 10' room	.100	L.F.	.001	.01	.04	.05
12' x 12' room	.083	L.F.	.001	.01	.03	.04
16' x 16' room	.063	L.F.	.001	.01	.03	.04
Painting, primer, & 1 coat	1.000	S.F.	.008	.12	.31	.43
& 2 coats	1.000	S.F.	.011	.19	.40	.59
Wallpaper, $7/double roll	1.000	S.F.	.013	.35	.48	.83
$17/double roll	1.000	S.F.	.015	.72	.57	1.29
$40/double roll	1.000	S.F.	.018	2.39	.71	3.10
Wallcovering, medium weight vinyl		S.F.	.017	.79	.64	1.43
Tile, ceramic adhesive thin set, 4 1/4" x 4 1/4" tiles	1.000	S.F.	.084	2.38	2.98	5.36
6" x 6" tiles	1.000	S.F.	.080	3.28	2.83	6.11
Pregrouted sheets	1.000	S.F.	.067	5.05	2.36	7.41
Trim, painted or stained, baseboard	.125	L.F.	.006	.34	.27	.61
Base shoe	.125	L.F.	.005	.17	.23	.40
Chair rail	.125	L.F.	.005	.21	.21	.42
Cornice molding	.125	L.F.	.004	.15	.18	.33
Cove base, vinyl	.125	L.F.	.003	.09	.13	.22
Paneling, not including furring or trim						
Plywood, prefinished, 1/4" thick, 4' x 8' sheets, vert. grooves						
Birch faced, minimum	1.000	S.F.	.032	.94	1.40	2.34
Average	1.000	S.F.	.038	1.42	1.66	3.08
Maximum	1.000	S.F.	.046	2.08	1.99	4.07
Mahogany, African	1.000	S.F.	.040	2.65	1.74	4.39
Philippine (lauan)	1.000	S.F.	.032	1.14	1.40	2.54
Oak or cherry, minimum	1.000	S.F.	.032	2.22	1.40	3.62
Maximum	1.000	S.F.	.040	3.41	1.74	5.15
Rosewood	1.000	S.F.	.050	4.84	2.18	7.02
Teak	1.000	S.F.	.040	3.41	1.74	5.15
Chestnut	1.000	S.F.	.043	5.05	1.86	6.91
Pecan	1.000	S.F.	.040	2.18	1.74	3.92
Walnut, minimum	1.000	S.F.	.032	2.92	1.40	4.32
Maximum	1.000	S.F.	.040	5.50	1.74	7.24

6 | INTERIORS   08 | Drywall & Thincoat Ceiling Systems

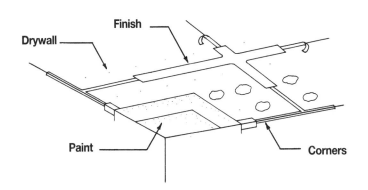

System Description	QUAN.	UNIT	LABOR HOURS	COST PER S.F. MAT.	COST PER S.F. INST.	COST PER S.F. TOTAL
1/2" SHEETROCK, TAPED & FINISHED						
Gypsum wallboard, 1/2" thick, standard	1.000	S.F.	.008	.35	.35	.70
Finish, taped & finished	1.000	S.F.	.008	.04	.35	.39
Corners, taped & finished, 12' x 12' room	.333	L.F.	.006	.03	.24	.27
Paint, primer & 2 coats	1.000	S.F.	.011	.19	.40	.59
TOTAL		S.F.	.033	.61	1.34	1.95
THINCOAT, SKIM COAT ON 1/2" GYPSUM WALLBOARD						
Gypsum wallboard, 1/2" thick, thincoat backer	1.000	S.F.	.008	.35	.35	.70
Thincoat plaster	1.000	S.F.	.011	.09	.44	.53
Corners, taped & finished, 12' x 12' room	.333	L.F.	.006	.03	.24	.27
Paint, primer & 2 coats	1.000	S.F.	.011	.19	.40	.59
TOTAL		S.F.	.036	.66	1.43	2.09
WATER-RESISTANT GYPSUM WALLBOARD, 1/2" THICK, TAPED & FINISHED						
Gypsum wallboard, 1/2" thick, water-resistant	1.000	S.F.	.008	.43	.35	.78
Finish, taped & finished	1.000	S.F.	.008	.04	.35	.39
Corners, taped & finished, 12' x 12' room	.333	L.F.	.006	.03	.24	.27
Paint, primer & 2 coats	1.000	S.F.	.011	.19	.40	.59
TOTAL		S.F.	.033	.69	1.34	2.03
5/8" GYPSUM WALLBOARD, TAPED & FINISHED						
Gypsum wallboard, 5/8" thick, standard	1.000	S.F.	.008	.43	.35	.78
Finish, taped & finished	1.000	S.F.	.008	.04	.35	.39
Corners, taped & finished, 12' x 12' room	.333	L.F.	.006	.03	.24	.27
Paint, primer & 2 coats	1.000	S.F.	.011	.19	.40	.59
TOTAL		S.F.	.033	.69	1.34	2.03

The costs in this system are based on a square foot of ceiling.

Description	QUAN.	UNIT	LABOR HOURS	COST PER S.F. MAT.	COST PER S.F. INST.	COST PER S.F. TOTAL

Drywall & Thincoat Ceilings

	QUAN.	UNIT	LABOR HOURS	COST PER S.F.		
				MAT.	INST.	TOTAL
Gypsum wallboard ceilings, 1/2" thick, standard	1.000	S.F.	.008	.35	.35	.70
Fire resistant	1.000	S.F.	.008	.42	.35	.77
Water resistant	1.000	S.F.	.008	.43	.35	.78
5/8" thick, standard	1.000	S.F.	.008	.43	.35	.78
Fire resistant	1.000	S.F.	.008	.42	.35	.77
Water resistant	1.000	S.F.	.008	.43	.35	.78
Gypsum wallboard backer for thincoat ceiling system, 1/2" thick	1.000	S.F.	.016	.77	.70	1.47
5/8" thick	1.000	S.F.	.016	.85	.70	1.55
Gypsum wallboard ceilings, taped & finished	1.000	S.F.	.008	.04	.35	.39
Texture spray	1.000	S.F.	.010	.06	.38	.44
Thincoat plaster	1.000	S.F.	.011	.09	.44	.53
Corners taped & finished, 4' x 4' room	1.000	L.F.	.015	.10	.73	.83
6' x 6' room	.667	L.F.	.010	.07	.49	.56
10' x 10' room	.400	L.F.	.006	.04	.29	.33
12' x 12' room	.333	L.F.	.005	.03	.24	.27
16' x 16' room	.250	L.F.	.003	.02	.14	.16
Thincoat system, 4' x 4' room	1.000	L.F.	.011	.09	.44	.53
6' x 6' room	.667	L.F.	.007	.06	.29	.35
10' x 10' room	.400	L.F.	.004	.04	.17	.21
12' x 12' room	.333	L.F.	.004	.03	.15	.18
16' x 16' room	.250	L.F.	.002	.02	.09	.11
Painting, primer & 1 coat	1.000	S.F.	.008	.12	.31	.43
& 2 coats	1.000	S.F.	.011	.19	.40	.59
Wallpaper, double roll, solid pattern, avg. workmanship	1.000	S.F.	.013	.35	.48	.83
Basic pattern, avg. workmanship	1.000	S.F.	.015	.72	.57	1.29
Basic pattern, quality workmanship	1.000	S.F.	.018	2.39	.71	3.10
Tile, ceramic adhesive thin set, 4 1/4" x 4 1/4" tiles	1.000	S.F.	.084	2.38	2.98	5.36
6" x 6" tiles	1.000	S.F.	.080	3.28	2.83	6.11
Pregrouted sheets	1.000	S.F.	.067	5.05	2.36	7.41

6 | INTERIORS — 12 | Plaster & Stucco Wall Systems

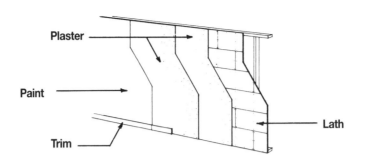

System Description	QUAN.	UNIT	LABOR HOURS	COST PER S.F. MAT.	COST PER S.F. INST.	COST PER S.F. TOTAL
PLASTER ON GYPSUM LATH						
Plaster, gypsum or perlite, 2 coats	1.000	S.F.	.053	.41	2.12	2.53
Lath, 3/8" gypsum	1.000	S.F.	.010	.55	.40	.95
Corners, expanded metal, 32 L.F. per 12' x 12' room	.083	L.F.	.002	.01	.07	.08
Painting, primer & 2 coats	1.000	S.F.	.011	.19	.40	.59
Paint trim, to 6" wide, primer + 1 coat enamel	.125	L.F.	.001	.01	.05	.06
Trim, baseboard	.125	L.F.	.005	.33	.22	.55
TOTAL		S.F.	.082	1.50	3.26	4.76
PLASTER ON METAL LATH						
Plaster, gypsum or perlite, 2 coats	1.000	S.F.	.053	.41	2.12	2.53
Lath, 2.5 Lb. diamond, metal	1.000	S.F.	.010	.32	.40	.72
Corners, expanded metal, 32 L.F. per 12' x 12' room	.083	L.F.	.002	.01	.07	.08
Painting, primer & 2 coats	1.000	S.F.	.011	.19	.40	.59
Paint trim, to 6" wide, primer + 1 coat enamel	.125	L.F.	.001	.01	.05	.06
Trim, baseboard	.125	L.F.	.005	.33	.22	.55
TOTAL		S.F.	.082	1.27	3.26	4.53
STUCCO ON METAL LATH						
Stucco, 2 coats	1.000	S.F.	.041	.25	1.63	1.88
Lath, 2.5 Lb. diamond, metal	1.000	S.F.	.010	.32	.40	.72
Corners, expanded metal, 32 L.F. per 12' x 12' room	.083	L.F.	.002	.01	.07	.08
Painting, primer & 2 coats	1.000	S.F.	.011	.19	.40	.59
Paint trim, to 6" wide, primer + 1 coat enamel	.125	L.F.	.001	.01	.05	.06
Trim, baseboard	.125	L.F.	.005	.33	.22	.55
TOTAL		S.F.	.070	1.11	2.77	3.88

The costs in these systems are based on a per square foot of wall area.
Do not deduct for openings.

Description	QUAN.	UNIT	LABOR HOURS	COST PER S.F. MAT.	COST PER S.F. INST.	COST PER S.F. TOTAL

Plaster & Stucco Wall Price Sheet	QUAN.	UNIT	LABOR HOURS	COST PER S.F.		
				MAT.	INST.	TOTAL
Plaster, gypsum or perlite, 2 coats	1.000	S.F.	.053	.41	2.12	2.53
3 coats	1.000	S.F.	.065	.58	2.56	3.14
Lath, gypsum, standard, 3/8" thick	1.000	S.F.	.010	.55	.40	.95
Fire resistant, 3/8" thick	1.000	S.F.	.013	.43	.49	.92
1/2" thick	1.000	S.F.	.014	.56	.53	1.09
Metal, diamond, 2.5 Lb.	1.000	S.F.	.010	.32	.40	.72
3.4 Lb.	1.000	S.F.	.012	.53	.46	.99
Rib, 2.75 Lb.	1.000	S.F.	.012	.39	.46	.85
3.4 Lb.	1.000	S.F.	.013	.56	.49	1.05
Corners, expanded metal, 32 L.F. per 4' x 4' room	.250	L.F.	.005	.04	.22	.26
6' x 6' room	.110	L.F.	.002	.02	.10	.12
10' x 10' room	.100	L.F.	.002	.02	.09	.11
12' x 12' room	.083	L.F.	.002	.01	.07	.08
16' x 16' room	.063	L.F.	.001	.01	.05	.06
Painting, primer & 1 coats	1.000	S.F.	.008	.12	.31	.43
Primer & 2 coats	1.000	S.F.	.011	.19	.40	.59
Wallpaper, low price double roll	1.000	S.F.	.013	.35	.48	.83
Medium price double roll	1.000	S.F.	.015	.72	.57	1.29
High price double roll	1.000	S.F.	.018	2.39	.71	3.10
Tile, ceramic thin set, 4-1/4" x 4-1/4" tiles	1.000	S.F.	.084	2.38	2.98	5.36
6" x 6" tiles	1.000	S.F.	.080	3.28	2.83	6.11
Pregrouted sheets	1.000	S.F.	.067	5.05	2.36	7.41
Trim, painted or stained, baseboard	.125	L.F.	.006	.34	.27	.61
Base shoe	.125	L.F.	.005	.17	.23	.40
Chair rail	.125	L.F.	.005	.21	.21	.42
Cornice molding	.125	L.F.	.004	.15	.18	.33
Cove base, vinyl	.125	L.F.	.003	.09	.13	.22
Paneling not including furring or trim						
Plywood, prefinished, 1/4" thick, 4' x 8' sheets, vert. grooves						
Birch faced, minimum	1.000	S.F.	.032	.94	1.40	2.34
Average	1.000	S.F.	.038	1.42	1.66	3.08
Maximum	1.000	S.F.	.046	2.08	1.99	4.07
Mahogany, African	1.000	S.F.	.040	2.65	1.74	4.39
Philippine (lauan)	1.000	S.F.	.032	1.14	1.40	2.54
Oak or cherry, minimum	1.000	S.F.	.032	2.22	1.40	3.62
Maximum	1.000	S.F.	.040	3.41	1.74	5.15
Rosewood	1.000	S.F.	.050	4.84	2.18	7.02
Teak	1.000	S.F.	.040	3.41	1.74	5.15
Chestnut	1.000	S.F.	.043	5.05	1.86	6.91
Pecan	1.000	S.F.	.040	2.18	1.74	3.92
Walnut, minimum	1.000	S.F.	.032	2.92	1.40	4.32
Maximum	1.000	S.F.	.040	5.50	1.74	7.24

6 | INTERIORS — 16 | Plaster & Stucco Ceiling Systems

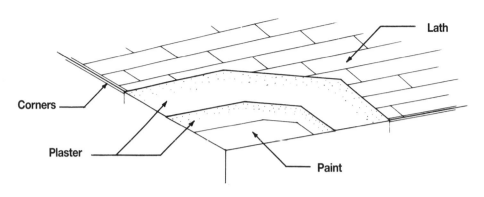

System Description	QUAN.	UNIT	LABOR HOURS	COST PER S.F. MAT.	COST PER S.F. INST.	COST PER S.F. TOTAL
PLASTER ON GYPSUM LATH						
Plaster, gypsum or perlite, 2 coats	1.000	S.F.	.061	.41	2.41	2.82
Gypsum lath, plain or perforated, nailed, 3/8" thick	1.000	S.F.	.010	.55	.40	.95
Gypsum lath, ceiling installation adder	1.000	S.F.	.004		.16	.16
Corners, expanded metal, 12' x 12' room	.330	L.F.	.007	.05	.29	.34
Painting, primer & 2 coats	1.000	S.F.	.011	.19	.40	.59
TOTAL		S.F.	.093	1.20	3.66	4.86
PLASTER ON METAL LATH						
Plaster, gypsum or perlite, 2 coats	1.000	S.F.	.061	.41	2.41	2.82
Lath, 2.5 Lb. diamond, metal	1.000	S.F.	.012	.32	.46	.78
Corners, expanded metal, 12' x 12' room	.330	L.F.	.007	.05	.29	.34
Painting, primer & 2 coats	1.000	S.F.	.011	.19	.40	.59
TOTAL		S.F.	.091	.97	3.56	4.53
STUCCO ON GYPSUM LATH						
Stucco, 2 coats	1.000	S.F.	.041	.25	1.63	1.88
Gypsum lath, plain or perforated, nailed, 3/8" thick	1.000	S.F.	.010	.55	.40	.95
Gypsum lath, ceiling installation adder	1.000	S.F.	.004		.16	.16
Corners, expanded metal, 12' x 12' room	.330	L.F.	.007	.05	.29	.34
Painting, primer & 2 coats	1.000	S.F.	.011	.19	.40	.59
TOTAL		S.F.	.073	1.04	2.88	3.92
STUCCO ON METAL LATH						
Stucco, 2 coats	1.000	S.F.	.041	.25	1.63	1.88
Lath, 2.5 Lb. diamond, metal	1.000	S.F.	.012	.32	.46	.78
Corners, expanded metal, 12' x 12' room	.330	L.F.	.007	.05	.29	.34
Painting, primer & 2 coats	1.000	S.F.	.011	.19	.40	.59
TOTAL		S.F.	.071	.81	2.78	3.59

The costs in these systems are based on a square foot of ceiling area.

Description	QUAN.	UNIT	LABOR HOURS	COST PER S.F. MAT.	COST PER S.F. INST.	COST PER S.F. TOTAL

Plaster & Stucco Ceiling Price Sheet

	QUAN.	UNIT	LABOR HOURS	COST PER S.F. MAT.	COST PER S.F. INST.	COST PER S.F. TOTAL
Plaster, gypsum or perlite, 2 coats	1.000	S.F.	.061	.41	2.41	2.82
3 coats	1.000	S.F.	.065	.58	2.56	3.14
Lath, gypsum, standard, 3/8" thick	1.000	S.F.	.014	.55	.56	1.11
1/2" thick	1.000	S.F.	.015	.52	.59	1.11
Fire resistant, 3/8" thick	1.000	S.F.	.017	.43	.65	1.08
1/2" thick	1.000	S.F.	.018	.56	.69	1.25
Metal, diamond, 2.5 Lb.	1.000	S.F.	.012	.32	.46	.78
3.4 Lb.	1.000	S.F.	.015	.53	.57	1.10
Rib, 2.75 Lb.	1.000	S.F.	.012	.39	.46	.85
3.4 Lb.	1.000	S.F.	.013	.56	.49	1.05
Corners expanded metal, 4' x 4' room	1.000	L.F.	.020	.15	.87	1.02
6' x 6' room	.667	L.F.	.013	.10	.58	.68
10' x 10' room	.400	L.F.	.008	.06	.35	.41
12' x 12' room	.333	L.F.	.007	.05	.29	.34
16' x 16' room	.250	L.F.	.004	.03	.16	.19
Painting, primer & 1 coat	1.000	S.F.	.008	.12	.31	.43
Primer & 2 coats	1.000	S.F.	.011	.19	.40	.59

6 | INTERIORS — 18 | Suspended Ceiling Systems

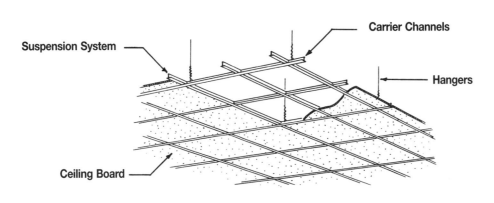

System Description	QUAN.	UNIT	LABOR HOURS	COST PER S.F. MAT.	COST PER S.F. INST.	COST PER S.F. TOTAL
2' X 2' GRID, FILM FACED FIBERGLASS, 5/8" THICK						
Suspension system, 2' x 2' grid, T bar	1.000	S.F.	.012	.75	.54	1.29
Ceiling board, film faced fiberglass, 5/8" thick	1.000	S.F.	.013	.65	.56	1.21
Carrier channels, 1-1/2" x 3/4"	1.000	S.F.	.017	.13	.74	.87
Hangers, #12 wire	1.000	S.F.	.002	.08	.07	.15
TOTAL		S.F.	.044	1.61	1.91	3.52
2' X 4' GRID, FILM FACED FIBERGLASS, 5/8" THICK						
Suspension system, 2' x 4' grid, T bar	1.000	S.F.	.010	.60	.44	1.04
Ceiling board, film faced fiberglass, 5/8" thick	1.000	S.F.	.013	.65	.56	1.21
Carrier channels, 1-1/2" x 3/4"	1.000	S.F.	.017	.13	.74	.87
Hangers, #12 wire	1.000	S.F.	.002	.08	.07	.15
TOTAL		S.F.	.042	1.46	1.81	3.27
2' X 2' GRID, MINERAL FIBER, REVEAL EDGE, 1" THICK						
Suspension system, 2' x 2' grid, T bar	1.000	S.F.	.012	.75	.54	1.29
Ceiling board, mineral fiber, reveal edge, 1" thick	1.000	S.F.	.013	1.44	.58	2.02
Carrier channels, 1-1/2" x 3/4"	1.000	S.F.	.017	.13	.74	.87
Hangers, #12 wire	1.000	S.F.	.002	.08	.07	.15
TOTAL		S.F.	.044	2.40	1.93	4.33
2' X 4' GRID, MINERAL FIBER, REVEAL EDGE, 1" THICK						
Suspension system, 2' x 4' grid, T bar	1.000	S.F.	.010	.60	.44	1.04
Ceiling board, mineral fiber, reveal edge, 1" thick	1.000	S.F.	.013	1.44	.58	2.02
Carrier channels, 1-1/2" x 3/4"	1.000	S.F.	.017	.13	.74	.87
Hangers, #12 wire	1.000	S.F.	.002	.08	.07	.15
TOTAL		S.F.	.042	2.25	1.83	4.08

Description	QUAN.	UNIT	LABOR HOURS	COST PER S.F. MAT.	COST PER S.F. INST.	COST PER S.F. TOTAL

Suspended Ceiling Price Sheet	QUAN.	UNIT	LABOR HOURS	COST PER S.F.		
				MAT.	INST.	TOTAL
Suspension systems, T bar, 2' x 2' grid	1.000	S.F.	.012	.75	.54	1.29
2' x 4' grid	1.000	S.F.	.010	.60	.44	1.04
Concealed Z bar, 12" module	1.000	S.F.	.015	.72	.67	1.39
Ceiling boards, fiberglass, film faced, 2' x 2' or 2' x 4', 5/8" thick	1.000	S.F.	.013	.65	.56	1.21
3/4" thick	1.000	S.F.	.013	1.50	.58	2.08
3" thick thermal R11	1.000	S.F.	.018	1.57	.78	2.35
Glass cloth faced, 3/4" thick	1.000	S.F.	.016	1.91	.70	2.61
1" thick	1.000	S.F.	.016	2.12	.72	2.84
1-1/2" thick, nubby face	1.000	S.F.	.017	2.73	.73	3.46
Mineral fiber boards, 5/8" thick, aluminum face 2' x 2'	1.000	S.F.	.013	1.80	.58	2.38
2' x 4'	1.000	S.F.	.012	1.22	.54	1.76
Standard faced, 2' x 2' or 2' x 4'	1.000	S.F.	.012	.74	.52	1.26
Plastic coated face, 2' x 2' or 2' x 4'	1.000	S.F.	.020	1.16	.87	2.03
Fire rated, 2 hour rating, 5/8" thick	1.000	S.F.	.012	1.10	.52	1.62
Tegular edge, 2' x 2' or 2' x 4', 5/8" thick, fine textured	1.000	S.F.	.013	1.36	.74	2.10
Rough textured	1.000	S.F.	.015	1.78	.74	2.52
3/4" thick, fine textured	1.000	S.F.	.016	1.95	.78	2.73
Rough textured	1.000	S.F.	.018	2.20	.78	2.98
Luminous panels, prismatic, acrylic	1.000	S.F.	.020	2.08	.87	2.95
Polystyrene	1.000	S.F.	.020	1.07	.87	1.94
Flat or ribbed, acrylic	1.000	S.F.	.020	3.62	.87	4.49
Polystyrene	1.000	S.F.	.020	2.48	.87	3.35
Drop pan, white, acrylic	1.000	S.F.	.020	5.30	.87	6.17
Polystyrene	1.000	S.F.	.020	4.43	.87	5.30
Carrier channels, 4'-0" on center, 3/4" x 1-1/2"	1.000	S.F.	.017	.13	.74	.87
1-1/2" x 3-1/2"	1.000	S.F.	.017	.34	.74	1.08
Hangers, #12 wire	1.000	S.F.	.002	.08	.07	.15

6 | INTERIORS   20 | Interior Door Systems

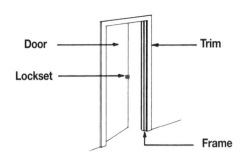

System Description	QUAN.	UNIT	LABOR HOURS	COST EACH MAT.	COST EACH INST.	COST EACH TOTAL
LAUAN, FLUSH DOOR, HOLLOW CORE						
Door, flush, lauan, hollow core, 2'-8" wide x 6'-8" high	1.000	Ea.	.889	35.50	39	74.50
Frame, pine, 4-5/8" jamb	17.000	L.F.	.725	117.30	31.62	148.92
Trim, stock pine, 11/16" x 2-1/2"	34.000	L.F.	1.133	50.66	49.30	99.96
Paint trim, to 6" wide, primer + 1 coat enamel	34.000	L.F.	.340	3.74	12.92	16.66
Butt hinges, chrome, 3-1/2" x 3-1/2"	1.500	Pr.		39.75		39.75
Lockset, passage	1.000	Ea.	.500	16.60	22	38.60
Prime door & frame, oil, brushwork	2.000	Face	1.600	4.74	61	65.74
Paint door and frame, oil, 2 coats	2.000	Face	2.667	8.50	102	110.50
TOTAL		Ea.	7.854	276.79	317.84	594.63
BIRCH, FLUSH DOOR, HOLLOW CORE						
Door, flush, birch, hollow core, 2'-8" wide x 6'-8" high	1.000	Ea.	.889	50.50	39	89.50
Frame, pine, 4-5/8" jamb	17.000	L.F.	.725	117.30	31.62	148.92
Trim, stock pine, 11/16" x 2-1/2"	34.000	L.F.	1.133	50.66	49.30	99.96
Butt hinges, chrome, 3-1/2" x 3-1/2"	1.500	Pr.		39.75		39.75
Lockset, passage	1.000	Ea.	.500	16.60	22	38.60
Prime door & frame, oil, brushwork	2.000	Face	1.600	4.74	61	65.74
Paint door and frame, oil, 2 coats	2.000	Face	2.667	8.50	102	110.50
TOTAL		Ea.	7.514	288.05	304.92	592.97
RAISED PANEL, SOLID, PINE DOOR						
Door, pine, raised panel, 2'-8" wide x 6'-8" high	1.000	Ea.	.889	167	39	206
Frame, pine, 4-5/8" jamb	17.000	L.F.	.725	117.30	31.62	148.92
Trim, stock pine, 11/16" x 2-1/2"	34.000	L.F.	1.133	50.66	49.30	99.96
Butt hinges, bronze, 3-1/2" x 3-1/2"	1.500	Pr.		45		45
Lockset, passage	1.000	Ea.	.500	16.60	22	38.60
Prime door & frame, oil, brushwork	2.000		1.600	4.74	61	65.74
Paint door and frame, oil, 2 coats	2.000		2.667	8.50	102	110.50
TOTAL		Ea.	7.514	409.80	304.92	714.72

The costs in these systems are based on a cost per each door.

Description	QUAN.	UNIT	LABOR HOURS	COST EACH MAT.	COST EACH INST.	COST EACH TOTAL

Interior Door Price Sheet	QUAN.	UNIT	LABOR HOURS	COST EACH		
				MAT.	INST.	TOTAL
Door, hollow core, lauan 1-3/8" thick, 6'-8" high x 1'-6" wide	1.000	Ea.	.889	31.50	39	70.50
2'-0" wide	1.000	Ea.	.889	30	39	69
2'-6" wide	1.000	Ea.	.889	33.50	39	72.50
2'-8" wide	1.000	Ea.	.889	35.50	39	74.50
3'-0" wide	1.000	Ea.	.941	37	41	78
Birch 1-3/8" thick, 6'-8" high x 1'-6" wide	1.000	Ea.	.889	39.50	39	78.50
2'-0" wide	1.000	Ea.	.889	44	39	83
2'-6" wide	1.000	Ea.	.889	49	39	88
2'-8" wide	1.000	Ea.	.889	50.50	39	89.50
3'-0" wide	1.000	Ea.	.941	55	41	96
Louvered pine 1-3/8" thick, 6'-8" high x 1'-6" wide	1.000	Ea.	.842	110	36.50	146.50
2'-0" wide	1.000	Ea.	.889	138	39	177
2'-6" wide	1.000	Ea.	.889	150	39	189
2'-8" wide	1.000	Ea.	.889	158	39	197
3'-0" wide	1.000	Ea.	.941	170	41	211
Paneled pine 1-3/8" thick, 6'-8" high x 1'-6" wide	1.000	Ea.	.842	120	36.50	156.50
2'-0" wide	1.000	Ea.	.889	138	39	177
2'-6" wide	1.000	Ea.	.889	154	39	193
2'-8" wide	1.000	Ea.	.889	167	39	206
3'-0" wide	1.000	Ea.	.941	177	41	218
Frame, pine, 1'-6" thru 2'-0" wide door, 3-5/8" deep	16.000	L.F.	.683	89.50	30	119.50
4-5/8" deep	16.000	L.F.	.683	110	30	140
5-5/8" deep	16.000	L.F.	.683	69	30	99
2'-6" thru 3'0" wide door, 3-5/8" deep	17.000	L.F.	.725	95	31.50	126.50
4-5/8" deep	17.000	L.F.	.725	117	31.50	148.50
5-5/8" deep	17.000	L.F.	.725	73.50	31.50	105
Trim, casing, painted, both sides, 1'-6" thru 2'-6" wide door	32.000	L.F.	1.855	49.50	76.50	126
2'-6" thru 3'-0" wide door	34.000	L.F.	1.971	52.50	81.50	134
Butt hinges 3-1/2" x 3-1/2", steel plated, chrome	1.500	Pr.		40		40
Bronze	1.500	Pr.		45		45
Locksets, passage, minimum	1.000	Ea.	.500	16.60	22	38.60
Maximum	1.000	Ea.	.575	19.10	25.50	44.60
Privacy, miniumum	1.000	Ea.	.625	21	27.50	48.50
Maximum	1.000	Ea.	.675	22.50	29.50	52
Paint 2 sides, primer & 2 cts., flush door, 1'-6" to 2'-0" wide	2.000	Face	5.547	15.80	212	227.80
2'-6" thru 3'-0" wide	2.000	Face	6.933	19.75	265	284.75
Louvered door, 1'-6" thru 2'-0" wide	2.000	Face	6.400	15.60	245	260.60
2'-6" thru 3'-0" wide	2.000	Face	8.000	19.50	305	324.50
Paneled door, 1'-6" thru 2'-0" wide	2.000	Face	6.400	15.60	245	260.60
2'-6" thru 3'-0" wide	2.000	Face	8.000	19.50	305	324.50

6 | INTERIORS — 24 | Closet Door Systems

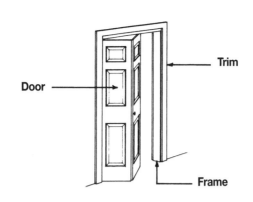

System Description	QUAN.	UNIT	LABOR HOURS	COST EACH MAT.	COST EACH INST.	COST EACH TOTAL
BI-PASSING, FLUSH, LAUAN, HOLLOW CORE, 4'-0" X 6'-8"						
Door, flush, lauan, hollow core, 4'-0" x 6'-8" opening	1.000	Ea.	1.333	182	58	240
Frame, pine, 4-5/8" jamb	18.000	L.F.	.768	124.20	33.48	157.68
Trim, stock pine, 11/16" x 2-1/2"	36.000	L.F.	1.200	53.64	52.20	105.84
Prime door & frame, oil, brushwork	2.000	Face	1.600	4.74	61	65.74
Paint door and frame, oil, 2 coats	2.000	Face	2.667	8.50	102	110.50
TOTAL		Ea.	7.568	373.08	306.68	679.76
BI-PASSING, FLUSH, BIRCH, HOLLOW CORE, 6'-0" X 6'-8"						
Door, flush, birch, hollow core, 6'-0" x 6'-8" opening	1.000	Ea.	1.600	273	70	343
Frame, pine, 4-5/8" jamb	19.000	L.F.	.811	131.10	35.34	166.44
Trim, stock pine, 11/16" x 2-1/2"	38.000	L.F.	1.267	56.62	55.10	111.72
Prime door & frame, oil, brushwork	2.000	Face	2.000	5.93	76.25	82.18
Paint door and frame, oil, 2 coats	2.000	Face	3.333	10.63	127.50	138.13
TOTAL		Ea.	9.011	477.28	364.19	841.47
BI-FOLD, PINE, PANELED, 3'-0" X 6'-8"						
Door, pine, paneled, 3'-0" x 6'-8" opening	1.000	Ea.	1.231	177	53.50	230.50
Frame, pine, 4-5/8" jamb	17.000	L.F.	.725	117.30	31.62	148.92
Trim, stock pine, 11/16" x 2-1/2"	34.000	L.F.	1.133	50.66	49.30	99.96
Prime door & frame, oil, brushwork	2.000	Face	1.600	4.74	61	65.74
Paint door and frame, oil, 2 coats	2.000	Face	2.667	8.50	102	110.50
TOTAL		Ea.	7.356	358.20	297.42	655.62
BI-FOLD, PINE, LOUVERED, 6'-0" X 6'-8"						
Door, pine, louvered, 6'-0" x 6'-8" opening	1.000	Ea.	1.600	270	70	340
Frame, pine, 4-5/8" jamb	19.000	L.F.	.811	131.10	35.34	166.44
Trim, stock pine, 11/16" x 2-1/2"	38.000	L.F.	1.267	56.62	55.10	111.72
Prime door & frame, oil, brushwork	2.500	Face	2.000	5.93	76.25	82.18
Paint door and frame, oil, 2 coats	2.500	Face	3.333	10.63	127.50	138.13
TOTAL		Ea.	9.011	474.28	364.19	838.47

The costs in this system are based on a cost per each door.

Description	QUAN.	UNIT	LABOR HOURS	COST EACH MAT.	COST EACH INST.	COST EACH TOTAL

Closet Door Price Sheet

	QUAN.	UNIT	LABOR HOURS	COST EACH MAT.	COST EACH INST.	TOTAL
Doors, bi-passing, pine, louvered, 4'-0" x 6'-8" opening	1.000	Ea.	1.333	450	58	508
6'-0" x 6'-8" opening	1.000	Ea.	1.600	550	70	620
Paneled, 4'-0" x 6'-8" opening	1.000	Ea.	1.333	530	58	588
6'-0" x 6'-8" opening	1.000	Ea.	1.600	630	70	700
Flush, birch, hollow core, 4'-0" x 6'-8" opening	1.000	Ea.	1.333	227	58	285
6'-0" x 6'-8" opening	1.000	Ea.	1.600	273	70	343
Flush, lauan, hollow core, 4'-0" x 6'-8" opening	1.000	Ea.	1.333	182	58	240
6'-0" x 6'-8" opening	1.000	Ea.	1.600	213	70	283
Bi-fold, pine, louvered, 3'-0" x 6'-8" opening	1.000	Ea.	1.231	177	53.50	230.50
6'-0" x 6'-8" opening	1.000	Ea.	1.600	270	70	340
Paneled, 3'-0" x 6'-8" opening	1.000	Ea.	1.231	177	53.50	230.50
6'-0" x 6'-8" opening	1.000	Ea.	1.600	270	70	340
Flush, birch, hollow core, 3'-0" x 6'-8" opening	1.000	Ea.	1.231	58	53.50	111.50
6'-0" x 6'-8" opening	1.000	Ea.	1.600	114	70	184
Flush, lauan, hollow core, 3'-0" x 6'8" opening	1.000	Ea.	1.231	229	53.50	282.50
6'-0" x 6'-8" opening	1.000	Ea.	1.600	435	70	505
Frame pine, 3'-0" door, 3-5/8" deep	17.000	L.F.	.725	95	31.50	126.50
4-5/8" deep	17.000	L.F.	.725	117	31.50	148.50
5-5/8" deep	17.000	L.F.	.725	73.50	31.50	105
4'-0" door, 3-5/8" deep	18.000	L.F.	.768	101	33.50	134.50
4-5/8" deep	18.000	L.F.	.768	124	33.50	157.50
5-5/8" deep	18.000	L.F.	.768	77.50	33.50	111
6'-0" door, 3-5/8" deep	19.000	L.F.	.811	106	35.50	141.50
4-5/8" deep	19.000	L.F.	.811	131	35.50	166.50
5-5/8" deep	19.000	L.F.	.811	82	35.50	117.50
Trim both sides, painted 3'-0" x 6'-8" door	34.000	L.F.	1.971	52.50	81.50	134
4'-0" x 6'-8" door	36.000	L.F.	2.086	56	86	142
6'-0" x 6'-8" door	38.000	L.F.	2.203	59	91	150
Paint 2 sides, primer & 2 cts., flush door & frame, 3' x 6'-8" opng	2.000	Face	2.914	9.95	122	131.95
4'-0" x 6'-8" opening	2.000	Face	3.886	13.25	163	176.25
6'-0" x 6'-8" opening	2.000	Face	4.857	16.55	204	220.55
Paneled door & frame, 3'-0" x 6'-8" opening	2.000	Face	6.000	14.65	230	244.65
4'-0" x 6'-8" opening	2.000	Face	8.000	19.50	305	324.50
6'-0" x 6'-8" opening	2.000	Face	10.000	24.50	385	409.50
Louvered door & frame, 3'-0" x 6'-8" opening	2.000	Face	6.000	14.65	230	244.65
4'-0" x 6'-8" opening	2.000	Face	8.000	19.50	305	324.50
6'-0" x 6'-8" opening	2.000	Face	10.000	24.50	385	409.50

6 | INTERIORS　60 | Carpet Systems

System Description	QUAN.	UNIT	LABOR HOURS	COST PER S.F. MAT.	COST PER S.F. INST.	COST PER S.F. TOTAL
Carpet, direct glue-down, nylon, level loop, 26 oz.	1.000	S.F.	.018	2.75	.72	3.47
32 oz.	1.000	S.F.	.018	3.48	.72	4.20
40 oz.	1.000	S.F.	.018	5.15	.72	5.87
Nylon, plush, 20 oz.	1.000	S.F.	.018	1.65	.72	2.37
24 oz.	1.000	S.F.	.018	1.74	.72	2.46
30 oz.	1.000	S.F.	.018	2.57	.72	3.29
42 oz.	1.000	S.F.	.022	3.89	.86	4.75
48 oz.	1.000	S.F.	.022	4.46	.86	5.32
54 oz.	1.000	S.F.	.022	4.89	.86	5.75
Olefin, 15 oz.	1.000	S.F.	.018	.92	.72	1.64
22 oz.	1.000	S.F.	.018	1.09	.72	1.81
Tile, foam backed, needle punch	1.000	S.F.	.014	3.71	.56	4.27
Tufted loop or shag	1.000	S.F.	.014	2.20	.56	2.76
Wool, 36 oz., level loop	1.000	S.F.	.018	11.90	.72	12.62
32 oz., patterned	1.000	S.F.	.020	11.75	.80	12.55
48 oz., patterned	1.000	S.F.	.020	12	.80	12.80
Padding, sponge rubber cushion, minimum	1.000	S.F.	.006	.48	.24	.72
Maximum	1.000	S.F.	.006	1.06	.24	1.30
Felt, 32 oz. to 56 oz., minimum	1.000	S.F.	.006	.48	.24	.72
Maximum	1.000	S.F.	.006	.86	.24	1.10
Bonded urethane, 3/8" thick, minimum	1.000	S.F.	.006	.51	.24	.75
Maximum	1.000	S.F.	.006	.87	.24	1.11
Prime urethane, 1/4" thick, minimum	1.000	S.F.	.006	.29	.24	.53
Maximum	1.000	S.F.	.006	.54	.24	.78
Stairs, for stairs, add to above carpet prices	1.000	Riser	.267		10.65	10.65
Underlayment plywood, 3/8" thick	1.000	S.F.	.011	.86	.47	1.33
1/2" thick	1.000	S.F.	.011	.99	.48	1.47
5/8" thick	1.000	S.F.	.011	1.20	.50	1.70
3/4" thick	1.000	S.F.	.012	1.22	.54	1.76
Particle board, 3/8" thick	1.000	S.F.	.011	.36	.47	.83
1/2" thick	1.000	S.F.	.011	.41	.48	.89
5/8" thick	1.000	S.F.	.011	.45	.50	.95
3/4" thick	1.000	S.F.	.012	.70	.54	1.24
Hardboard, 4' x 4', 0.215" thick	1.000	S.F.	.011	.43	.47	.90

6 | INTERIORS — 64 | Flooring Systems

System Description	QUAN.	UNIT	LABOR HOURS	COST PER S.F. MAT.	COST PER S.F. INST.	COST PER S.F. TOTAL
Resilient flooring, asphalt tile on concrete, 1/8" thick						
Color group B	1.000	S.F.	.020	1.22	.80	2.02
Color group C & D	1.000	S.F.	.020	1.34	.80	2.14
Asphalt tile on wood subfloor, 1/8" thick						
Color group B	1.000	S.F.	.020	1.44	.80	2.24
Color group C & D	1.000	S.F.	.020	1.56	.80	2.36
Vinyl composition tile, 12" x 12", 1/16" thick	1.000	S.F.	.016	1.07	.64	1.71
Embossed	1.000	S.F.	.016	1.93	.64	2.57
Marbleized	1.000	S.F.	.016	1.93	.64	2.57
Plain	1.000	S.F.	.016	2.48	.64	3.12
.080" thick, embossed	1.000	S.F.	.016	1.29	.64	1.93
Marbleized	1.000	S.F.	.016	2.20	.64	2.84
Plain	1.000	S.F.	.016	2.20	.64	2.84
1/8" thick, marbleized	1.000	S.F.	.016	1.53	.64	2.17
Plain	1.000	S.F.	.016	2.28	.64	2.92
Vinyl tile, 12" x 12", .050" thick, minimum	1.000	S.F.	.016	2.42	.64	3.06
Maximum	1.000	S.F.	.016	5.50	.64	6.14
1/8" thick, minimum	1.000	S.F.	.016	3.96	.64	4.60
Maximum	1.000	S.F.	.016	5.85	.64	6.49
1/8" thick, solid colors	1.000	S.F.	.016	5.65	.64	6.29
Florentine pattern	1.000	S.F.	.016	5.70	.64	6.34
Marbleized or travertine pattern	1.000	S.F.	.016	12.10	.64	12.74
Vinyl sheet goods, backed, .070" thick, minimum	1.000	S.F.	.032	3.03	1.28	4.31
Maximum	1.000	S.F.	.040	3.33	1.60	4.93
.093" thick, minimum	1.000	S.F.	.035	3.30	1.39	4.69
Maximum	1.000	S.F.	.040	4.68	1.60	6.28
.125" thick, minimum	1.000	S.F.	.035	3.69	1.39	5.08
Maximum	1.000	S.F.	.040	5.50	1.60	7.10
Wood, oak, finished in place, 25/32" x 2-1/2" clear	1.000	S.F.	.074	4.24	2.91	7.15
Select	1.000	S.F.	.074	3.94	2.91	6.85
No. 1 common	1.000	S.F.	.074	5.10	2.91	8.01
Prefinished, oak, 2-1/2" wide	1.000	S.F.	.047	7.05	2.05	9.10
3-1/4" wide	1.000	S.F.	.043	9	1.89	10.89
Ranch plank, oak, random width	1.000	S.F.	.055	8.70	2.41	11.11
Parquet, 5/16" thick, finished in place, oak, minimum	1.000	S.F.	.077	5.40	3.04	8.44
Maximum	1.000	S.F.	.107	6.25	4.35	10.60
Teak, minimum	1.000	S.F.	.077	6	3.04	9.04
Maximum	1.000	S.F.	.107	9.85	4.35	14.20
Sleepers, treated, 16" O.C., 1" x 2"	1.000	S.F.	.007	.22	.30	.52
1" x 3"	1.000	S.F.	.008	.34	.35	.69
2" x 4"	1.000	S.F.	.011	.58	.47	1.05
2" x 6"	1.000	S.F.	.012	.91	.54	1.45
Subfloor, plywood, 1/2" thick	1.000	S.F.	.011	.52	.47	.99
5/8" thick	1.000	S.F.	.012	.65	.52	1.17
3/4" thick	1.000	S.F.	.013	.78	.56	1.34
Ceramic tile, color group 2, 1" x 1"	1.000	S.F.	.087	5.05	3.09	8.14
2" x 2" or 2" x 1"	1.000	S.F.	.084	5.40	2.98	8.38
Color group 1, 8" x 8"		S.F.	.064	3.70	2.26	5.96
12" x 12"		S.F.	.049	4.64	1.74	6.38
16" x 16"		S.F.	.029	6.45	1.03	7.48

6 | INTERIORS 90 | Stairways

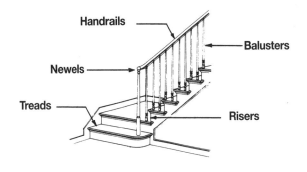

System Description	QUAN.	UNIT	LABOR HOURS	COST EACH MAT.	COST EACH INST.	COST EACH TOTAL
7 RISERS, OAK TREADS, BOX STAIRS						
Treads, oak, 1-1/4" x 10" wide, 3' long	6.000	Ea.	2.667	390	116.40	506.40
Risers, 3/4" thick, beech	7.000	Ea.	2.625	136.50	114.45	250.95
Balusters, birch, 30" high	12.000	Ea.	3.429	90	149.40	239.40
Newels, 3-1/4" wide	2.000	Ea.	2.286	84	100	184
Handrails, oak laminated	7.000	L.F.	.933	248.50	40.60	289.10
Stringers, 2" x 10", 3 each	21.000	L.F.	.306	8.61	13.23	21.84
TOTAL		Ea.	12.246	957.61	534.08	1,491.69
14 RISERS, OAK TREADS, BOX STAIRS						
Treads, oak, 1-1/4" x 10" wide, 3' long	13.000	Ea.	5.778	845	252.20	1,097.20
Risers, 3/4" thick, beech	14.000	Ea.	5.250	273	228.90	501.90
Balusters, birch, 30" high	26.000	Ea.	7.428	195	323.70	518.70
Newels, 3-1/4" wide	2.000	Ea.	2.286	84	100	184
Handrails, oak, laminated	14.000	L.F.	1.867	497	81.20	578.20
Stringers, 2" x 10", 3 each	42.000	L.F.	5.169	55.02	224.70	279.72
TOTAL		Ea.	27.778	1,949.02	1,210.70	3,159.72
14 RISERS, PINE TREADS, BOX STAIRS						
Treads, pine, 9-1/2" x 3/4" thick	13.000	Ea.	5.778	188.50	252.20	440.70
Risers, 3/4" thick, pine	14.000	Ea.	5.091	143.64	222.60	366.24
Balusters, pine, 30" high	26.000	Ea.	7.428	107.90	323.70	431.60
Newels, 3-1/4" wide	2.000	Ea.	2.286	84	100	184
Handrails, oak, laminated	14.000	L.F.	1.867	497	81.20	578.20
Stringers, 2" x 10", 3 each	42.000	L.F.	5.169	55.02	224.70	279.72
TOTAL		Ea.	27.619	1,076.06	1,204.40	2,280.46

Description	QUAN.	UNIT	LABOR HOURS	COST EACH MAT.	COST EACH INST.	COST EACH TOTAL

6 | INTERIORS — 90 | Stairways

Stairway Price Sheet	QUAN.	UNIT	LABOR HOURS	COST EACH MAT.	COST EACH INST.	COST EACH TOTAL
Treads, oak, 1-1/16" x 9-1/2", 3' long, 7 riser stair	6.000	Ea.	2.667	390	116	506
14 riser stair	13.000	Ea.	5.778	845	252	1,097
1-1/16" x 11-1/2", 3' long, 7 riser stair	6.000	Ea.	2.667	390	116	506
14 riser stair	13.000	Ea.	5.778	845	252	1,097
Pine, 3/4" x 9-1/2", 3' long, 7 riser stair	6.000	Ea.	2.667	87	116	203
14 riser stair	13.000	Ea.	5.778	189	252	441
3/4" x 11-1/4", 3' long, 7 riser stair	6.000	Ea.	2.667	132	116	248
14 riser stair	13.000	Ea.	5.778	286	252	538
Risers, oak, 3/4" x 7-1/2" high, 7 riser stair	7.000	Ea.	2.625	144	114	258
14 riser stair	14.000	Ea.	5.250	288	229	517
Beech, 3/4" x 7-1/2" high, 7 riser stair	7.000	Ea.	2.625	137	114	251
14 riser stair	14.000	Ea.	5.250	273	229	502
Baluster, turned, 30" high, pine, 7 riser stair	12.000	Ea.	3.429	50	149	199
14 riser stair	26.000	Ea.	7.428	108	325	433
30" birch, 7 riser stair	12.000	Ea.	3.429	90	149	239
14 riser stair	26.000	Ea.	7.428	195	325	520
42" pine, 7 riser stair	12.000	Ea.	3.556	66	155	221
14 riser stair	26.000	Ea.	7.704	143	335	478
42" birch, 7 riser stair	12.000	Ea.	3.556	144	155	299
14 riser stair	26.000	Ea.	7.704	310	335	645
Newels, 3-1/4" wide, starting, 7 riser stair	2.000	Ea.	2.286	84	100	184
14 riser stair	2.000	Ea.	2.286	84	100	184
Landing, 7 riser stair	2.000	Ea.	3.200	234	140	374
14 riser stair	2.000	Ea.	3.200	234	140	374
Handrails, oak, laminated, 7 riser stair	7.000	L.F.	.933	249	40.50	289.50
14 riser stair	14.000	L.F.	1.867	495	81	576
Stringers, fir, 2" x 10" 7 riser stair	21.000	L.F.	2.585	27.50	112	139.50
14 riser stair	42.000	L.F.	5.169	55	225	280
2" x 12", 7 riser stair	21.000	L.F.	2.585	33	112	145
14 riser stair	42.000	L.F.	5.169	66	225	291

Special Stairways	QUAN.	UNIT	LABOR HOURS	COST EACH MAT.	COST EACH INST.	COST EACH TOTAL
Basement stairs, open risers	1.000	Flight	4.000	700	174	874
Spiral stairs, oak, 4'-6" diameter, prefabricated, 9' high	1.000	Flight	10.667	5,125	465	5,590
Aluminum, 5'-0" diameter stock unit	1.000	Flight	9.956	7,275	585	7,860
Custom unit	1.000	Flight	9.956	13,800	585	14,385
Cast iron, 4'-0" diameter, minimum	1.000	Flight	9.956	6,500	585	7,085
Maximum	1.000	Flight	17.920	8,900	1,050	9,950
Steel, industrial, pre-erected, 3'-6" wide, bar rail	1.000	Flight	7.724	6,650	690	7,340
Picket rail	1.000	Flight	7.724	7,425	690	8,115

Division 7
Specialties

7 | SPECIALTIES — 08 | Kitchen Systems

System Description	QUAN.	UNIT	LABOR HOURS	COST PER L.F. MAT.	COST PER L.F. INST.	COST PER L.F. TOTAL
KITCHEN, ECONOMY GRADE						
Top cabinets, economy grade	1.000	L.F.	.171	36.80	7.52	44.32
Bottom cabinets, economy grade	1.000	L.F.	.256	55.20	11.28	66.48
Square edge, plastic face countertop	1.000	L.F.	.267	26.50	11.65	38.15
Blocking, wood, 2" x 4"	1.000	L.F.	.032	.41	1.40	1.81
Soffit, framing, wood, 2" x 4"	4.000	L.F.	.071	1.64	3.12	4.76
Soffit drywall	2.000	S.F.	.047	.98	2.06	3.04
Drywall painting	2.000	S.F.	.013	.12	.58	.70
TOTAL		L.F.	.857	121.65	37.61	159.26
AVERAGE GRADE						
Top cabinets, average grade	1.000	L.F.	.213	46	9.40	55.40
Bottom cabinets, average grade	1.000	L.F.	.320	69	14.10	83.10
Solid surface countertop, solid color	1.000	L.F.	.800	66	35	101
Blocking, wood, 2" x 4"	1.000	L.F.	.032	.41	1.40	1.81
Soffit framing, wood, 2" x 4"	4.000	L.F.	.071	1.64	3.12	4.76
Soffit drywall	2.000	S.F.	.047	.98	2.06	3.04
Drywall painting	2.000	S.F.	.013	.12	.58	.70
TOTAL		L.F.	1.496	184.15	65.66	249.81
CUSTOM GRADE						
Top cabinets, custom grade	1.000	L.F.	.256	110	11.20	121.20
Bottom cabinets, custom grade	1.000	L.F.	.384	165	16.80	181.80
Solid surface countertop, premium patterned color	1.000	L.F.	1.067	115	46.50	161.50
Blocking, wood, 2" x 4"	1.000	L.F.	.032	.41	1.40	1.81
Soffit framing, wood, 2" x 4"	4.000	L.F.	.071	1.64	3.12	4.76
Soffit drywall	2.000	S.F.	.047	.98	2.06	3.04
Drywall painting	2.000	S.F.	.013	.12	.58	.70
TOTAL		L.F.	1.870	393.15	81.66	474.81

Description	QUAN.	UNIT	LABOR HOURS	COST PER L.F. MAT.	COST PER L.F. INST.	COST PER L.F. TOTAL

Kitchen Price Sheet	QUAN.	UNIT	LABOR HOURS	COST PER L.F.		
				MAT.	INST.	TOTAL
Top cabinets, economy grade	1.000	L.F.	.171	37	7.50	44.50
Average grade	1.000	L.F.	.213	46	9.40	55.40
Custom grade	1.000	L.F.	.256	110	11.20	121.20
Bottom cabinets, economy grade	1.000	L.F.	.256	55	11.30	66.30
Average grade	1.000	L.F.	.320	69	14.10	83.10
Custom grade	1.000	L.F.	.384	165	16.80	181.80
Counter top, laminated plastic, 7/8" thick, no splash	1.000	L.F.	.267	21	11.65	32.65
With backsplash	1.000	L.F.	.267	27.50	11.65	39.15
1-1/4" thick, no splash	1.000	L.F.	.286	31.50	12.45	43.95
With backsplash	1.000	L.F.	.286	30	12.45	42.45
Post formed, laminated plastic	1.000	L.F.	.267	10.80	11.65	22.45
Ceramic tile, with backsplash			.427	15.60	7	22.60
Marble, with backsplash, minimum	1.000	L.F.	.471	38.50	21	59.50
Maximum	1.000	L.F.	.615	96	27.50	123.50
Maple, solid laminated, no backsplash	1.000	L.F.	.286	63	12.45	75.45
With backsplash	1.000	L.F.	.286	74.50	12.45	86.95
Solid Surface, with backsplash, minimum	1.000	L.F.	.842	72.50	36.50	109
Maximum	1.000	L.F.	.067	115	46.50	161.50
Blocking, wood, 2" x 4"	1.000	L.F.	.032	.41	1.40	1.81
2" x 6"	1.000	L.F.	.036	.63	1.57	2.20
2" x 8"	1.000	L.F.	.040	.92	1.74	2.66
Soffit framing, wood, 2" x 3"	4.000	L.F.	.064	1.32	2.80	4.12
2" x 4"	4.000	L.F.	.071	1.64	3.12	4.76
Soffit, drywall, painted	2.000	S.F.	.060	1.10	2.64	3.74
Paneling, standard	2.000	S.F.	.064	1.88	2.80	4.68
Deluxe	2.000	S.F.	.091	4.16	3.98	8.14
Sinks, porcelain on cast iron, single bowl, 21" x 24"	1.000	Ea.	10.334	500	450	950
21" x 30"	1.000	Ea.	10.334	485	450	935
Double bowl, 20" x 32"	1.000	Ea.	10.810	585	470	1,055
Stainless steel, single bowl, 16" x 20"	1.000	Ea.	10.334	645	450	1,095
22" x 25"	1.000	Ea.	10.334	690	450	1,140
Double bowl, 20" x 32"	1.000	Ea.	10.810	405	470	875

7 | SPECIALTIES   12 | Appliance Systems

Kitchen Price Sheet	QUAN.	UNIT	LABOR HOURS	COST PER L.F. MAT.	COST PER L.F. INST.	COST PER L.F. TOTAL
Range, free standing, minimum	1.000	Ea.	3.600	400	147	547
Maximum	1.000	Ea.	6.000	1,800	223	2,023
Built-in, minimum	1.000	Ea.	3.333	620	160	780
Maximum	1.000	Ea.	10.000	1,700	445	2,145
Counter top range, 4-burner, minimum	1.000	Ea.	3.333	340	160	500
Maximum	1.000	Ea.	4.667	750	224	974
Compactor, built-in, minimum	1.000	Ea.	2.215	510	99.50	609.50
Maximum	1.000	Ea.	3.282	600	146	746
Dishwasher, built-in, minimum	1.000	Ea.	6.735	455	325	780
Maximum	1.000	Ea.	9.235	500	450	950
Garbage disposer, minimum	1.000	Ea.	2.810	122	137	259
Maximum	1.000	Ea.	2.810	242	137	379
Microwave oven, minimum	1.000	Ea.	2.615	120	126	246
Maximum	1.000	Ea.	4.615	485	222	707
Range hood, ducted, minimum	1.000	Ea.	4.658	91	211	302
Maximum	1.000	Ea.	5.991	695	270	965
Ductless, minimum	1.000	Ea.	2.615	88	119	207
Maximum	1.000	Ea.	3.948	695	178	873
Refrigerator, 16 cu.ft., minimum	1.000	Ea.	2.000	565	64	629
Maximum	1.000	Ea.	3.200	900	102	1,002
16 cu.ft. with icemaker, minimum	1.000	Ea.	4.210	740	165	905
Maximum	1.000	Ea.	5.410	1,075	203	1,278
19 cu.ft., minimum	1.000	Ea.	2.667	550	84.50	634.50
Maximum	1.000	Ea.	4.667	965	148	1,113
19 cu.ft. with icemaker, minimum	1.000	Ea.	5.143	755	194	949
Maximum	1.000	Ea.	7.143	1,175	257	1,432
Sinks, porcelain on cast iron single bowl, 21" x 24"	1.000	Ea.	10.334	500	450	950
21" x 30"	1.000	Ea.	10.334	485	450	935
Double bowl, 20" x 32"	1.000	Ea.	10.810	585	470	1,055
Stainless steel, single bowl 16" x 20"	1.000	Ea.	10.334	645	450	1,095
22" x 25"	1.000	Ea.	10.334	690	450	1,140
Double bowl, 20" x 32"	1.000	Ea.	10.810	405	470	875
Water heater, electric, 30 gallon	1.000	Ea.	3.636	425	177	602
40 gallon	1.000	Ea.	4.000	450	194	644
Gas, 30 gallon	1.000	Ea.	4.000	690	194	884
75 gallon	1.000	Ea.	5.333	965	259	1,224
Wall, packaged terminal heater/air conditioner cabinet, wall sleeve, louver, electric heat, thermostat, manual changeover, 208V		Ea.				
6000 BTUH cooling, 8800 BTU heating	1.000	Ea.	2.667	1,100	118	1,218
9000 BTUH cooling, 13,900 BTU heating	1.000	Ea.	3.200	1,175	141	1,316
12,000 BTUH cooling, 13,900 BTU heating	1.000	Ea.	4.000	1,275	177	1,452
15,000 BTUH cooling, 13,900 BTU heating	1.000	Ea.	5.333	1,525	236	1,761

7 | SPECIALTIES   16 | Bath Accessories

System Description	QUAN.	UNIT	LABOR HOURS	COST EACH		
				MAT.	INST.	TOTAL
Curtain rods, stainless, 1" diameter, 3' long	1.000	Ea.	.615	36.50	27	63.50
5' long	1.000	Ea.	.615	36.50	27	63.50
Grab bar, 1" diameter, 12" long	1.000	Ea.	.283	21.50	12.35	33.85
36" long	1.000	Ea.	.340	22	14.85	36.85
1-1/4" diameter, 12" long	1.000	Ea.	.333	25.50	14.55	40.05
36" long	1.000	Ea.	.400	26	17.45	43.45
1-1/2" diameter, 12" long	1.000	Ea.	.383	29.50	16.75	46.25
36" long	1.000	Ea.	.460	30	20	50
Mirror, 18" x 24"	1.000	Ea.	.400	59	17.45	76.45
72" x 24"	1.000	Ea.	1.333	330	58	388
Medicine chest with mirror, 18" x 24"	1.000	Ea.	.400	215	17.45	232.45
36" x 24"	1.000	Ea.	.600	325	26	351
Toilet tissue dispenser, surface mounted, minimum	1.000	Ea.	.267	14.85	11.65	26.50
Maximum	1.000	Ea.	.400	22.50	17.50	40
Flush mounted, minimum	1.000	Ea.	.293	16.35	12.80	29.15
Maximum	1.000	Ea.	.427	24	18.65	42.65
Towel bar, 18" long, minimum	1.000	Ea.	.278	34.50	12.10	46.60
Maximum	1.000	Ea.	.348	43	15.15	58.15
24" long, minimum	1.000	Ea.	.313	38.50	13.65	52.15
Maximum	1.000	Ea.	.383	47.50	16.65	64.15
36" long, minimum	1.000	Ea.	.381	96	16.60	112.60
Maximum	1.000	Ea.	.419	106	18.25	124.25

7 | SPECIALTIES 24 | Masonry Fireplace Systems

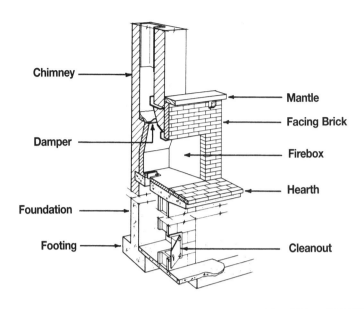

System Description	QUAN.	UNIT	LABOR HOURS	COST EACH		
				MAT.	INST.	TOTAL
MASONRY FIREPLACE						
Footing, 8" thick, concrete, 4' x 7'	.700	C.Y.	2.110	110.60	79.73	190.33
Foundation, concrete block, 32" x 60" x 4' deep	1.000	Ea.	5.275	150	210.60	360.60
Fireplace, brick firebox, 30" x 29" opening	1.000	Ea.	40.000	530	1,550	2,080
Damper, cast iron, 30" opening	1.000	Ea.	1.333	87.50	59	146.50
Facing brick, standard size brick, 6' x 5'	30.000	S.F.	5.217	127.20	208.50	335.70
Hearth, standard size brick, 3' x 6'	1.000	Ea.	8.000	195	310	505
Chimney, standard size brick, 8" x 12" flue, one story house	12.000	V.L.F.	12.000	384	468	852
Mantle, 4" x 8", wood	6.000	L.F.	1.333	36	58.20	94.20
Cleanout, cast iron, 8" x 8"	1.000	Ea.	.667	37	29.50	66.50
TOTAL		Ea.	75.935	1,657.30	2,973.53	4,630.83

The costs in this system are on a cost each basis.

Description	QUAN.	UNIT	LABOR HOURS	COST EACH		
				MAT.	INST.	TOTAL

7 | SPECIALTIES 24 | Masonry Fireplace Systems

Masonry Fireplace Price Sheet	QUAN.	UNIT	LABOR HOURS	COST EACH MAT.	COST EACH INST.	COST EACH TOTAL
Footing 8" thick, 3' x 6'	.440	C.Y.	1.326	69.50	50	119.50
4' x 7'	.700	C.Y.	2.110	111	79.50	190.50
5' x 8'	1.000	C.Y.	3.014	158	114	272
1' thick, 3' x 6'	.670	C.Y.	2.020	106	76	182
4' x 7'	1.030	C.Y.	3.105	163	117	280
5' x 8'	1.480	C.Y.	4.461	234	168	402
Foundation-concrete block, 24" x 48", 4' deep	1.000	Ea.	4.267	120	168	288
8' deep	1.000	Ea.	8.533	240	335	575
24" x 60", 4' deep	1.000	Ea.	4.978	140	197	337
8' deep	1.000	Ea.	9.956	280	395	675
32" x 48", 4' deep	1.000	Ea.	4.711	133	186	319
8' deep	1.000	Ea.	9.422	265	370	635
32" x 60", 4' deep	1.000	Ea.	5.333	150	211	361
8' deep	1.000	Ea.	10.845	305	430	735
32" x 72", 4' deep	1.000	Ea.	6.133	173	242	415
8' deep	1.000	Ea.	12.267	345	485	830
Fireplace, brick firebox 30" x 29" opening	1.000	Ea.	40.000	530	1,550	2,080
48" x 30" opening	1.000	Ea.	60.000	795	2,325	3,120
Steel fire box with registers, 25" opening	1.000	Ea.	26.667	950	1,050	2,000
48" opening	1.000	Ea.	44.000	1,575	1,725	3,300
Damper, cast iron, 30" opening	1.000	Ea.	1.333	87.50	59	146.50
36" opening	1.000	Ea.	1.556	102	69	171
Steel, 30" opening	1.000	Ea.	1.333	78.50	59	137.50
36" opening	1.000	Ea.	1.556	91.50	69	160.50
Facing for fireplace, standard size brick, 6' x 5'	30.000	S.F.	5.217	127	209	336
7' x 5'	35.000	S.F.	6.087	148	243	391
8' x 6'	48.000	S.F.	8.348	204	335	539
Fieldstone, 6' x 5'	30.000	S.F.	5.217	545	209	754
7' x 5'	35.000	S.F.	6.087	635	243	878
8' x 6'	48.000	S.F.	8.348	870	335	1,205
Sheetrock on metal, studs, 6' x 5'	30.000	S.F.	.980	26.50	43	69.50
7' x 5'	35.000	S.F.	1.143	31	50	81
8' x 6'	48.000	S.F.	1.568	42.50	68.50	111
Hearth, standard size brick, 3' x 6'	1.000	Ea.	8.000	195	310	505
3' x 7'	1.000	Ea.	9.280	226	360	586
3' x 8'	1.000	Ea.	10.640	259	410	669
Stone, 3' x 6'	1.000	Ea.	8.000	210	310	520
3' x 7'	1.000	Ea.	9.280	244	360	604
3' x 8'	1.000	Ea.	10.640	280	410	690
Chimney, standard size brick, 8" x 12" flue, one story house	12.000	V.L.F.	12.000	385	470	855
Two story house	20.000	V.L.F.	20.000	640	780	1,420
Mantle wood, beams, 4" x 8"	6.000	L.F.	1.333	36	58	94
4" x 10"	6.000	L.F.	1.371	43	59.50	102.50
Ornate, prefabricated, 6' x 3'-6" opening, minimum	1.000	Ea.	1.600	162	70	232
Maximum	1.000	Ea.	1.600	202	70	272
Cleanout, door and frame, cast iron, 8" x 8"	1.000	Ea.	.667	37	29.50	66.50
12" x 12"	1.000	Ea.	.800	45	35.50	80.50

7 | SPECIALTIES — 30 | Prefabricated Fireplace Systems

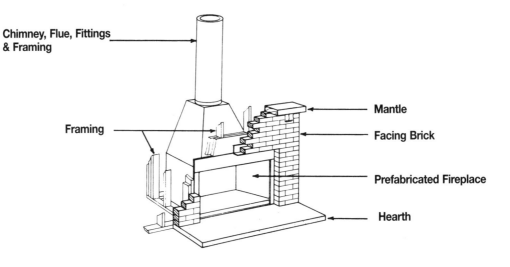

System Description	QUAN.	UNIT	LABOR HOURS	COST EACH MAT.	COST EACH INST.	COST EACH TOTAL
PREFABRICATED FIREPLACE						
Prefabricated fireplace, metal, minimum	1.000	Ea.	6.154	1,175	268	1,443
Framing, 2" x 4" studs, 6' x 5'	35.000	L.F.	.509	14.35	22.05	36.40
Fire resistant gypsum drywall, unfinished	40.000	S.F.	.320	16.80	14	30.80
Drywall finishing adder	40.000	S.F.	.320	1.60	14	15.60
Facing, brick, standard size brick, 6' x 5'	30.000	S.F.	5.217	127.20	208.50	335.70
Hearth, standard size brick, 3' x 6'	1.000	Ea.	8.000	195	310	505
Chimney, one story house, framing, 2" x 4" studs	80.000	L.F.	1.164	32.80	50.40	83.20
Sheathing, plywood, 5/8" thick	32.000	S.F.	.758	84.80	32.96	117.76
Flue, 10" metal, insulated pipe	12.000	V.L.F.	4.000	306	174	480
Fittings, ceiling support	1.000	Ea.	.667	149	29	178
Fittings, joist shield	1.000	Ea.	.667	85.50	29	114.50
Fittings, roof flashing	1.000	Ea.	.667	173	29	202
Mantle beam, wood, 4" x 8"	6.000	L.F.	1.333	36	58.20	94.20
TOTAL		Ea.	29.776	2,397.05	1,239.11	3,636.16

The costs in this system are on a cost each basis.

Description	QUAN.	UNIT	LABOR HOURS	COST EACH MAT.	COST EACH INST.	COST EACH TOTAL

7 | SPECIALTIES 30 | Prefabricated Fireplace Systems

Prefabricated Fireplace Price Sheet	QUAN.	UNIT	LABOR HOURS	COST EACH MAT.	COST EACH INST.	TOTAL
Prefabricated fireplace, minimum	1.000	Ea.	6.154	1,175	268	1,443
Average	1.000	Ea.	8.000	1,700	350	2,050
Maximum	1.000	Ea.	8.889	4,200	390	4,590
Framing, 2" x 4" studs, fireplace, 6' x 5'	35.000	L.F.	.509	14.35	22	36.35
7' x 5'	40.000	L.F.	.582	16.40	25	41.40
8' x 6'	45.000	L.F.	.655	18.45	28.50	46.95
Sheetrock, 1/2" thick, fireplace, 6' x 5'	40.000	S.F.	.640	18.40	28	46.40
7' x 5'	45.000	S.F.	.720	20.50	31.50	52
8' x 6'	50.000	S.F.	.800	23	35	58
Facing for fireplace, brick, 6' x 5'	30.000	S.F.	5.217	127	209	336
7' x 5'	35.000	S.F.	6.087	148	243	391
8' x 6'	48.000	S.F.	8.348	204	335	539
Fieldstone, 6' x 5'	30.000	S.F.	5.217	585	209	794
7' x 5'	35.000	S.F.	6.087	685	243	928
8' x 6'	48.000	S.F.	8.348	940	335	1,275
Hearth, standard size brick, 3' x 6'	1.000	Ea.	8.000	195	310	505
3' x 7'	1.000	Ea.	9.280	226	360	586
3' x 8'	1.000	Ea.	10.640	259	410	669
Stone, 3' x 6'	1.000	Ea.	8.000	210	310	520
3' x 7'	1.000	Ea.	9.280	244	360	604
3' x 8'	1.000	Ea.	10.640	280	410	690
Chimney, framing, 2" x 4", one story house	80.000	L.F.	1.164	33	50.50	83.50
Two story house	120.000	L.F.	1.746	49	75.50	124.50
Sheathing, plywood, 5/8" thick	32.000	S.F.	.758	85	33	118
Stucco on plywood	32.000	S.F.	1.125	41	47.50	88.50
Flue, 10" metal pipe, insulated, one story house	12.000	V.L.F.	4.000	305	174	479
Two story house	20.000	V.L.F.	6.667	510	290	800
Fittings, ceiling support	1.000	Ea.	.667	149	29	178
Fittings joist sheild, one story house	1.000	Ea.	.667	85.50	29	114.50
Two story house	2.000	Ea.	1.333	171	58	229
Fittings roof flashing	1.000	Ea.	.667	173	29	202
Mantle, wood beam, 4" x 8"	6.000	L.F.	1.333	36	58	94
4" x 10"	6.000	L.F.	1.371	43	59.50	102.50
Ornate prefabricated, 6' x 3'-6" opening, minimum	1.000	Ea.	1.600	162	70	232
Maximum	1.000	Ea.	1.600	202	70	272

7 | SPECIALTIES   32 | Greenhouse Systems

System Description	QUAN.	UNIT	LABOR HOURS	COST EACH MAT.	COST EACH INST.	COST EACH TOTAL
Economy, lean to, shell only, not including 2' stub wall, fndtn, flrs, heat						
4' x 16'	1.000	Ea.	26.212	2,525	1,150	3,675
4' x 24'	1.000	Ea.	30.259	2,925	1,325	4,250
6' x 10'	1.000	Ea.	16.552	2,100	725	2,825
6' x 16'	1.000	Ea.	23.034	2,925	1,000	3,925
6' x 24'	1.000	Ea.	29.793	3,775	1,300	5,075
8' x 10'	1.000	Ea.	22.069	2,800	965	3,765
8' x 16'	1.000	Ea.	38.400	4,875	1,675	6,550
8' x 24'	1.000	Ea.	49.655	6,300	2,175	8,475
Free standing, 8' x 8'	1.000	Ea.	17.356	3,300	755	4,055
8' x 16'	1.000	Ea.	30.211	5,725	1,325	7,050
8' x 24'	1.000	Ea.	39.051	7,425	1,700	9,125
10' x 10'	1.000	Ea.	18.824	3,950	820	4,770
10' x 16'	1.000	Ea.	24.095	5,050	1,050	6,100
10' x 24'	1.000	Ea.	31.624	6,625	1,375	8,000
14' x 10'	1.000	Ea.	20.741	4,900	905	5,805
14' x 16'	1.000	Ea.	24.889	5,875	1,075	6,950
14' x 24'	1.000	Ea.	33.349	7,875	1,450	9,325
Standard, lean to, shell only, not incl. 2' stub wall, fndtn, flrs, heat 4'x10'	1.000	Ea.	28.235	2,725	1,225	3,950
4' x 16'	1.000	Ea.	39.341	3,800	1,725	5,525
4' x 24'	1.000	Ea.	45.412	4,400	1,975	6,375
6' x 10'	1.000	Ea.	24.827	3,150	1,075	4,225
6' x 16'	1.000	Ea.	34.538	4,375	1,500	5,875
6' x 24'	1.000	Ea.	44.689	5,675	1,950	7,625
8' x 10'	1.000	Ea.	33.103	4,200	1,450	5,650
8' x 16'	1.000	Ea.	57.600	7,300	2,525	9,825
8' x 24'	1.000	Ea.	74.482	9,450	3,250	12,700
Free standing, 8' x 8'	1.000	Ea.	26.034	4,950	1,125	6,075
8' x 16'	1.000	Ea.	45.316	8,600	1,975	10,575
8' x 24'	1.000	Ea.	58.577	11,100	2,550	13,650
10' x 10'	1.000	Ea.	28.236	5,925	1,225	7,150
10' x 16'	1.000	Ea.	36.142	7,575	1,575	9,150
10' x 24'	1.000	Ea.	47.436	9,950	2,075	12,025
14' x 10'	1.000	Ea.	31.112	7,350	1,350	8,700
14' x 16'	1.000	Ea.	37.334	8,825	1,625	10,450
14' x 24'	1.000	Ea.	50.030	11,800	2,175	13,975
Deluxe, lean to, shell only, not incl. 2' stub wall, fndtn, flrs or heat, 4'x10'	1.000	Ea.	20.645	4,675	900	5,575
4' x 16'	1.000	Ea.	33.032	7,500	1,450	8,950
4' x 24'	1.000	Ea.	49.548	11,200	2,150	13,350
6' x 10'	1.000	Ea.	30.968	7,025	1,350	8,375
6' x 16'	1.000	Ea.	49.548	11,200	2,150	13,350
6' x 24'	1.000	Ea.	74.323	16,800	3,250	20,050
8' x 10'	1.000	Ea.	41.290	9,350	1,800	11,150
8' x 16'	1.000	Ea.	66.065	15,000	2,875	17,875
8' x 24'	1.000	Ea.	99.097	22,500	4,325	26,825
Freestanding, 8' x 8'	1.000	Ea.	18.618	6,475	815	7,290
8' x 16'	1.000	Ea.	37.236	12,900	1,625	14,525
8' x 24'	1.000	Ea.	55.855	19,400	2,450	21,850
10' x 10'	1.000	Ea.	29.091	10,100	1,275	11,375
10' x 16'	1.000	Ea.	46.546	16,200	2,025	18,225
10' x 24'	1.000	Ea.	69.818	24,200	3,050	27,250
14' x 10'	1.000	Ea.	40.727	14,100	1,775	15,875
14' x 16'	1.000	Ea.	65.164	22,600	2,850	25,450
14' x 24'	1.000	Ea.	97.746	33,900	4,275	38,175

7 | SPECIALTIES   36 | Swimming Pool Systems

System Description	QUAN.	UNIT	LABOR HOURS	COST EACH MAT.	COST EACH INST.	COST EACH TOTAL
Swimming pools, vinyl lined, metal sides, sand bottom, 12' x 28'	1.000	Ea.	50.177	3,625	2,225	5,850
12' x 32'	1.000	Ea.	55.366	4,000	2,450	6,450
12' x 36'	1.000	Ea.	60.061	4,350	2,675	7,025
16' x 32'	1.000	Ea.	66.798	4,825	2,975	7,800
16' x 36'	1.000	Ea.	71.190	5,150	3,150	8,300
16' x 40'	1.000	Ea.	74.703	5,400	3,300	8,700
20' x 36'	1.000	Ea.	77.860	5,625	3,475	9,100
20' x 40'	1.000	Ea.	82.135	5,950	3,650	9,600
20' x 44'	1.000	Ea.	90.348	6,525	4,025	10,550
24' x 40'	1.000	Ea.	98.562	7,125	4,375	11,500
24' x 44'	1.000	Ea.	108.418	7,850	4,825	12,675
24' x 48'	1.000	Ea.	118.274	8,550	5,250	13,800
Vinyl lined, concrete sides, 12' x 28'	1.000	Ea.	79.447	5,750	3,525	9,275
12' x 32'	1.000	Ea.	88.818	6,425	3,925	10,350
12' x 36'	1.000	Ea.	97.656	7,075	4,325	11,400
16' x 32'	1.000	Ea.	111.393	8,050	4,950	13,000
16' x 36'	1.000	Ea.	121.354	8,775	5,400	14,175
16' x 40'	1.000	Ea.	130.445	9,425	5,800	15,225
28' x 36'	1.000	Ea.	140.585	10,200	6,225	16,425
20' x 40'	1.000	Ea.	149.336	10,800	6,625	17,425
20' x 44'	1.000	Ea.	164.270	11,900	7,300	19,200
24' x 40'	1.000	Ea.	179.203	13,000	7,950	20,950
24' x 44'	1.000	Ea.	197.124	14,300	8,750	23,050
24' x 48'	1.000	Ea.	215.044	15,600	9,550	25,150
Gunite, bottom and sides, 12' x 28'	1.000	Ea.	129.767	8,400	5,750	14,150
12' x 32'	1.000	Ea.	142.164	9,200	6,300	15,500
12' x 36'	1.000	Ea.	153.028	9,900	6,750	16,650
16' x 32'	1.000	Ea.	167.743	10,900	7,400	18,300
16' x 36'	1.000	Ea.	176.421	11,400	7,800	19,200
16' x 40'	1.000	Ea.	182.368	11,800	8,075	19,875
20' x 36'	1.000	Ea.	187.949	12,200	8,325	20,525
20' x 40'	1.000	Ea.	179.200	16,000	7,925	23,925
20' x 44'	1.000	Ea.	197.120	17,600	8,700	26,300
24' x 40'	1.000	Ea.	215.040	19,200	9,500	28,700
24' x 44'	1.000	Ea.	273.244	17,700	12,100	29,800
24' x 48'	1.000	Ea.	298.077	19,300	13,200	32,500

7 | SPECIALTIES 40 | Wood Deck Systems

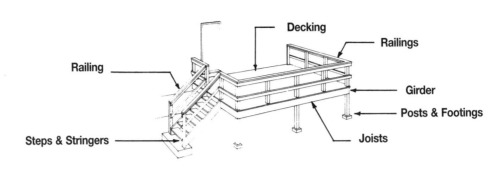

System Description	QUAN.	UNIT	LABOR HOURS	COST PER S.F.		
				MAT.	INST.	TOTAL
8' X 12' DECK, PRESSURE TREATED LUMBER, JOISTS 16" O.C.						
Decking, 2" x 6" lumber	2.080	L.F.	.027	1.31	1.16	2.47
Lumber preservative	2.080	L.F.		.31		.31
Joists, 2" x 8", 16" O.C.	1.000	L.F.	.015	.92	.63	1.55
Lumber preservative	1.000	L.F.		.20		.20
Girder, 2" x 10"	.125	L.F.	.002	.16	.10	.26
Lumber preservative	.125	L.F.		.03		.03
Hand excavation for footings	.250	L.F.	.006		.19	.19
Concrete footings	.250	L.F.	.006	.32	.23	.55
4" x 4" Posts	.250	L.F.	.010	.46	.45	.91
Lumber preservative	.250	L.F.		.05		.05
Framing, pressure treated wood stairs, 3' wide, 8 closed risers	1.000	Set	.080	1.22	3.50	4.72
Railings, 2" x 4"	1.000	L.F.	.026	.41	1.13	1.54
Lumber preservative	1.000	L.F.		.10		.10
TOTAL		S.F.	.172	5.49	7.39	12.88
12' X 16' DECK, PRESSURE TREATED LUMBER, JOISTS 24" O.C.						
Decking, 2" x 6"	2.080	L.F.	.027	1.31	1.16	2.47
Lumber preservative	2.080	L.F.		.31		.31
Joists, 2" x 10", 24" O.C.	.800	L.F.	.014	1.05	.62	1.67
Lumber preservative	.800	L.F.		.20		.20
Girder, 2" x 10"	.083	L.F.	.001	.11	.06	.17
Lumber preservative	.083	L.F.		.02		.02
Hand excavation for footings	.122	L.F.	.006		.19	.19
Concrete footings	.122	L.F.	.006	.32	.23	.55
4" x 4" Posts	.122	L.F.	.005	.22	.22	.44
Lumber preservative	.122	L.F.		.02		.02
Framing, pressure treated wood stairs, 3' wide, 8 closed risers	1.000	Set	.040	.61	1.75	2.36
Railings, 2" x 4"	.670	L.F.	.017	.27	.76	1.03
Lumber preservative	.670	L.F.		.07		.07
TOTAL		S.F.	.116	4.51	4.99	9.50
12' X 24' DECK, REDWOOD OR CEDAR, JOISTS 16" O.C.						
Decking, 2" x 6" redwood	2.080	L.F.	.027	7.78	1.16	8.94
Joists, 2" x 10", 16" O.C.	1.000	L.F.	.018	8	.78	8.78
Girder, 2" x 10"	.083	L.F.	.001	.66	.06	.72
Hand excavation for footings	.111	L.F.	.006		.19	.19
Concrete footings	.111	L.F.	.006	.32	.23	.55
Lumber preservative	.111	L.F.		.02		.02
Post, 4" x 4", including concrete footing	.111	L.F.	.009	3.50	.39	3.89
Framing, redwood or cedar stairs, 3' wide, 8 closed risers	1.000	Set	.028	2.07	1.23	3.30
Railings, 2" x 4"	.540	L.F.	.005	1.35	.20	1.55
TOTAL		S.F.	.100	23.70	4.24	27.94

The costs in this system are on a square foot basis.

7 | SPECIALTIES 40 | Wood Deck Systems

Wood Deck Price Sheet	QUAN.	UNIT	LABOR HOURS	COST PER S.F. MAT.	INST.	TOTAL
Decking, treated lumber, 1" x 4"	3.430	L.F.	.031	2.78	1.37	4.15
1" x 6"	2.180	L.F.	.033	2.92	1.44	4.36
2" x 4"	3.200	L.F.	.041	1.64	1.79	3.43
2" x 6"	2.080	L.F.	.027	1.62	1.16	2.78
Redwood or cedar,, 1" x 4"	3.430	L.F.	.035	3.26	1.51	4.77
1" x 6"	2.180	L.F.	.036	3.40	1.58	4.98
2" x 4"	3.200	L.F.	.028	8.25	1.23	9.48
2" x 6"	2.080	L.F.	.027	7.80	1.16	8.96
Joists for deck, treated lumber, 2" x 8", 16" O.C.	1.000	L.F.	.015	1.12	.63	1.75
24" O.C.	.800	L.F.	.012	.90	.50	1.40
2" x 10", 16" O.C.	1.000	L.F.	.018	1.56	.78	2.34
24" O.C.	.800	L.F.	.014	1.25	.62	1.87
Redwood or cedar, 2" x 8", 16" O.C.	1.000	L.F.	.015	4.98	.63	5.61
24" O.C.	.800	L.F.	.012	3.98	.50	4.48
2" x 10", 16" O.C.	1.000	L.F.	.018	8	.78	8.78
24" O.C.	.800	L.F.	.014	6.40	.62	7.02
Girder for joists, treated lumber, 2" x 10", 8' x 12' deck	.125	L.F.	.002	.19	.10	.29
12' x 16' deck	.083	L.F.	.001	.13	.06	.19
12' x 24' deck	.083	L.F.	.001	.13	.06	.19
Redwood or cedar, 2" x 10", 8' x 12' deck	.125	L.F.	.002	1	.10	1.10
12' x 16' deck	.083	L.F.	.001	.66	.06	.72
12' x 24' deck	.083	L.F.	.001	.66	.06	.72
Posts, 4" x 4", including concrete footing, 8' x 12' deck	.250	S.F.	.022	.83	.87	1.70
12' x 16' deck	.122	L.F.	.017	.56	.64	1.20
12' x 24' deck	.111	L.F.	.017	.54	.62	1.16
Stairs 2" x 10" stringers, treated lumber, 8' x 12' deck	1.000	Set	.020	1.22	3.50	4.72
12' x 16' deck	1.000	Set	.012	.61	1.75	2.36
12' x 24' deck	1.000	Set	.008	.43	1.23	1.66
Redwood or cedar, 8' x 12' deck	1.000	Set	.040	5.90	3.50	9.40
12' x 16' deck	1.000	Set	.020	2.95	1.75	4.70
12' x 24' deck	1.000	Set	.012	2.07	1.23	3.30
Railings 2" x 4", treated lumber, 8' x 12' deck	1.000	L.F.	.026	.51	1.13	1.64
12' x 16' deck	.670	L.F.	.017	.34	.76	1.10
12' x 24' deck	.540	L.F.	.014	.27	.61	.88
Redwood or cedar, 8' x 12' deck	1.000	L.F.	.009	2.51	.38	2.89
12' x 16' deck	.670	L.F.	.006	1.65	.25	1.90
12' x 24' deck	.540	L.F.	.005	1.35	.20	1.55

Division 8
Mechanical

8 | MECHANICAL — 04 | Two Fixture Lavatory Systems

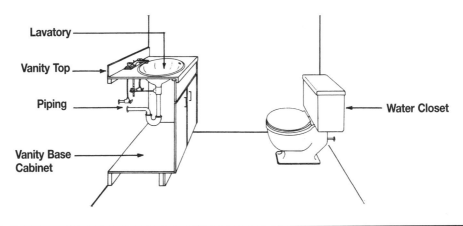

System Description	QUAN.	UNIT	LABOR HOURS	COST EACH MAT.	COST EACH INST.	COST EACH TOTAL
LAVATORY INSTALLED WITH VANITY, PLUMBING IN 2 WALLS						
Water closet, floor mounted, 2 piece, close coupled, white	1.000	Ea.	3.019	201	132	333
Rough-in, vent, 2" diameter DWV piping	1.000	Ea.	.955	24.80	41.80	66.60
Waste, 4" diameter DWV piping	1.000	Ea.	.828	31.95	36.15	68.10
Supply, 1/2" diameter type "L" copper supply piping	1.000	Ea.	.593	22.56	28.80	51.36
Lavatory, 20" x 18", P.E. cast iron white	1.000	Ea.	2.500	242	109	351
Rough-in, vent, 1-1/2" diameter DWV piping	1.000	Ea.	.901	24	39.40	63.40
Waste, 2" diameter DWV piping	1.000	Ea.	.955	24.80	41.80	66.60
Supply, 1/2" diameter type "L" copper supply piping	1.000	Ea.	.988	37.60	48	85.60
Piping, supply, 1/2" diameter type "L" copper supply piping	10.000	L.F.	.988	37.60	48	85.60
Waste, 4" diameter DWV piping	7.000	L.F.	1.931	74.55	84.35	158.90
Vent, 2" diameter DWV piping	12.000	L.F.	2.866	74.40	125.40	199.80
Vanity base cabinet, 2 door, 30" wide	1.000	Ea.	1.000	249	43.50	292.50
Vanity top, plastic & laminated, square edge	2.670	L.F.	.712	89.45	31.11	120.56
TOTAL		Ea.	18.236	1,133.71	809.31	1,943.02
LAVATORY WITH WALL-HUNG LAVATORY, PLUMBING IN 2 WALLS						
Water closet, floor mounted, 2 piece close coupled, white	1.000	Ea.	3.019	201	132	333
Rough-in, vent, 2" diameter DWV piping	1.000	Ea.	.955	24.80	41.80	66.60
Waste, 4" diameter DWV piping	1.000	Ea.	.828	31.95	36.15	68.10
Supply, 1/2" diameter type "L" copper supply piping	1.000	Ea.	.593	22.56	28.80	51.36
Lavatory, 20" x 18", P.E. cast iron, wall hung, white	1.000	Ea.	2.000	305	87.50	392.50
Rough-in, vent, 1-1/2" diameter DWV piping	1.000	Ea.	.901	24	39.40	63.40
Waste, 2" diameter DWV piping	1.000	Ea.	.955	24.80	41.80	66.60
Supply, 1/2" diameter type "L" copper supply piping	1.000	Ea.	.988	37.60	48	85.60
Piping, supply, 1/2" diameter type "L" copper supply piping	10.000	L.F.	.988	37.60	48	85.60
Waste, 4" diameter DWV piping	7.000	L.F.	1.931	74.55	84.35	158.90
Vent, 2" diameter DWV piping	12.000	L.F.	2.866	74.40	125.40	199.80
Carrier, steel for studs, no arms	1.000	Ea.	1.143	56	55.50	111.50
TOTAL		Ea.	17.167	914.26	768.70	1,682.96

Description	QUAN.	UNIT	LABOR HOURS	COST EACH MAT.	COST EACH INST.	COST EACH TOTAL

Two Fixture Lavatory Price Sheet	QUAN.	UNIT	LABOR HOURS	COST EACH		
				MAT.	INST.	TOTAL
Water closet, close coupled standard 2 piece, white	1.000	Ea.	3.019	201	132	333
Color	1.000	Ea.	3.019	238	132	370
One piece elongated bowl, white	1.000	Ea.	3.019	585	132	717
Color	1.000	Ea.	3.019	720	132	852
Low profile, one piece elongated bowl, white	1.000	Ea.	3.019	810	132	942
Color	1.000	Ea.	3.019	1,050	132	1,182
Rough-in for water closet						
1/2" copper supply, 4" cast iron waste, 2" cast iron vent	1.000	Ea.	2.376	79.50	107	186.50
4" PVC waste, 2" PVC vent	1.000	Ea.	2.678	45.50	120	165.50
4" copper waste, 2" copper vent	1.000	Ea.	2.520	245	117	362
3" cast iron waste, 1-1/2" cast iron vent	1.000	Ea.	2.244	71.50	101	172.50
3" PVC waste, 1-1/2" PVC vent	1.000	Ea.	2.388	40	112	152
3" copper waste, 1-1/2" copper vent	1.000	Ea.	2.524	229	113	342
1/2" PVC supply, 4" PVC waste, 2" PVC vent	1.000	Ea.	2.974	38.50	134	172.50
3" PVC waste, 1-1/2" PVC vent	1.000	Ea.	2.684	33	126	159
1/2" steel supply, 4" cast iron waste, 2" cast iron vent	1.000	Ea.	2.545	73	115	188
4" cast iron waste, 2" steel vent	1.000	Ea.	2.590	85.50	117	202.50
4" PVC waste, 2" PVC vent	1.000	Ea.	2.847	39	128	167
Lavatory, vanity top mounted, P.E. on cast iron 20" x 18" white	1.000	Ea.	2.500	242	109	351
Color	1.000	Ea.	2.500	273	109	382
Steel, enameled 10" x 17" white	1.000	Ea.	2.759	160	121	281
Color	1.000	Ea.	2.500	165	109	274
Vitreous china 20" x 16", white	1.000	Ea.	2.963	291	130	421
Color	1.000	Ea.	2.963	291	130	421
Wall hung, P.E. on cast iron, 20" x 18", white	1.000	Ea.	2.000	305	87.50	392.50
Color	1.000	Ea.	2.000	320	87.50	407.50
Vitreous china 19" x 17", white	1.000	Ea.	2.286	197	100	297
Color	1.000	Ea.	2.286	220	100	320
Rough-in supply waste and vent for lavatory						
1/2" copper supply, 2" cast iron waste, 1-1/2" cast iron vent	1.000	Ea.	2.844	86.50	129	215.50
2" PVC waste, 1-1/2" PVC vent	1.000	Ea.	2.962	51.50	139	190.50
2" copper waste, 1-1/2" copper vent	1.000	Ea.	2.308	156	112	268
1-1/2" PVC waste, 1-1/4" PVC vent	1.000	Ea.	2.639	51	128	179
1-1/2" copper waste, 1-1/4" copper vent	1.000	Ea.	2.114	129	103	232
1/2" PVC supply, 2" PVC waste, 1-1/2" PVC vent	1.000	Ea.	3.456	39.50	163	202.50
1-1/2" PVC waste, 1-1/4" PVC vent	1.000	Ea.	3.133	39	152	191
1/2" steel supply, 2" cast iron waste, 1-1/2" cast iron vent	1.000	Ea.	3.126	75.50	143	218.50
2" cast iron waste, 2" steel vent	1.000	Ea.	3.225	89	147	236
2" PVC waste, 1-1/2" PVC vent	1.000	Ea.	3.244	41	152	193
1-1/2" PVC waste, 1-1/4" PVC vent	1.000	Ea.	2.921	40.50	142	182.50
Piping, supply, 1/2" copper, type "L"	10.000	L.F.	.988	37.50	48	85.50
1/2" steel	10.000	L.F.	1.270	27	61.50	88.50
1/2" PVC	10.000	L.F.	1.482	25.50	72	97.50
Waste, 4" cast iron	7.000	L.F.	1.931	74.50	84.50	159
4" copper	7.000	L.F.	2.800	360	123	483
4" PVC	7.000	L.F.	2.333	35.50	102	137.50
Vent, 2" cast iron	12.000	L.F.	2.866	74.50	125	199.50
2" copper	12.000	L.F.	2.182	204	106	310
2" PVC	12.000	L.F.	3.254	23	142	165
2" steel	12.000	Ea.	3.000	112	131	243
Vanity base cabinet, 2 door, 24" x 30"	1.000	Ea.	1.000	249	43.50	292.50
24" x 36"	1.000	Ea.	1.200	330	52.50	382.50
Vanity top, laminated plastic, square edge 25" x 32"	2.670	L.F.	.712	89.50	31	120.50
25" x 38"	3.170	L.F.	.845	106	37	143
Post formed, laminated plastic, 25" x 32"	2.670	L.F.	.712	29	31	60
25" x 38"	3.170	L.F.	.845	34	37	71
Cultured marble, 25" x 32" with bowl	1.000	Ea.	2.500	200	109	309
25" x 38" with bowl	1.000	Ea.	2.500	237	109	346
Carrier for lavatory, steel for studs	1.000	Ea.	1.143	56	55.50	111.50
Wood 2" x 8" blocking	1.330	L.F.	.053	1.22	2.31	3.53

8 | MECHANICAL   12 | Three Fixture Bathroom Systems

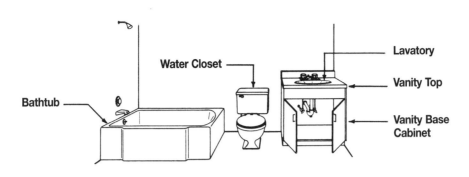

System Description	QUAN.	UNIT	LABOR HOURS	COST EACH MAT.	COST EACH INST.	COST EACH TOTAL
BATHROOM INSTALLED WITH VANITY						
Water closet, floor mounted, 2 piece, close coupled, white	1.000	Ea.	3.019	201	132	333
Rough-in, waste, 4" diameter DWV piping	1.000	Ea.	.828	31.95	36.15	68.10
Vent, 2" diameter DWV piping	1.000	Ea.	.955	24.80	41.80	66.60
Supply, 1/2" diameter type "L" copper supply piping	1.000	Ea.	.593	22.56	28.80	51.36
Lavatory, 20" x 18", P.E. cast iron with accessories, white	1.000	Ea.	2.500	242	109	351
Rough-in, supply, 1/2" diameter type "L" copper supply piping	1.000	Ea.	.988	37.60	48	85.60
Waste, 1-1/2" diameter DWV piping	1.000	Ea.	1.803	48	78.80	126.80
Bathtub, P.E. cast iron, 5' long with accessories, white	1.000	Ea.	3.636	830	159	989
Rough-in, waste, 4" diameter DWV piping	1.000	Ea.	.828	31.95	36.15	68.10
Vent, 1-1/2" diameter DWV piping	1.000	Ea.	.593	50.80	28.80	79.60
Supply, 1/2" diameter type "L" copper supply piping	1.000	Ea.	.988	37.60	48	85.60
Piping, supply, 1/2" diameter type "L" copper supply piping	20.000	L.F.	1.975	75.20	96	171.20
Waste, 4" diameter DWV piping	9.000	L.F.	2.483	95.85	108.45	204.30
Vent, 2" diameter DWV piping	6.000	L.F.	1.500	56.10	65.70	121.80
Vanity base cabinet, 2 door, 30" wide	1.000	Ea.	1.000	249	43.50	292.50
Vanity top, plastic laminated square edge	2.670	L.F.	.712	70.76	31.11	101.87
TOTAL		Ea.	24.401	2,105.17	1,091.26	3,196.43
BATHROOM WITH WALL HUNG LAVATORY						
Water closet, floor mounted, 2 piece, close coupled, white	1.000	Ea.	3.019	201	132	333
Rough-in, vent, 2" diameter DWV piping	1.000	Ea.	.955	24.80	41.80	66.60
Waste, 4" diameter DWV piping	1.000	Ea.	.828	31.95	36.15	68.10
Supply, 1/2" diameter type "L" copper supply piping	1.000	Ea.	.593	22.56	28.80	51.36
Lavatory, 20" x 18" P.E. cast iron, wall hung, white	1.000	Ea.	2.000	305	87.50	392.50
Rough-in, waste, 1-1/2" diameter DWV piping	1.000	Ea.	1.803	48	78.80	126.80
Supply, 1/2" diameter type "L" copper supply piping	1.000	Ea.	.988	37.60	48	85.60
Bathtub, P.E. cast iron, 5' long with accessories, white	1.000	Ea.	3.636	830	159	989
Rough-in, waste, 4" diameter DWV piping	1.000	Ea.	.828	31.95	36.15	68.10
Supply, 1/2" diameter type "L" copper supply piping	1.000	Ea.	.988	37.60	48	85.60
Vent, 1-1/2" diameter DWV piping	1.000	Ea.	1.482	127	72	199
Piping, supply, 1/2" diameter type "L" copper supply piping	20.000	L.F.	1.975	75.20	96	171.20
Waste, 4" diameter DWV piping	9.000	L.F.	2.483	95.85	108.45	204.30
Vent, 2" diameter DWV piping	6.000	L.F.	1.500	56.10	65.70	121.80
Carrier, steel, for studs, no arms	1.000	Ea.	1.143	56	55.50	111.50
TOTAL		Ea.	24.221	1,980.61	1,093.85	3,074.46

The costs in this system are on a cost each basis. All necessary piping is included.

Three Fixture Bathroom Price Sheet	QUAN.	UNIT	LABOR HOURS	COST EACH MAT.	COST EACH INST.	COST EACH TOTAL
Water closet, close coupled standard 2 piece, white	1.000	Ea.	3.019	201	132	333
Color	1.000	Ea.	3.019	238	132	370
One piece, elongated bowl, white	1.000	Ea.	3.019	585	132	717
Color	1.000	Ea.	3.019	720	132	852
Low profile, one piece elongated bowl, white	1.000	Ea.	3.019	810	132	942
Color	1.000	Ea.	3.019	1,050	132	1,182
Rough-in, for water closet						
1/2" copper supply, 4" cast iron waste, 2" cast iron vent	1.000	Ea.	2.376	79.50	107	186.50
4" PVC/DWV waste, 2" PVC vent	1.000	Ea.	2.678	45.50	120	165.50
4" copper waste, 2" copper vent	1.000	Ea.	2.520	245	117	362
3" cast iron waste, 1-1/2" cast iron vent	1.000	Ea.	2.244	71.50	101	172.50
3" PVC waste, 1-1/2" PVC vent	1.000	Ea.	2.388	40	112	152
3" copper waste, 1-1/2" copper vent	1.000	Ea.	2.014	162	94	256
1/2" PVC supply, 4" PVC waste, 2" PVC vent	1.000	Ea.	2.974	38.50	134	172.50
3" PVC waste, 1-1/2" PVC supply	1.000	Ea.	2.684	33	126	159
1/2" steel supply, 4" cast iron waste, 2" cast iron vent	1.000	Ea.	2.545	73	115	188
4" cast iron waste, 2" steel vent	1.000	Ea.	2.590	85.50	117	202.50
4" PVC waste, 2" PVC vent	1.000	Ea.	2.847	39	128	167
Lavatory, wall hung, P.E. cast iron 20" x 18", white	1.000	Ea.	2.000	305	87.50	392.50
Color	1.000	Ea.	2.000	320	87.50	407.50
Vitreous china 19" x 17", white	1.000	Ea.	2.286	197	100	297
Color	1.000	Ea.	2.286	220	100	320
Lavatory, for vanity top, P.E. cast iron 20" x 18", white	1.000	Ea.	2.500	242	109	351
Color	1.000	Ea.	2.500	273	109	382
Steel, enameled 20" x 17", white	1.000	Ea.	2.759	160	121	281
Color	1.000	Ea.	2.500	165	109	274
Vitreous china 20" x 16", white	1.000	Ea.	2.963	291	130	421
Color	1.000	Ea.	2.963	291	130	421
Rough-in, for lavatory						
1/2" copper supply, 1-1/2" C.I. waste, 1-1/2" C.I. vent	1.000	Ea.	2.791	85.50	127	212.50
1-1/2" PVC waste, 1-1/4" PVC vent	1.000	Ea.	2.639	51	128	179
1/2" steel supply, 1-1/4" cast iron waste, 1-1/4" steel vent	1.000	Ea.	2.890	74.50	132	206.50
1-1/4" PVC@ waste, 1-1/4" PVC vent	1.000	Ea.	2.794	41	136	177
1/2" PVC supply, 1-1/2" PVC waste, 1-1/2" PVC vent	1.000	Ea.	3.260	38.50	158	196.50
Bathtub, P.E. cast iron, 5' long corner with fittings, white	1.000	Ea.	3.636	830	159	989
Color	1.000	Ea.	3.636	1,025	159	1,184
Rough-in, for bathtub						
1/2" copper supply, 4" cast iron waste, 1-1/2" copper vent	1.000	Ea.	2.409	120	113	233
4" PVC waste, 1-1/2" PVC vent	1.000	Ea.	2.877	59	135	194
1/2" steel supply, 4" cast iron waste, 1-1/2" steel vent	1.000	Ea.	2.898	87	133	220
4" PVC waste, 1-1/2" PVC vent	1.000	Ea.	3.159	48.50	149	197.50
1/2" PVC supply, 4" PVC waste, 1-1/2" PVC vent	1.000	Ea.	3.371	47	159	206
Piping, supply 1/2" copper	20.000	L.F.	1.975	75	96	171
1/2" steel	20.000	L.F.	2.540	53.50	123	176.50
1/2" PVC	20.000	L.F.	2.963	51.50	144	195.50
Piping, waste, 4" cast iron no hub	9.000	L.F.	2.483	96	108	204
4" PVC/DWV	9.000	L.F.	3.000	45.50	131	176.50
4" copper/DWV	9.000	L.F.	3.600	465	158	623
Piping, vent 2" cast iron no hub	6.000	L.F.	1.433	37	62.50	99.50
2" copper/DWV	6.000	L.F.	1.091	102	53	155
2" PVC/DWV	6.000	L.F.	1.627	11.50	71	82.50
2" steel, galvanized	6.000	L.F.	1.500	56	65.50	121.50
Vanity base cabinet, 2 door, 24" x 30"	1.000	Ea.	1.000	249	43.50	292.50
24" x 36"	1.000	Ea.	1.200	330	52.50	382.50
Vanity top, laminated plastic square edge 25" x 32"	2.670	L.F.	.712	71	31	102
25" x 38"	3.160	L.F.	.843	83.50	37	120.50
Cultured marble, 25" x 32", with bowl	1.000	Ea.	2.500	200	109	309
25" x 38", with bowl	1.000	Ea.	2.500	237	109	346
Carrier, for lavatory, steel for studs, no arms	1.000	Ea.	1.143	56	55.50	111.50
Wood, 2" x 8" blocking	1.300	L.F.	.052	1.20	2.26	3.46

8 | MECHANICAL 16 | Three Fixture Bathroom Systems

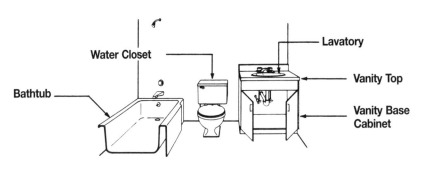

System Description	QUAN.	UNIT	LABOR HOURS	COST EACH MAT.	COST EACH INST.	COST EACH TOTAL
BATHROOM WITH LAVATORY INSTALLED IN VANITY						
Water closet, floor mounted, 2 piece, close coupled, white	1.000	Ea.	3.019	201	132	333
Rough-in, waste, 4" diameter DWV piping	1.000	Ea.	.828	31.95	36.15	68.10
Vent, 2" diameter DWV piping	1.000	Ea.	.955	24.80	41.80	66.60
Supply, 1/2" diameter type "L" copper supply piping	1.000	Ea.	.593	22.56	28.80	51.36
Lavatory, 20" x 18", P.E. cast iron with accessories, white	1.000	Ea.	2.500	242	109	351
Rough-in, waste, 1-1/2" diameter DWV piping	1.000	Ea.	1.803	48	78.80	126.80
Supply, 1/2" diameter type "L" copper supply piping	1.000	Ea.	.988	37.60	48	85.60
Bathtub, P.E. cast iron 5' long with accessories, white	1.000	Ea.	3.636	830	159	989
Rough-in, waste, 4" diameter DWV piping	1.000	Ea.	.828	31.95	36.15	68.10
Vent, 1-1/2" diameter DWV piping	1.000	Ea.	.593	50.80	28.80	79.60
Supply, 1/2" diameter type "L" copper supply piping	1.000	Ea.	.988	37.60	48	85.60
Piping, supply, 1/2" diameter type "L" copper supply piping	10.000	L.F.	.988	37.60	48	85.60
Waste, 4" diameter DWV piping	6.000	L.F.	1.655	63.90	72.30	136.20
Vent, 2" diameter DWV piping	6.000	L.F.	1.500	56.10	65.70	121.80
Vanity base cabinet, 2 door, 30" wide	1.000	Ea.	1.000	249	43.50	292.50
Vanity top, plastic laminated square edge	2.670	L.F.	.712	70.76	31.11	101.87
TOTAL		Ea.	22.586	2,035.62	1,007.11	3,042.73
BATHROOM WITH WALL HUNG LAVATORY						
Water closet, floor mounted, 2 piece, close coupled, white	1.000	Ea.	3.019	201	132	333
Rough-in, vent, 2" diameter DWV piping	1.000	Ea.	.955	24.80	41.80	66.60
Waste, 4" diameter DWV piping	1.000	Ea.	.828	31.95	36.15	68.10
Supply, 1/2" diameter type "L" copper supply piping	1.000	Ea.	.593	22.56	28.80	51.36
Lavatory, 20" x 18" P.E. cast iron, wall hung, white	1.000	Ea.	2.000	305	87.50	392.50
Rough-in, waste, 1-1/2" diameter DWV piping	1.000	Ea.	1.803	48	78.80	126.80
Supply, 1/2" diameter type "L" copper supply piping	1.000	Ea.	.988	37.60	48	85.60
Bathtub, P.E. cast iron, 5' long with accessories, white	1.000	Ea.	3.636	830	159	989
Rough-in, waste, 4" diameter DWV piping	1.000	Ea.	.828	31.95	36.15	68.10
Supply, 1/2" diameter type "L" copper supply piping	1.000	Ea.	.988	37.60	48	85.60
Vent, 1-1/2" diameter DWV piping	1.000	Ea.	.593	50.80	28.80	79.60
Piping, supply, 1/2" diameter type "L" copper supply piping	10.000	L.F.	.988	37.60	48	85.60
Waste, 4" diameter DWV piping	6.000	L.F.	1.655	63.90	72.30	136.20
Vent, 2" diameter DWV piping	6.000	L.F.	1.500	56.10	65.70	121.80
Carrier, steel, for studs, no arms	1.000	Ea.	1.143	56	55.50	111.50
TOTAL		Ea.	21.517	1,834.86	966.50	2,801.36

The costs in this system are on a cost each basis. All necessary piping is included.

Three Fixture Bathroom Price Sheet

	QUAN.	UNIT	LABOR HOURS	COST EACH MAT.	COST EACH INST.	COST EACH TOTAL
Water closet, close coupled standard 2 piece, white	1.000	Ea.	3.019	201	132	333
Color	1.000	Ea.	3.019	238	132	370
One piece elongated bowl, white	1.000	Ea.	3.019	585	132	717
Color	1.000	Ea.	3.019	720	132	852
Low profile, one piece elongated bowl, white	1.000	Ea.	3.019	810	132	942
Color	1.000	Ea.	3.019	1,050	132	1,182
Rough-in for water closet						
1/2" copper supply, 4" cast iron waste, 2" cast iron vent	1.000	Ea.	2.376	79.50	107	186.50
4" PVC/DWV waste, 2" PVC vent	1.000	Ea.	2.678	45.50	120	165.50
4" carrier waste, 2" copper vent	1.000	Ea.	2.520	245	117	362
3" cast iron waste, 1-1/2" cast iron vent	1.000	Ea.	2.244	71.50	101	172.50
3" PVC waste, 1-1/2" PVC vent	1.000	Ea.	2.388	40	112	152
3" copper waste, 1-1/2" copper vent	1.000	Ea.	2.014	162	94	256
1/2" PVC supply, 4" PVC waste, 2" PVC vent	1.000	Ea.	2.974	38.50	134	172.50
3" PVC waste, 1-1/2" PVC supply	1.000	Ea.	2.684	33	126	159
1/2" steel supply, 4" cast iron waste, 2" cast iron vent	1.000	Ea.	2.545	73	115	188
4" cast iron waste, 2" steel vent	1.000	Ea.	2.590	85.50	117	202.50
4" PVC waste, 2" PVC vent	1.000	Ea.	2.847	39	128	167
Lavatory, wall hung, PE cast iron 20" x 18", white	1.000	Ea.	2.000	305	87.50	392.50
Color	1.000	Ea.	2.000	320	87.50	407.50
Vitreous china 19" x 17", white	1.000	Ea.	2.286	197	100	297
Color	1.000	Ea.	2.286	220	100	320
Lavatory, for vanity top, PE cast iron 20" x 18", white	1.000	Ea.	2.500	242	109	351
Color	1.000	Ea.	2.500	273	109	382
Steel enameled 20" x 17", white	1.000	Ea.	2.759	160	121	281
Color	1.000	Ea.	2.500	165	109	274
Vitreous china 20" x 16", white	1.000	Ea.	2.963	291	130	421
Color	1.000	Ea.	2.963	291	130	421
Rough-in for lavatory						
1/2" copper supply, 1-1/2" cast iron waste, 1-1/2" cast iron vent	1.000	Ea.	2.791	85.50	127	212.50
1-1/2" PVC waste, 1-1/4" PVC vent	1.000	Ea.	2.639	51	128	179
1/2" steel supply, 1-1/4" cast iron waste, 1-1/4" steel vent	1.000	Ea.	2.890	74.50	132	206.50
1-1/4" PVC waste, 1-1/4" PVC vent	1.000	Ea.	2.794	41	136	177
1/2" PVC supply, 1-1/2" PVC waste, 1-1/2" PVC vent	1.000	Ea.	3.260	38.50	158	196.50
Bathtub, PE cast iron, 5' long corner with fittings, white	1.000	Ea.	3.636	830	159	989
Color	1.000	Ea.	3.636	1,025	159	1,184
Rough-in for bathtub						
1/2" copper supply, 4" cast iron waste, 1-1/2" copper vent	1.000	Ea.	2.409	120	113	233
4" PVC waste, 1/2" PVC vent	1.000	Ea.	2.877	59	135	194
1/2" steel supply, 4" cast iron waste, 1-1/2" steel vent	1.000	Ea.	2.898	87	133	220
4" PVC waste, 1-1/2" PVC vent	1.000	Ea.	3.159	48.50	149	197.50
1/2" PVC supply, 4" PVC waste, 1-1/2" PVC vent	1.000	Ea.	3.371	47	159	206
Piping supply, 1/2" copper	10.000	L.F.	.988	37.50	48	85.50
1/2" steel	10.000	L.F.	1.270	27	61.50	88.50
1/2" PVC	10.000	L.F.	1.482	25.50	72	97.50
Piping waste, 4" cast iron no hub	6.000	L.F.	1.655	64	72.50	136.50
4" PVC/DWV	6.000	L.F.	2.000	30.50	87.50	118
4" copper/DWV	6.000	L.F.	2.400	310	105	415
Piping vent 2" cast iron no hub	6.000	L.F.	1.433	37	62.50	99.50
2" copper/DWV	6.000	L.F.	1.091	102	53	155
2" PVC/DWV	6.000	L.F.	1.627	11.50	71	82.50
2" steel, galvanized	6.000	L.F.	1.500	56	65.50	121.50
Vanity base cabinet, 2 door, 24" x 30"	1.000	Ea.	1.000	249	43.50	292.50
24" x 36"	1.000	Ea.	1.200	330	52.50	382.50
Vanity top, laminated plastic square edge 25" x 32"	2.670	L.F.	.712	71	31	102
25" x 38"	3.160	L.F.	.843	83.50	37	120.50
Cultured marble, 25" x 32", with bowl	1.000	Ea.	2.500	200	109	309
25" x 38", with bowl	1.000	Ea.	2.500	237	109	346
Carrier, for lavatory, steel for studs, no arms	1.000	Ea.	1.143	56	55.50	111.50
Wood, 2" x 8" blocking	1.300	L.F.	.052	1.20	2.26	3.46

8 | MECHANICAL — 20 | Three Fixture Bathroom Systems

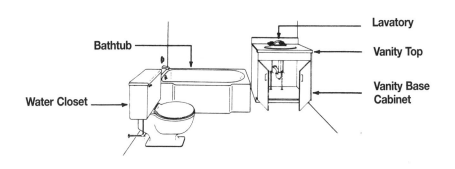

System Description	QUAN.	UNIT	LABOR HOURS	COST EACH MAT.	COST EACH INST.	COST EACH TOTAL
BATHROOM WITH LAVATORY INSTALLED IN VANITY						
Water closet, floor mounted, 2 piece, close coupled, white	1.000	Ea.	3.019	201	132	333
Rough-in, vent, 2" diameter DWV piping	1.000	Ea.	.955	24.80	41.80	66.60
Waste, 4" diameter DWV piping	1.000	Ea.	.828	31.95	36.15	68.10
Supply, 1/2" diameter type "L" copper supply piping	1.000	Ea.	.593	22.56	28.80	51.36
Lavatory, 20" x 18", PE cast iron with accessories, white	1.000	Ea.	2.500	242	109	351
Rough-in, vent, 1-1/2" diameter DWV piping	1.000	Ea.	1.803	48	78.80	126.80
Supply, 1/2" diameter type "L" copper supply piping	1.000	Ea.	.988	37.60	48	85.60
Bathtub, P.E. cast iron, 5' long with accessories, white	1.000	Ea.	3.636	830	159	989
Rough-in, waste, 4" diameter DWV piping	1.000	Ea.	.828	31.95	36.15	68.10
Supply, 1/2" diameter type "L" copper supply piping	1.000	Ea.	.988	37.60	48	85.60
Vent, 1-1/2" diameter DWV piping	1.000	Ea.	.593	50.80	28.80	79.60
Piping, supply, 1/2" diameter type "L" copper supply piping	32.000	L.F.	3.161	120.32	153.60	273.92
Waste, 4" diameter DWV piping	12.000	L.F.	3.310	127.80	144.60	272.40
Vent, 2" diameter DWV piping	6.000	L.F.	1.500	56.10	65.70	121.80
Vanity base cabinet, 2 door, 30" wide	1.000	Ea.	1.000	249	43.50	292.50
Vanity top, plastic laminated square edge	2.670	L.F.	.712	70.76	31.11	101.87
TOTAL		Ea.	26.414	2,182.24	1,185.01	3,367.25
BATHROOM WITH WALL HUNG LAVATORY						
Water closet, floor mounted, 2 piece, close coupled, white	1.000	Ea.	3.019	201	132	333
Rough-in, vent, 2" diameter DWV piping	1.000	Ea.	.955	24.80	41.80	66.60
Waste, 4" diameter DWV piping	1.000	Ea.	.828	31.95	36.15	68.10
Supply, 1/2" diameter type "L" copper supply piping	1.000	Ea.	.593	22.56	28.80	51.36
Lavatory, 20" x 18" P.E. cast iron, wall hung, white	1.000	Ea.	2.000	305	87.50	392.50
Rough-in, waste, 1-1/2" diameter DWV piping	1.000	Ea.	1.803	48	78.80	126.80
Supply, 1/2" diameter type "L" copper supply piping	1.000	Ea.	.988	37.60	48	85.60
Bathtub, P.E. cast iron, 5' long with accessories, white	1.000	Ea.	3.636	830	159	989
Rough-in, waste, 4" diameter DWV piping	1.000	Ea.	.828	31.95	36.15	68.10
Supply, 1/2" diameter type "L" copper supply piping	1.000	Ea.	.988	37.60	48	85.60
Vent, 1-1/2" diameter DWV piping	1.000	Ea.	.593	50.80	28.80	79.60
Piping, supply, 1/2" diameter type "L" copper supply piping	32.000	L.F.	3.161	120.32	153.60	273.92
Waste, 4" diameter DWV piping	12.000	L.F.	3.310	127.80	144.60	272.40
Vent, 2" diameter DWV piping	6.000	L.F.	1.500	56.10	65.70	121.80
Carrier steel, for studs, no arms	1.000	Ea.	1.143	56	55.50	111.50
TOTAL		Ea.	25.345	1,981.48	1,144.40	3,125.88

The costs in this system are on a cost each basis. All necessary piping is included.

Three Fixture Bathroom Price Sheet	QUAN.	UNIT	LABOR HOURS	COST EACH		
				MAT.	INST.	TOTAL
Water closet, close coupled, standard 2 piece, white	1.000	Ea.	3.019	201	132	333
Color	1.000	Ea.	3.019	238	132	370
One piece, elongated bowl, white	1.000	Ea.	3.019	585	132	717
Color	1.000	Ea.	3.019	720	132	852
Low profile, one piece, elongated bowl, white	1.000	Ea.	3.019	810	132	942
Color	1.000	Ea.	3.019	1,050	132	1,182
Rough-in, for water closet						
1/2" copper supply, 4" cast iron waste, 2" cast iron vent	1.000	Ea.	2.376	79.50	107	186.50
4" PVC/DWV waste, 2" PVC vent	1.000	Ea.	2.678	45.50	120	165.50
4" copper waste, 2" copper vent	1.000	Ea.	2.520	245	117	362
3" cast iron waste, 1-1/2" cast iron vent	1.000	Ea.	2.244	71.50	101	172.50
3" PVC waste, 1-1/2" PVC vent	1.000	Ea.	2.388	40	112	152
3" copper waste, 1-1/2" copper vent	1.000	Ea.	2.014	162	94	256
1/2" PVC supply, 4" PVC waste, 2" PVC vent	1.000	Ea.	2.974	38.50	134	172.50
3" PVC waste, 1-1/2" PVC supply	1.000	Ea.	2.684	33	126	159
1/2" steel supply, 4" cast iron waste, 2" cast iron vent	1.000	Ea.	2.545	73	115	188
4" cast iron waste, 2" steel vent	1.000	Ea.	2.590	85.50	117	202.50
4" PVC waste, 2" PVC vent	1.000	Ea.	2.847	39	128	167
Lavatory wall hung, P.E. cast iron, 20" x 18", white	1.000	Ea.	2.000	305	87.50	392.50
Color	1.000	Ea.	2.000	320	87.50	407.50
Vitreous china, 19" x 17", white	1.000	Ea.	2.286	197	100	297
Color	1.000	Ea.	2.286	220	100	320
Lavatory, for vanity top, P.E., cast iron, 20" x 18", white	1.000	Ea.	2.500	242	109	351
Color	1.000	Ea.	2.500	273	109	382
Steel, enameled, 20" x 17", white	1.000	Ea.	2.759	160	121	281
Color	1.000	Ea.	2.500	165	109	274
Vitreous china, 20" x 16", white	1.000	Ea.	2.963	291	130	421
Color	1.000	Ea.	2.963	291	130	421
Rough-in, for lavatory						
1/2" copper supply, 1-1/2" C.I. waste, 1-1/2" C.I. vent	1.000	Ea.	2.791	85.50	127	212.50
1-1/2" PVC waste, 1-1/4" PVC vent	1.000	Ea.	2.639	51	128	179
1/2" steel supply, 1-1/4" cast iron waste, 1-1/4" steel vent	1.000	Ea.	2.890	74.50	132	206.50
1-1/4" PVC waste, 1-1/4" PVC vent	1.000	Ea.	2.794	41	136	177
1/2" PVC supply, 1-1/2" PVC waste, 1-1/2" PVC vent	1.000	Ea.	3.260	38.50	158	196.50
Bathtub, P.E. cast iron, 5' long corner with fittings, white	1.000	Ea.	3.636	830	159	989
Color	1.000	Ea.	3.636	1,025	159	1,184
Rough-in, for bathtub						
1/2" copper supply, 4" cast iron waste, 1-1/2" copper vent	1.000	Ea.	2.409	120	113	233
4" PVC waste, 1/2" PVC vent	1.000	Ea.	2.877	59	135	194
1/2" steel supply, 4" cast iron waste, 1-1/2" steel vent	1.000	Ea.	2.898	87	133	220
4" PVC waste, 1-1/2" PVC vent	1.000	Ea.	3.159	48.50	149	197.50
1/2" PVC supply, 4" PVC waste, 1-1/2" PVC vent	1.000	Ea.	3.371	47	159	206
Piping, supply, 1/2" copper	32.000	L.F.	3.161	120	154	274
1/2" steel	32.000	L.F.	4.063	86	197	283
1/2" PVC	32.000	L.F.	4.741	82	230	312
Piping, waste, 4" cast iron no hub	12.000	L.F.	3.310	128	145	273
4" PVC/DWV	12.000	L.F.	4.000	60.50	175	235.50
4" copper/DWV	12.000	L.F.	4.800	620	210	830
Piping, vent, 2" cast iron no hub	6.000	L.F.	1.433	37	62.50	99.50
2" copper/DWV	6.000	L.F.	1.091	102	53	155
2" PVC/DWV	6.000	L.F.	1.627	11.50	71	82.50
2" steel, galvanized	6.000	L.F.	1.500	56	65.50	121.50
Vanity base cabinet, 2 door, 24" x 30"	1.000	Ea.	1.000	249	43.50	292.50
24" x 36"	1.000	Ea.	1.200	330	52.50	382.50
Vanity top, laminated plastic square edge, 25" x 32"	2.670	L.F.	.712	71	31	102
25" x 38"	3.160	L.F.	.843	83.50	37	120.50
Cultured marble, 25" x 32", with bowl	1.000	Ea.	2.500	200	109	309
25" x 38", with bowl	1.000	Ea.	2.500	237	109	346
Carrier, for lavatory, steel for studs, no arms	1.000	Ea.	1.143	56	55.50	111.50
Wood, 2" x 8" blocking	1.300	L.F.	.052	1.20	2.26	3.46

8 | MECHANICAL — 24 | Three Fixture Bathroom Systems

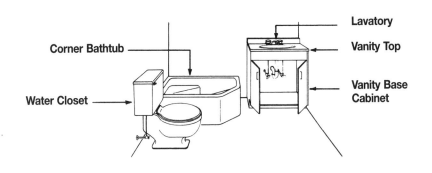

System Description	QUAN.	UNIT	LABOR HOURS	COST EACH MAT.	COST EACH INST.	COST EACH TOTAL
BATHROOM WITH LAVATORY INSTALLED IN VANITY						
Water closet, floor mounted, 2 piece, close coupled, white	1.000	Ea.	3.019	201	132	333
Rough-in, vent, 2" diameter DWV piping	1.000	Ea.	.955	24.80	41.80	66.60
Waste, 4" diameter DWV piping	1.000	Ea.	.828	31.95	36.15	68.10
Supply, 1/2" diameter type "L" copper supply piping	1.000	Ea.	.593	22.56	28.80	51.36
Lavatory, 20" x 18", P.E. cast iron with fittings, white	1.000	Ea.	2.500	242	109	351
Rough-in, waste, 1-1/2" diameter DWV piping	1.000	Ea.	1.803	48	78.80	126.80
Supply, 1/2" diameter type "L" copper supply piping	1.000	Ea.	.988	37.60	48	85.60
Bathtub, P.E. cast iron, corner with fittings, white	1.000	Ea.	3.636	1,825	159	1,984
Rough-in, waste, 4" diameter DWV piping	1.000	Ea.	.828	31.95	36.15	68.10
Supply, 1/2" diameter type "L" copper supply piping	1.000	Ea.	.988	37.60	48	85.60
Vent, 1-1/2" diameter DWV piping	1.000	Ea.	.593	50.80	28.80	79.60
Piping, supply, 1/2" diameter type "L" copper supply piping	32.000	L.F.	3.161	120.32	153.60	273.92
Waste, 4" diameter DWV piping	12.000	L.F.	3.310	127.80	144.60	272.40
Vent, 2" diameter DWV piping	6.000	L.F.	1.500	56.10	65.70	121.80
Vanity base cabinet, 2 door, 30" wide	1.000	Ea.	1.000	249	43.50	292.50
Vanity top, plastic laminated, square edge	2.670	L.F.	.712	89.45	31.11	120.56
TOTAL		Ea.	26.414	3,195.93	1,185.01	4,380.94
BATHROOM WITH WALL HUNG LAVATORY						
Water closet, floor mounted, 2 piece, close coupled, white	1.000	Ea.	3.019	201	132	333
Rough-in, vent, 2" diameter DWV piping	1.000	Ea.	.955	24.80	41.80	66.60
Waste, 4" diameter DWV piping	1.000	Ea.	.828	31.95	36.15	68.10
Supply, 1/2" diameter type "L" copper supply piping	1.000	Ea.	.593	22.56	28.80	51.36
Lavatory, 20" x 18", P.E. cast iron, with fittings, white	1.000	Ea.	2.000	305	87.50	392.50
Rough-in, waste, 1-1/2" diameter DWV piping	1.000	Ea.	1.803	48	78.80	126.80
Supply, 1/2" diameter type "L" copper supply piping	1.000	Ea.	.988	37.60	48	85.60
Bathtub, P.E. cast iron, corner, with fittings, white	1.000	Ea.	3.636	1,825	159	1,984
Rough-in, waste, 4" diameter DWV piping	1.000	Ea.	.828	31.95	36.15	68.10
Supply, 1/2" diameter type "L" copper supply piping	1.000	Ea.	.988	37.60	48	85.60
Vent, 1-1/2" diameter DWV piping	1.000	Ea.	.593	50.80	28.80	79.60
Piping, supply, 1/2" diameter type "L" copper supply piping	32.000	L.F.	3.161	120.32	153.60	273.92
Waste, 4" diameter DWV piping	12.000	L.F.	3.310	127.80	144.60	272.40
Vent, 2" diameter DWV piping	6.000	L.F.	1.500	56.10	65.70	121.80
Carrier, steel, for studs, no arms	1.000	Ea.	1.143	56	55.50	111.50
TOTAL		Ea.	25.345	2,976.48	1,144.40	4,120.88

The costs in this system are on a cost each basis. All necessary piping is included.

Three Fixture Bathroom Price Sheet	QUAN.	UNIT	LABOR HOURS	COST EACH MAT.	COST EACH INST.	COST EACH TOTAL
Water closet, close coupled, standard 2 piece, white	1.000	Ea.	3.019	201	132	333
Color	1.000	Ea.	3.019	238	132	370
One piece elongated bowl, white	1.000	Ea.	3.019	585	132	717
Color	1.000	Ea.	3.019	720	132	852
Low profile, one piece elongated bowl, white	1.000	Ea.	3.019	810	132	942
Color	1.000	Ea.	3.019	1,050	132	1,182
Rough-in, for water closet						
1/2" copper supply, 4" cast iron waste, 2" cast iron vent	1.000	Ea.	2.376	79.50	107	186.50
4" PVC/DWV waste, 2" PVC vent	1.000	Ea.	2.678	45.50	120	165.50
4" copper waste, 2" copper vent	1.000	Ea.	2.520	245	117	362
3" cast iron waste, 1-1/2" cast iron vent	1.000	Ea.	2.244	71.50	101	172.50
3" PVC waste, 1-1/2" PVC vent	1.000	Ea.	2.388	40	112	152
3" copper waste, 1-1/2" copper vent	1.000	Ea.	2.014	162	94	256
1/2" PVC supply, 4" PVC waste, 2" PVC vent	1.000	Ea.	2.974	38.50	134	172.50
3" PVC waste, 1-1/2" PVC supply	1.000	Ea.	2.684	33	126	159
1/2" steel supply, 4" cast iron waste, 2" cast iron vent	1.000	Ea.	2.545	73	115	188
4" cast iron waste, 2" steel vent	1.000	Ea.	2.590	85.50	117	202.50
4" PVC waste, 2" PVC vent	1.000	Ea.	2.847	39	128	167
Lavatory, wall hung P.E. cast iron 20" x 18", white	1.000	Ea.	2.000	305	87.50	392.50
Color	1.000	Ea.	2.000	320	87.50	407.50
Vitreous china 19" x 17", white	1.000	Ea.	2.286	197	100	297
Color	1.000	Ea.	2.286	220	100	320
Lavatory, for vanity top, P.E., cast iron, 20" x 18", white	1.000	Ea.	2.500	242	109	351
Color	1.000	Ea.	2.500	273	109	382
Steel enameled 20" x 17", white	1.000	Ea.	2.759	160	121	281
Color	1.000	Ea.	2.500	165	109	274
Vitreous china 20" x 16", white	1.000	Ea.	2.963	291	130	421
Color	1.000	Ea.	2.963	291	130	421
Rough-in, for lavatory						
1/2" copper supply, 1-1/2" cast iron waste, 1-1/2" cast iron vent	1.000	Ea.	2.791	85.50	127	212.50
1-1/2" PVC waste, 1-1/4" PVC vent	1.000	Ea.	2.639	51	128	179
1/2" steel supply, 1-1/4" cast iron waste, 1-1/4" steel vent	1.000	Ea.	2.890	74.50	132	206.50
1-1/4" PVC waste, 1-1/4" PVC vent	1.000	Ea.	2.794	41	136	177
1/2" PVC supply, 1-1/2" PVC waste, 1-1/2" PVC vent	1.000	Ea.	3.260	38.50	158	196.50
Bathtub, P.E. cast iron, corner with fittings, white	1.000	Ea.	3.636	1,825	159	1,984
Color	1.000	Ea.	4.000	2,075	175	2,250
Rough-in, for bathtub						
1/2" copper supply, 4" cast iron waste, 1-1/2" copper vent	1.000	Ea.	2.409	120	113	233
4" PVC waste, 1-1/2" PVC vent	1.000	Ea.	2.877	59	135	194
1/2" steel supply, 4" cast iron waste, 1-1/2" steel vent	1.000	Ea.	2.898	87	133	220
4" PVC waste, 1-1/2" PVC vent	1.000	Ea.	3.159	48.50	149	197.50
1/2" PVC supply, 4" PVC waste, 1-1/2" PVC vent	1.000	Ea.	3.371	47	159	206
Piping, supply, 1/2" copper	32.000	L.F.	3.161	120	154	274
1/2" steel	32.000	L.F.	4.063	86	197	283
1/2" PVC	32.000	L.F.	4.741	82	230	312
Piping, waste, 4" cast iron, no hub	12.000	L.F.	3.310	128	145	273
4" PVC/DWV	12.000	L.F.	4.000	60.50	175	235.50
4" copper/DWV	12.000	L.F.	4.800	620	210	830
Piping, vent 2" cast iron, no hub	6.000	L.F.	1.433	37	62.50	99.50
2" copper/DWV	6.000	L.F.	1.091	102	53	155
2" PVC/DWV	6.000	L.F.	1.627	11.50	71	82.50
2" steel, galvanized	6.000	L.F.	1.500	56	65.50	121.50
Vanity base cabinet, 2 door, 24" x 30"	1.000	Ea.	1.000	249	43.50	292.50
24" x 36"	1.000	Ea.	1.200	330	52.50	382.50
Vanity top, laminated plastic square edge 25" x 32"	2.670	L.F.	.712	89.50	31	120.50
25" x 38"	3.160	L.F.	.843	106	37	143
Cultured marble, 25" x 32", with bowl	1.000	Ea.	2.500	200	109	309
25" x 38", with bowl	1.000	Ea.	2.500	237	109	346
Carrier, for lavatory, steel for studs, no arms	1.000	Ea.	1.143	56	55.50	111.50
Wood, 2" x 8" blocking	1.300	L.F.	.053	1.22	2.31	3.53

8 | MECHANICAL — 28 | Three Fixture Bathroom Systems

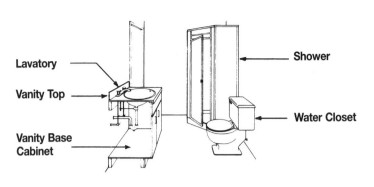

System Description	QUAN.	UNIT	LABOR HOURS	MAT.	INST.	TOTAL
BATHROOM WITH SHOWER, LAVATORY INSTALLED IN VANITY						
Water closet, floor mounted, 2 piece, close coupled, white	1.000	Ea.	3.019	201	132	333
Rough-in, vent, 2" diameter DWV piping	1.000	Ea.	.955	24.80	41.80	66.60
Waste, 4" diameter DWV piping	1.000	Ea.	.828	31.95	36.15	68.10
Supply, 1/2" diameter type "L" copper supply piping	1.000	Ea.	.593	22.56	28.80	51.36
Lavatory, 20" x 18" P.E. cast iron with fittings, white	1.000	Ea.	2.500	242	109	351
Rough-in, waste, 1-1/2" diameter DWV piping	1.000	Ea.	1.803	48	78.80	126.80
Supply, 1/2" diameter type "L" copper supply piping	1.000	Ea.	.988	37.60	48	85.60
Shower, steel enameled, stone base, corner, white	1.000	Ea.	3.200	380	140	520
Shower mixing valve	1.000	Ea.	1.333	120	65	185
Shower door	1.000	Ea.	1.000	225	48.50	273.50
Rough-in, vent, 1-1/2" diameter DWV piping	1.000	Ea.	.225	6	9.85	15.85
Waste, 2" diameter DWV piping	1.000	Ea.	1.433	37.20	62.70	99.90
Supply, 1/2" diameter type "L" copper supply piping	1.000	Ea.	1.580	60.16	76.80	136.96
Piping, supply, 1/2" diameter type "L" copper supply piping	36.000	L.F.	4.148	157.92	201.60	359.52
Waste, 4" diameter DWV piping	7.000	L.F.	2.759	106.50	120.50	227
Vent, 2" diameter DWV piping	6.000	L.F.	2.250	84.15	98.55	182.70
Vanity base 2 door, 30" wide	1.000	Ea.	1.000	249	43.50	292.50
Vanity top, plastic laminated, square edge	2.170	L.F.	.712	73.43	31.11	104.54
TOTAL		Ea.	30.326	2,107.27	1,372.66	3,479.93
BATHROOM WITH SHOWER, WALL HUNG LAVATORY						
Water closet, floor mounted, close coupled	1.000	Ea.	3.019	201	132	333
Rough-in, vent, 2" diameter DWV piping	1.000	Ea.	.955	24.80	41.80	66.60
Waste, 4" diameter DWV piping	1.000	Ea.	.828	31.95	36.15	68.10
Supply, 1/2" diameter type "L" copper supply piping	1.000	Ea.	.593	22.56	28.80	51.36
Lavatory, 20" x 18" P.E. cast iron with fittings, white	1.000	Ea.	2.000	305	87.50	392.50
Rough-in, waste, 1-1/2" diameter DWV piping	1.000	Ea.	1.803	48	78.80	126.80
Supply, 1/2" diameter type "L" copper supply piping	1.000	Ea.	.988	37.60	48	85.60
Shower, steel enameled, stone base, white	1.000	Ea.	3.200	380	140	520
Mixing valve	1.000	Ea.	1.333	120	65	185
Shower door	1.000	Ea.	1.000	225	48.50	273.50
Rough-in, vent, 1-1/2" diameter DWV piping	1.000	Ea.	.225	6	9.85	15.85
Waste, 2" diameter DWV piping	1.000	Ea.	1.433	37.20	62.70	99.90
Supply, 1/2" diameter type "L" copper supply piping	1.000	Ea.	1.580	60.16	76.80	136.96
Piping, supply, 1/2" diameter type "L" copper supply piping	36.000	L.F.	4.148	157.92	201.60	359.52
Waste, 4" diameter DWV piping	7.000	L.F.	2.759	106.50	120.50	227
Vent, 2" diameter DWV piping	6.000	L.F.	2.250	84.15	98.55	182.70
Carrier, steel, for studs, no arms	1.000	Ea.	1.143	56	55.50	111.50
TOTAL		Ea.	29.257	1,903.84	1,332.05	3,235.89

The costs in this system are on a cost each basis. All necessary piping is included.

Three Fixture Bathroom Price Sheet	QUAN.	UNIT	LABOR HOURS	COST EACH		
				MAT.	INST.	TOTAL
Water closet, close coupled, standard 2 piece, white	1.000	Ea.	3.019	201	132	333
Color	1.000	Ea.	3.019	238	132	370
One piece elongated bowl, white	1.000	Ea.	3.019	585	132	717
Color	1.000	Ea.	3.019	720	132	852
Low profile, one piece elongated bowl, white	1.000	Ea.	3.019	810	132	942
Color	1.000	Ea.	3.019	1,050	132	1,182
Rough-in, for water closet						
1/2" copper supply, 4" cast iron waste, 2" cast iron vent	1.000	Ea.	2.376	79.50	107	186.50
4" PVC/DWV waste, 2" PVC vent	1.000	Ea.	2.678	45.50	120	165.50
4" copper waste, 2" copper vent	1.000	Ea.	2.520	245	117	362
3" cast iron waste, 1-1/2" cast iron vent	1.000	Ea.	2.244	71.50	101	172.50
3" PVC waste, 1-1/2" PVC vent	1.000	Ea.	2.388	40	112	152
3" copper waste, 1-1/2" copper vent	1.000	Ea.	2.014	162	94	256
1/2" PVC supply, 4" PVC waste, 2" PVC vent	1.000	Ea.	2.974	38.50	134	172.50
3" PVC waste, 1-1/2" PVC supply	1.000	Ea.	2.684	33	126	159
1/2" steel supply, 4" cast iron waste, 2" cast iron vent	1.000	Ea.	2.545	73	115	188
4" cast iron waste, 2" steel vent	1.000	Ea.	2.590	85.50	117	202.50
4" PVC waste, 2" PVC vent	1.000	Ea.	2.847	39	128	167
Lavatory, wall hung, P.E. cast iron 20" x 18", white	1.000	Ea.	2.000	305	87.50	392.50
Color	1.000	Ea.	2.000	320	87.50	407.50
Vitreous china 19" x 17", white	1.000	Ea.	2.286	197	100	297
Color	1.000	Ea.	2.286	220	100	320
Lavatory, for vanity top, P.E. cast iron 20" x 18", white	1.000	Ea.	2.500	242	109	351
Color	1.000	Ea.	2.500	273	109	382
Steel enameled 20" x 17", white	1.000	Ea.	2.759	160	121	281
Color	1.000	Ea.	2.500	165	109	274
Vitreous china 20" x 16", white	1.000	Ea.	2.963	291	130	421
Color	1.000	Ea.	2.963	291	130	421
Rough-in, for lavatory						
1/2" copper supply, 1-1/2" cast iron waste, 1-1/2" cast iron vent	1.000	Ea.	2.791	85.50	127	212.50
1-1/2" PVC waste, 1-1/2" PVC vent	1.000	Ea.	2.639	51	128	179
1/2" steel supply, 1-1/4" cast iron waste, 1-1/4" steel vent	1.000	Ea.	2.890	74.50	132	206.50
1-1/4" PVC waste, 1-1/4" PVC vent	1.000	Ea.	2.921	40.50	142	182.50
1/2" PVC supply, 1-1/2" PVC waste, 1-1/2" PVC vent	1.000	Ea.	3.260	38.50	158	196.50
Shower, steel enameled stone base, 32" x 32", white	1.000	Ea.	8.000	380	140	520
Color	1.000	Ea.	7.822	900	128	1,028
36" x 36" white	1.000	Ea.	8.889	960	146	1,106
Color	1.000	Ea.	8.889	1,025	146	1,171
Rough-in, for shower						
1/2" copper supply, 4" cast iron waste, 1-1/2" copper vent	1.000	Ea.	3.238	103	149	252
4" PVC waste, 1-1/2" PVC vent	1.000	Ea.	3.429	73.50	159	232.50
1/2" steel supply, 4" cast iron waste, 1-1/2" steel vent	1.000	Ea.	3.665	87	170	257
4" PVC waste, 1-1/2" PVC vent	1.000	Ea.	3.881	56	180	236
1/2" PVC supply, 4" PVC waste, 1-1/2" PVC vent	1.000	Ea.	4.219	54	197	251
Piping, supply, 1/2" copper	36.000	L.F.	4.148	158	202	360
1/2" steel	36.000	L.F.	5.333	113	258	371
1/2" PVC	36.000	L.F.	6.222	108	300	408
Piping, waste, 4" cast iron no hub	7.000	L.F.	2.759	107	121	228
4" PVC/DWV	7.000	L.F.	3.333	50.50	146	196.50
4" copper/DWV	7.000	L.F.	4.000	515	175	690
Piping, vent, 2" cast iron no hub	6.000	L.F.	2.149	56	94	150
2" copper/DWV	6.000	L.F.	1.636	153	79.50	232.50
2" PVC/DWV	6.000	L.F.	2.441	17.30	107	124.30
2" steel, galvanized	6.000	L.F.	2.250	84	98.50	182.50
Vanity base cabinet, 2 door, 24" x 30"	1.000	Ea.	1.000	249	43.50	292.50
24" x 36"	1.000	Ea.	1.200	330	52.50	382.50
Vanity top, laminated plastic square edge, 25" x 32"	2.170	L.F.	.712	73.50	31	104.50
25" x 38"	2.670	L.F.	.845	87	37	124
Carrier, for lavatory, steel for studs, no arms	1.000	Ea.	1.143	56	55.50	111.50
Wood, 2" x 8" blocking	1.300	L.F.	.052	1.20	2.26	3.46

8 | MECHANICAL 32 | Three Fixture Bathroom Systems

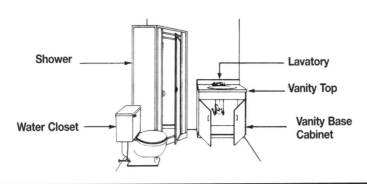

System Description	QUAN.	UNIT	LABOR HOURS	COST EACH MAT.	COST EACH INST.	COST EACH TOTAL
BATHROOM WITH LAVATORY INSTALLED IN VANITY						
Water closet, floor mounted, 2 piece, close coupled, white	1.000	Ea.	3.019	201	132	333
Rough-in, vent, 2" diameter DWV piping	1.000	Ea.	.955	24.80	41.80	66.60
Waste, 4" diameter DWV piping	1.000	Ea.	.828	31.95	36.15	68.10
Supply, 1/2" diameter type "L" copper supply piping	1.000	Ea.	.593	22.56	28.80	51.36
Lavatory, 20" x 18", P.E. cast iron with fittings, white	1.000	Ea.	2.500	242	109	351
Rough-in, waste, 1-1/2" diameter DWV piping	1.000	Ea.	1.803	48	78.80	126.80
Supply, 1/2" diameter type "L" copper supply piping	1.000	Ea.	.988	37.60	48	85.60
Shower, steel enameled, stone base, corner, white	1.000	Ea.	3.200	380	140	520
Mixing valve	1.000	Ea.	1.333	120	65	185
Shower door	1.000	Ea.	1.000	225	48.50	273.50
Rough-in, vent, 1-1/2" diameter DWV piping	1.000	Ea.	.225	6	9.85	15.85
Waste, 2" diameter DWV piping	1.000	Ea.	1.433	37.20	62.70	99.90
Supply, 1/2" diameter type "L" copper supply piping	1.000	Ea.	1.580	60.16	76.80	136.96
Piping, supply, 1/2" diameter type "L" copper supply piping	36.000	L.F.	3.556	135.36	172.80	308.16
Waste, 4" diameter DWV piping	7.000	L.F.	1.931	74.55	84.35	158.90
Vent, 2" diameter DWV piping	6.000	L.F.	1.500	56.10	65.70	121.80
Vanity base, 2 door, 30" wide	1.000	Ea.	1.000	249	43.50	292.50
Vanity top, plastic laminated, square edge	2.670	L.F.	.712	70.76	31.11	101.87
TOTAL		Ea.	28.156	2,022.04	1,274.86	3,296.90
BATHROOM, WITH WALL HUNG LAVATORY						
Water closet, floor mounted, 2 piece, close coupled, white	1.000	Ea.	3.019	201	132	333
Rough-in, vent, 2" diameter DWV piping	1.000	Ea.	.955	24.80	41.80	66.60
Waste, 4" diameter DWV piping	1.000	Ea.	.828	31.95	36.15	68.10
Supply, 1/2" diameter type "L" copper supply piping	1.000	Ea.	.593	22.56	28.80	51.36
Lavatory, wall hung, 20" x 18" P.E. cast iron with fittings, white	1.000	Ea.	2.000	305	87.50	392.50
Rough-in, waste, 1-1/2" diameter DWV piping	1.000	Ea.	1.803	48	78.80	126.80
Supply, 1/2" diameter type "L" copper supply piping	1.000	Ea.	.988	37.60	48	85.60
Shower, steel enameled, stone base, corner, white	1.000	Ea.	3.200	380	140	520
Mixing valve	1.000	Ea.	1.333	120	65	185
Shower door	1.000	Ea.	1.000	225	48.50	273.50
Rough-in, waste, 1-1/2" diameter DWV piping	1.000	Ea.	.225	6	9.85	15.85
Waste, 2" diameter DWV piping	1.000	Ea.	1.433	37.20	62.70	99.90
Supply, 1/2" diameter type "L" copper supply piping	1.000	Ea.	1.580	60.16	76.80	136.96
Piping, supply, 1/2" diameter type "L" copper supply piping	36.000	L.F.	3.556	135.36	172.80	308.16
Waste, 4" diameter DWV piping	7.000	L.F.	1.931	74.55	84.35	158.90
Vent, 2" diameter DWV piping	6.000	L.F.	1.500	56.10	65.70	121.80
Carrier, steel, for studs, no arms	1.000	Ea.	1.143	56	55.50	111.50
TOTAL		Ea.	27.087	1,821.28	1,234.25	3,055.53

The costs in this system are on a cost each basis. All necessary piping is included.

Three Fixture Bathroom Price Sheet

	QUAN.	UNIT	LABOR HOURS	COST EACH MAT.	COST EACH INST.	COST EACH TOTAL
Water closet, close coupled, standard 2 piece, white	1.000	Ea.	3.019	201	132	333
Color	1.000	Ea.	3.019	238	132	370
One piece elongated bowl, white	1.000	Ea.	3.019	585	132	717
Color	1.000	Ea.	3.019	720	132	852
Low profile one piece elongated bowl, white	1.000	Ea.	3.019	810	132	942
Color	1.000	Ea.	3.623	1,050	132	1,182
Rough-in, for water closet						
1/2" copper supply, 4" cast iron waste, 2" cast iron vent	1.000	Ea.	2.376	79.50	107	186.50
4" P.V.C./DWV waste, 2" PVC vent	1.000	Ea.	2.678	45.50	120	165.50
4" copper waste, 2" copper vent	1.000	Ea.	2.520	245	117	362
3" cast iron waste, 1-1/2" cast iron vent	1.000	Ea.	2.244	71.50	101	172.50
3" PVC waste, 1-1/2" PVC vent	1.000	Ea.	2.388	40	112	152
3" copper waste, 1-1/2" copper vent	1.000	Ea.	2.014	162	94	256
1/2" P.V.C. supply, 4" P.V.C. waste, 2" P.V.C. vent	1.000	Ea.	2.974	38.50	134	172.50
3" P.V.C. waste, 1-1/2" P.V.C. vent	1.000	Ea.	2.684	33	126	159
1/2" steel supply, 4" cast iron waste, 2" cast iron vent	1.000	Ea.	2.545	73	115	188
4" cast iron waste, 2" steel vent	1.000	Ea.	2.590	85.50	117	202.50
4" P.V.C. waste, 2" P.V.C. vent	1.000	Ea.	2.847	39	128	167
Lavatory, wall hung P.E. cast iron 20" x 18", white	1.000	Ea.	2.000	305	87.50	392.50
Color	1.000	Ea.	2.000	320	87.50	407.50
Vitreous china 19" x 17", white	1.000	Ea.	2.286	197	100	297
Color	1.000	Ea.	2.286	220	100	320
Lavatory, for vanity top P.E. cast iron 20" x 18", white	1.000	Ea.	2.500	242	109	351
Color	1.000	Ea.	2.500	273	109	382
Steel enameled 20" x 17", white	1.000	Ea.	2.759	160	121	281
Color	1.000	Ea.	2.500	165	109	274
Vitreous china 20" x 16", white	1.000	Ea.	2.963	291	130	421
Color	1.000	Ea.	2.963	291	130	421
Rough-in, for lavatory						
1/2" copper supply, 1-1/2" cast iron waste, 1-1/2" cast iron vent	1.000	Ea.	2.791	85.50	127	212.50
1-1/2" P.V.C. waste, 1-1/2" P.V.C. vent	1.000	Ea.	2.639	51	128	179
1/2" steel supply, 1-1/2" cast iron waste, 1-1/4" steel vent	1.000	Ea.	2.890	74.50	132	206.50
1-1/2" P.V.C. waste, 1-1/4" P.V.C. vent	1.000	Ea.	2.921	40.50	142	182.50
1/2" P.V.C. supply, 1-1/2" P.V.C. waste, 1-1/2" P.V.C. vent	1.000	Ea.	3.260	38.50	158	196.50
Shower, steel enameled stone base, 32" x 32", white	1.000	Ea.	8.000	380	140	520
Color	1.000	Ea.	7.822	900	128	1,028
36" x 36", white	1.000	Ea.	8.889	960	146	1,106
Color	1.000	Ea.	8.889	1,025	146	1,171
Rough-in, for shower						
1/2" copper supply, 2" cast iron waste, 1-1/2" copper vent	1.000	Ea.	3.161	110	147	257
2" P.V.C. waste, 1-1/2" P.V.C. vent	1.000	Ea.	3.429	73.50	159	232.50
1/2" steel supply, 2" cast iron waste, 1-1/2" steel vent	1.000	Ea.	3.887	114	179	293
2" P.V.C. waste, 1-1/2" P.V.C. vent	1.000	Ea.	3.881	56	180	236
1/2" P.V.C. supply, 2" P.V.C. waste, 1-1/2" P.V.C. vent	1.000	Ea.	4.219	54	197	251
Piping, supply, 1/2" copper	36.000	L.F.	3.556	135	173	308
1/2" steel	36.000	L.F.	4.571	96.50	221	317.50
1/2" P.V.C.	36.000	L.F.	5.333	92.50	259	351.50
Waste, 4" cast iron, no hub	7.000	L.F.	1.931	74.50	84.50	159
4" P.V.C./DWV	7.000	L.F.	2.333	35.50	102	137.50
4" copper/DWV	7.000	L.F.	2.800	360	123	483
Vent, 2" cast iron, no hub	6.000	L.F.	1.091	102	53	155
2" copper/DWV	6.000	L.F.	1.091	102	53	155
2" P.V.C./DWV	6.000	L.F.	1.627	11.50	71	82.50
2" steel, galvanized	6.000	L.F.	1.500	56	65.50	121.50
Vanity base cabinet, 2 door, 24" x 30"	1.000	Ea.	1.000	249	43.50	292.50
24" x 36"	1.000	Ea.	1.200	330	52.50	382.50
Vanity top, laminated plastic square edge, 25" x 32"	2.670	L.F.	.712	71	31	102
25" x 38"	3.170	L.F.	.845	84	37	121
Carrier, for lavatory, steel, for studs, no arms	1.000	Ea.	1.143	56	55.50	111.50
Wood, 2" x 8" blocking	1.300	L.F.	.052	1.20	2.26	3.46

8 | MECHANICAL   36 | Four Fixture Bathroom Systems

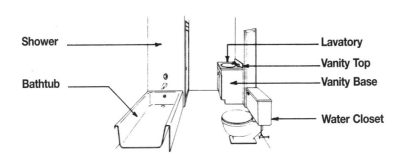

System Description	QUAN.	UNIT	LABOR HOURS	COST EACH MAT.	COST EACH INST.	COST EACH TOTAL
BATHROOM WITH LAVATORY INSTALLED IN VANITY						
Water closet, floor mounted, 2 piece, close coupled, white	1.000	Ea.	3.019	201	132	333
Rough-in, vent, 2" diameter DWV piping	1.000	Ea.	.955	24.80	41.80	66.60
Waste, 4" diameter DWV piping	1.000	Ea.	.828	31.95	36.15	68.10
Supply, 1/2" diameter type "L" copper supply piping	1.000	Ea.	.593	22.56	28.80	51.36
Lavatory, 20" x 18" P.E. cast iron with fittings, white	1.000	Ea.	2.500	242	109	351
Shower, steel, enameled, stone base, corner, white	1.000	Ea.	3.333	845	146	991
Mixing valve	1.000	Ea.	1.333	120	65	185
Shower door	1.000	Ea.	1.000	225	48.50	273.50
Rough-in, waste, 1-1/2" diameter DWV piping	2.000	Ea.	4.507	120	197	317
Supply, 1/2" diameter type "L" copper supply piping	2.000	Ea.	3.161	120.32	153.60	273.92
Bathtub, P.E. cast iron, 5' long with fittings, white	1.000	Ea.	3.636	830	159	989
Rough-in, waste, 4" diameter DWV piping	1.000	Ea.	.828	31.95	36.15	68.10
Supply, 1/2" diameter type "L" copper supply piping	1.000	Ea.	.988	37.60	48	85.60
Vent, 1-1/2" diameter DWV piping	1.000	Ea.	.593	50.80	28.80	79.60
Piping, supply, 1/2" diameter type "L" copper supply piping	42.000	L.F.	4.148	157.92	201.60	359.52
Waste, 4" diameter DWV piping	10.000	L.F.	2.759	106.50	120.50	227
Vent, 2" diameter DWV piping	13.000	L.F.	3.250	121.55	142.35	263.90
Vanity base, 2 doors, 30" wide	1.000	Ea.	1.000	249	43.50	292.50
Vanity top, plastic laminated, square edge	2.670	L.F.	.712	70.76	31.11	101.87
TOTAL		Ea.	39.143	3,608.71	1,768.86	5,377.57
BATHROOM WITH WALL HUNG LAVATORY						
Water closet, floor mounted, 2 piece, close coupled, white	1.000	Ea.	3.019	201	132	333
Rough-in, vent, 2" diameter DWV piping	1.000	Ea.	.955	24.80	41.80	66.60
Waste, 4" diameter DWV piping	1.000	Ea.	.828	31.95	36.15	68.10
Supply, 1/2" diameter type "L" copper supply piping	1.000	Ea.	.593	22.56	28.80	51.36
Lavatory, 20" x 18" P.E. cast iron with fittings, white	1.000	Ea.	2.000	305	87.50	392.50
Shower, steel enameled, stone base, corner, white	1.000	Ea.	3.333	845	146	991
Mixing valve	1.000	Ea.	1.333	120	65	185
Shower door	1.000	Ea.	1.000	225	48.50	273.50
Rough-in, waste, 1-1/2" diameter DWV piping	2.000	Ea.	4.507	120	197	317
Supply, 1/2" diameter type "L" copper supply piping	2.000	Ea.	3.161	120.32	153.60	273.92
Bathtub, P.E. cast iron, 5' long with fittings, white	1.000	Ea.	3.636	830	159	989
Rough-in, waste, 4" diameter DWV piping	1.000	Ea.	.828	31.95	36.15	68.10
Supply, 1/2" diameter type "L" copper supply piping	1.000	Ea.	.988	37.60	48	85.60
Vent, 1-1/2" diameter copper DWV piping	1.000	Ea.	.593	50.80	28.80	79.60
Piping, supply, 1/2" diameter type "L" copper supply piping	42.000	L.F.	4.148	157.92	201.60	359.52
Waste, 4" diameter DWV piping	10.000	L.F.	2.759	106.50	120.50	227
Vent, 2" diameter DWV piping	13.000	L.F.	3.250	121.55	142.35	263.90
Carrier, steel, for studs, no arms	1.000	Ea.	1.143	56	55.50	111.50
TOTAL		Ea.	38.074	3,407.95	1,728.25	5,136.20

The costs in this system are on a cost each basis. All necessary piping is included.

Four Fixture Bathroom Price Sheet	QUAN.	UNIT	LABOR HOURS	COST EACH		
				MAT.	INST.	TOTAL
Water closet, close coupled, standard 2 piece, white	1.000	Ea.	3.019	201	132	333
Color	1.000	Ea.	3.019	238	132	370
One piece elongated bowl, white	1.000	Ea.	3.019	585	132	717
Color	1.000	Ea.	3.019	720	132	852
Low profile, one piece elongated bowl, white	1.000	Ea.	3.019	810	132	942
Color	1.000	Ea.	3.019	1,050	132	1,182
1/2" copper supply, 4" cast iron waste, 2" cast iron vent	1.000	Ea.	2.376	79.50	107	186.50
4" PVC/DWV waste, 2" PVC vent	1.000	Ea.	2.678	45.50	120	165.50
4" copper waste, 2" copper vent	1.000	Ea.	2.520	245	117	362
3" cast iron waste, 1-1/2" cast iron vent	1.000	Ea.	2.244	71.50	101	172.50
3" P.V.C. waste, 1-1/2" P.V.C. vent	1.000	Ea.	2.388	40	112	152
3" copper waste, 1-1/2" copper vent	1.000	Ea.	2.014	162	94	256
1/2" P.V.C. supply, 4" P.V.C. waste, 2" P.V.C. vent	1.000	Ea.	2.974	38.50	134	172.50
3" P.V.C. waste, 1-1/2" P.V.C. vent	1.000	Ea.	2.684	33	126	159
1/2" steel supply, 4" cast iron waste, 2" cast iron vent	1.000	Ea.	2.545	73	115	188
4" cast iron waste, 2" steel vent	1.000	Ea.	2.590	85.50	117	202.50
4" P.V.C. waste, 2" P.V.C. vent	1.000	Ea.	2.847	39	128	167
Lavatory, wall hung P.E. cast iron 20" x 18", white	1.000	Ea.	2.000	305	87.50	392.50
Color	1.000	Ea.	2.000	320	87.50	407.50
Vitreous china 19" x 17", white	1.000	Ea.	2.286	197	100	297
Color	1.000	Ea.	2.286	220	100	320
Lavatory for vanity top, P.E. cast iron 20" x 18", white	1.000	Ea.	2.500	242	109	351
Color	1.000	Ea.	2.500	273	109	382
Steel enameled, 20" x 17", white	1.000	Ea.	2.759	160	121	281
Color	1.000	Ea.	2.500	165	109	274
Vitreous china 20" x 16", white	1.000	Ea.	2.963	291	130	421
Color	1.000	Ea.	2.963	291	130	421
Shower, steel enameled stone base, 36" square, white	1.000	Ea.	8.889	845	146	991
Color	1.000	Ea.	8.889	950	146	1,096
Rough-in, for lavatory or shower						
1/2" copper supply, 1-1/2" cast iron waste, 1-1/2" cast iron vent	1.000	Ea.	3.834	120	175	295
1-1/2" P.V.C. waste, 1-1/4" P.V.C. vent	1.000	Ea.	3.675	77	179	256
1/2" steel supply, 1-1/4" cast iron waste, 1-1/4" steel vent	1.000	Ea.	4.103	102	189	291
1-1/4" P.V.C. waste, 1-1/4" P.V.C. vent	1.000	Ea.	3.937	61	191	252
1/2" P.V.C. supply, 1-1/2" P.V.C. waste, 1-1/2" P.V.C. vent	1.000	Ea.	4.592	57	223	280
Bathtub, P.E. cast iron, 5' long with fittings, white	1.000	Ea.	3.636	830	159	989
Color	1.000	Ea.	3.636	1,025	159	1,184
Steel, enameled 5' long with fittings, white	1.000	Ea.	2.909	345	127	472
Color	1.000	Ea.	2.909	345	127	472
Rough-in, for bathtub						
1/2" copper supply, 4" cast iron waste, 1-1/2" copper vent	1.000	Ea.	2.409	120	113	233
4" P.V.C. waste, 1-1/2" P.V.C. vent	1.000	Ea.	2.877	59	135	194
1/2" steel supply, 4" cast iron waste, 1-1/2" steel vent	1.000	Ea.	2.898	87	133	220
4" P.V.C. waste, 1-1/2" P.V.C. vent	1.000	Ea.	3.159	48.50	149	197.50
1/2" P.V.C. supply, 4" P.V.C. waste, 1-1/2" P.V.C. vent	1.000	Ea.	3.371	47	159	206
Piping, supply, 1/2" copper	42.000	L.F.	4.148	158	202	360
1/2" steel	42.000	L.F.	5.333	113	258	371
1/2" P.V.C.	42.000	L.F.	6.222	108	300	408
Waste, 4" cast iron, no hub	10.000	L.F.	2.759	107	121	228
4" P.V.C./DWV	10.000	L.F.	3.333	50.50	146	196.50
4" copper/DWV	10.000	Ea.	4.000	515	175	690
Vent 2" cast iron, no hub	13.000	L.F.	3.105	80.50	136	216.50
2" copper/DWV	13.000	L.F.	2.364	221	115	336
2" P.V.C./DWV	13.000	L.F.	3.525	25	154	179
2" steel, galvanized	13.000	L.F.	3.250	122	142	264
Vanity base cabinet, 2 doors, 30" wide	1.000	Ea.	1.000	249	43.50	292.50
Vanity top, plastic laminated, square edge	2.670	L.F.	.712	71	31	102
Carrier, steel for studs, no arms	1.000	Ea.	1.143	56	55.50	111.50
Wood, 2" x 8" blocking	1.300	L.F.	.052	1.20	2.26	3.46

8 | MECHANICAL — 40 | Four Fixture Bathroom Systems

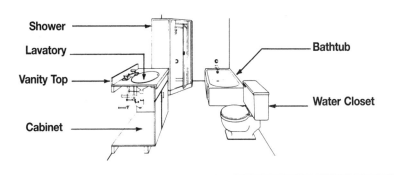

System Description	QUAN.	UNIT	LABOR HOURS	COST EACH MAT.	COST EACH INST.	COST EACH TOTAL
BATHROOM WITH LAVATORY INSTALLED IN VANITY						
Water closet, floor mounted, 2 piece, close coupled, white	1.000	Ea.	3.019	201	132	333
Rough-in, vent, 2" diameter DWV piping	1.000	Ea.	.955	24.80	41.80	66.60
Waste, 4" diameter DWV piping	1.000	Ea.	.828	31.95	36.15	68.10
Supply, 1/2" diameter type "L" copper supply piping	1.000	Ea.	.593	22.56	28.80	51.36
Lavatory, 20" x 18" P.E. cast iron with fittings, white	1.000	Ea.	2.500	242	109	351
Shower, steel, enameled, stone base, corner, white	1.000	Ea.	3.333	845	146	991
Mixing valve	1.000	Ea.	1.333	120	65	185
Shower door	1.000	Ea.	1.000	225	48.50	273.50
Rough-in, waste, 1-1/2" diameter DWV piping	2.000	Ea.	4.507	120	197	317
Supply, 1/2" diameter type "L" copper supply piping	2.000	Ea.	3.161	120.32	153.60	273.92
Bathtub, P.E. cast iron, 5' long with fittings, white	1.000	Ea.	3.636	830	159	989
Rough-in, waste, 4" diameter DWV piping	1.000	Ea.	.828	31.95	36.15	68.10
Supply, 1/2" diameter type "L" copper supply piping	1.000	Ea.	.988	37.60	48	85.60
Vent, 1-1/2" diameter DWV piping	1.000	Ea.	.593	50.80	28.80	79.60
Piping, supply, 1/2" diameter type "L" copper supply piping	42.000	L.F.	4.939	188	240	428
Waste, 4" diameter DWV piping	10.000	L.F.	4.138	159.75	180.75	340.50
Vent, 2" diameter DWV piping	13.000	L.F.	4.500	168.30	197.10	365.40
Vanity base, 2 doors, 30" wide	1.000	Ea.	1.000	249	43.50	292.50
Vanity top, plastic laminated, square edge	2.670	L.F.	.712	73.43	31.11	104.54
TOTAL		Ea.	42.563	3,741.46	1,922.26	5,663.72
BATHROOM WITH WALL HUNG LAVATORY						
Water closet, floor mounted, 2 piece, close coupled, white	1.000	Ea.	3.019	201	132	333
Rough-in, vent, 2" diameter DWV piping	1.000	Ea.	.955	24.80	41.80	66.60
Waste, 4" diameter DWV piping	1.000	Ea.	.828	31.95	36.15	68.10
Supply, 1/2" diameter type "L" copper supply piping	1.000	Ea.	.593	22.56	28.80	51.36
Lavatory, 20" x 18" P.E. cast iron with fittings, white	1.000	Ea.	2.000	305	87.50	392.50
Shower, steel enameled, stone base, corner, white	1.000	Ea.	3.333	845	146	991
Mixing valve	1.000	Ea.	1.333	120	65	185
Shower door	1.000	Ea.	1.000	225	48.50	273.50
Rough-in, waste, 1-1/2" diameter DWV piping	2.000	Ea.	4.507	120	197	317
Supply, 1/2" diameter type "L" copper supply piping	2.000	Ea.	3.161	120.32	153.60	273.92
Bathtub, P.E. cast iron, 5" long with fittings, white	1.000	Ea.	3.636	830	159	989
Rough-in, waste, 4" diameter DWV piping	1.000	Ea.	.828	31.95	36.15	68.10
Supply, 1/2" diameter type "L" copper supply piping	1.000	Ea.	.988	37.60	48	85.60
Vent, 1-1/2" diameter DWV piping	1.000	Ea.	.593	50.80	28.80	79.60
Piping, supply, 1/2" diameter type "L" copper supply piping	42.000	L.F.	4.939	188	240	428
Waste, 4" diameter DWV piping	10.000	L.F.	4.138	159.75	180.75	340.50
Vent, 2" diameter DWV piping	13.000	L.F.	4.500	168.30	197.10	365.40
Carrier, steel for studs, no arms	1.000	Ea.	1.143	56	55.50	111.50
TOTAL		Ea.	41.494	3,538.03	1,881.65	5,419.68

The costs in this system are on a cost each basis. All necessary piping is included.

Four Fixture Bathroom Price Sheet	QUAN.	UNIT	LABOR HOURS	COST EACH MAT.	COST EACH INST.	COST EACH TOTAL
Water closet, close coupled, standard 2 piece, white	1.000	Ea.	3.019	201	132	333
Color	1.000	Ea.	3.019	238	132	370
One piece, elongated bowl, white	1.000	Ea.	3.019	585	132	717
Color	1.000	Ea.	3.019	720	132	852
Low profile, one piece elongated bowl, white	1.000	Ea.	3.019	810	132	942
Color	1.000	Ea.	3.019	1,050	132	1,182
Rough-in, for water closet						
1/2" copper supply, 4" cast iron waste, 2" cast iron vent	1.000	Ea.	2.376	79.50	107	186.50
4" PVC/DWV waste, 2" PVC vent	1.000	Ea.	2.678	45.50	120	165.50
4" copper waste, 2" copper vent	1.000	Ea.	2.520	245	117	362
3" cast iron waste, 1-1/2" cast iron vent	1.000	Ea.	2.244	71.50	101	172.50
3" PVC waste, 1-1/2" PVC vent	1.000	Ea.	2.388	40	112	152
3" PVC waste, 1-1/2" PVC vent	1.000	Ea.	2.014	162	94	256
1/2" PVC supply, 4" PVC waste, 2" PVC vent	1.000	Ea.	2.974	38.50	134	172.50
3" PVC waste, 1-1/2" PVC vent	1.000	Ea.	2.684	33	126	159
1/2" steel supply, 4" cast iron waste, 2" cast iron vent	1.000	Ea.	2.545	73	115	188
4" cast iron waste, 2" steel vent	1.000	Ea.	2.590	85.50	117	202.50
4" PVC waste, 2" PVC vent	1.000	Ea.	2.847	39	128	167
Lavatory wall hung, P.E. cast iron 20" x 18", white	1.000	Ea.	2.000	305	87.50	392.50
Color	1.000	Ea.	2.000	320	87.50	407.50
Vitreous china 19" x 17", white	1.000	Ea.	2.286	197	100	297
Color	1.000	Ea.	2.286	220	100	320
Lavatory for vanity top, P.E. cast iron, 20" x 18", white	1.000	Ea.	2.500	242	109	351
Color	1.000	Ea.	2.500	273	109	382
Steel, enameled 20" x 17", white	1.000	Ea.	2.759	160	121	281
Color	1.000	Ea.	2.500	165	109	274
Vitreous china 20" x 16", white	1.000	Ea.	2.963	291	130	421
Color	1.000	Ea.	2.963	291	130	421
Shower, steel enameled, stone base 36" square, white	1.000	Ea.	8.889	845	146	991
Color	1.000	Ea.	8.889	950	146	1,096
Rough-in, for lavatory and shower						
1/2" copper supply, 1-1/2" cast iron waste, 1-1/2" cast iron vent	1.000	Ea.	7.668	240	350	590
1-1/2" PVC waste, 1-1/4" PVC vent	1.000	Ea.	7.352	154	355	509
1/2" steel supply, 1-1/4" cast iron waste, 1-1/4" steel vent	1.000	Ea.	8.205	205	380	585
1-1/4" PVC waste, 1-1/4" PVC vent	1.000	Ea.	7.873	122	380	502
1/2" PVC supply, 1-1/2" PVC waste, 1-1/2" PVC vent	1.000	Ea.	9.185	114	445	559
Bathtub, P.E. cast iron, 5' long with fittings, white	1.000	Ea.	3.636	830	159	989
Color	1.000	Ea.	3.636	1,025	159	1,184
Steel enameled, 5' long with fittings, white	1.000	Ea.	2.909	345	127	472
Color	1.000	Ea.	2.909	345	127	472
Rough-in, for bathtub						
1/2" copper supply, 4" cast iron waste, 1-1/2" copper vent	1.000	Ea.	2.409	120	113	233
4" PVC waste, 1-1/2" PVC vent	1.000	Ea.	2.877	59	135	194
1/2" steel supply, 4" cast iron waste, 1-1/2" steel vent	1.000	Ea.	2.898	87	133	220
4" PVC waste, 1-1/2" PVC vent	1.000	Ea.	3.159	48.50	149	197.50
1/2" PVC supply, 4" PVC waste, 1-1/2" PVC vent	1.000	Ea.	3.371	47	159	206
Piping supply, 1/2" copper	42.000	L.F.	4.148	158	202	360
1/2" steel	42.000	L.F.	5.333	113	258	371
1/2" PVC	42.000	L.F.	6.222	108	300	408
Piping, waste, 4" cast iron, no hub	10.000	L.F.	3.586	138	157	295
4" PVC/DWV	10.000	L.F.	4.333	65.50	190	255.50
4" copper/DWV	10.000	L.F.	5.200	670	228	898
Piping, vent, 2" cast iron, no hub	13.000	L.F.	3.105	80.50	136	216.50
2" copper/DWV	13.000	L.F.	2.364	221	115	336
2" PVC/DWV	13.000	L.F.	3.525	25	154	179
2" steel, galvanized	13.000	L.F.	3.250	122	142	264
Vanity base cabinet, 2 doors, 30" wide	1.000	Ea.	1.000	249	43.50	292.50
Vanity top, plastic laminated, square edge	3.160	L.F.	.843	83.50	37	120.50
Carrier, steel, for studs, no arms	1.000	Ea.	1.143	56	55.50	111.50
Wood, 2" x 8" blocking	1.300	L.F.	.052	1.20	2.26	3.46

8 | MECHANICAL | 44 | Five Fixture Bathroom Systems

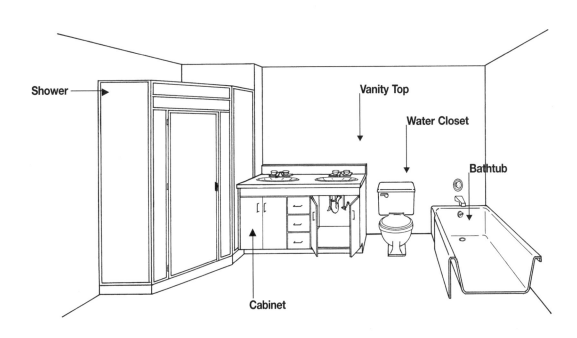

System Description	QUAN.	UNIT	LABOR HOURS	COST EACH MAT.	COST EACH INST.	COST EACH TOTAL
BATHROOM WITH SHOWER, BATHTUB, LAVATORIES IN VANITY						
Water closet, floor mounted, 1 piece combination, white	1.000	Ea.	3.019	810	132	942
Rough-in, vent, 2" diameter DWV piping	1.000	Ea.	.955	24.80	41.80	66.60
Waste, 4" diameter DWV piping	1.000	Ea.	.828	31.95	36.15	68.10
Supply, 1/2" diameter type "L" copper supply piping	1.000	Ea.	.593	22.56	28.80	51.36
Lavatory, 20" x 16", vitreous china oval, with fittings, white	2.000	Ea.	5.926	582	260	842
Shower, steel enameled, stone base, corner, white	1.000	Ea.	3.333	845	146	991
Mixing valve	1.000	Ea.	1.333	120	65	185
Shower door	1.000	Ea.	1.000	225	48.50	273.50
Rough-in, waste, 1-1/2" diameter DWV piping	3.000	Ea.	5.408	144	236.40	380.40
Supply, 1/2" diameter type "L" copper supply piping	3.000	Ea.	2.963	112.80	144	256.80
Bathtub, P.E. cast iron, 5' long with fittings, white	1.000	Ea.	3.636	830	159	989
Rough-in, waste, 4" diameter DWV piping	1.000	Ea.	1.103	42.60	48.20	90.80
Supply, 1/2" diameter type "L" copper supply piping	1.000	Ea.	.988	37.60	48	85.60
Vent, 1-1/2" diameter copper DWV piping	1.000	Ea.	.593	50.80	28.80	79.60
Piping, supply, 1/2" diameter type "L" copper supply piping	42.000	L.F.	4.148	157.92	201.60	359.52
Waste, 4" diameter DWV piping	10.000	L.F.	2.759	106.50	120.50	227
Vent, 2" diameter DWV piping	13.000	L.F.	3.250	121.55	142.35	263.90
Vanity base, 2 door, 24" x 48"	1.000	Ea.	1.400	395	61	456
Vanity top, plastic laminated, square edge	4.170	L.F.	1.112	110.51	48.58	159.09
TOTAL		Ea.	44.347	4,770.59	1,996.68	6,767.27

The costs in this system are on a cost each basis. All necessary piping is included

Description	QUAN.	UNIT	LABOR HOURS	COST EACH MAT.	COST EACH INST.	COST EACH TOTAL

Five Fixture Bathroom Price Sheet	QUAN.	UNIT	LABOR HOURS	COST EACH MAT.	COST EACH INST.	COST EACH TOTAL
Water closet, close coupled, standard 2 piece, white	1.000	Ea.	3.019	201	132	333
Color	1.000	Ea.	3.019	238	132	370
One piece elongated bowl, white	1.000	Ea.	3.019	585	132	717
Color	1.000	Ea.	3.019	720	132	852
Low profile, one piece elongated bowl, white	1.000	Ea.	3.019	810	132	942
Color	1.000	Ea.	3.019	1,050	132	1,182
Rough-in, supply, waste and vent for water closet						
1/2" copper supply, 4" cast iron waste, 2" cast iron vent	1.000	Ea.	2.376	79.50	107	186.50
4" P.V.C./DWV waste, 2" P.V.C. vent	1.000	Ea.	2.678	45.50	120	165.50
4" copper waste, 2" copper vent	1.000	Ea.	2.520	245	117	362
3" cast iron waste, 1-1/2" cast iron vent	1.000	Ea.	2.244	71.50	101	172.50
3" P.V.C. waste, 1-1/2" P.V.C. vent	1.000	Ea.	2.388	40	112	152
3" copper waste, 1-1/2" copper vent	1.000	Ea.	2.014	162	94	256
1/2" P.V.C. supply, 4" P.V.C. waste, 2" P.V.C. vent	1.000	Ea.	2.974	38.50	134	172.50
3" P.V.C. waste, 1-1/2" P.V.C. supply	1.000	Ea.	2.684	33	126	159
1/2" steel supply, 4" cast iron waste, 2" cast iron vent	1.000	Ea.	2.545	73	115	188
4" cast iron waste, 2" steel vent	1.000	Ea.	2.590	85.50	117	202.50
4" P.V.C. waste, 2" P.V.C. vent	1.000	Ea.	2.847	39	128	167
Lavatory, wall hung, P.E. cast iron 20" x 18", white	2.000	Ea.	4.000	610	175	785
Color	2.000	Ea.	4.000	640	175	815
Vitreous china, 19" x 17", white	2.000	Ea.	4.571	395	200	595
Color	2.000	Ea.	4.571	440	200	640
Lavatory, for vanity top, P.E. cast iron, 20" x 18", white	2.000	Ea.	5.000	485	218	703
Color	2.000	Ea.	5.000	545	218	763
Steel enameled 20" x 17", white	2.000	Ea.	5.517	320	242	562
Color	2.000	Ea.	5.000	330	218	548
Vitreous china 20" x 16", white	2.000	Ea.	5.926	580	260	840
Color	2.000	Ea.	5.926	580	260	840
Shower, steel enameled, stone base 36" square, white	1.000	Ea.	8.889	845	146	991
Color	1.000	Ea.	8.889	950	146	1,096
Rough-in, for lavatory or shower						
1/2" copper supply, 1-1/2" cast iron waste, 1-1/2" cast iron vent	3.000	Ea.	8.371	257	380	637
1-1/2" P.V.C. waste, 1-1/4" P.V.C. vent	3.000	Ea.	7.916	153	385	538
1/2" steel supply, 1-1/4" cast iron waste, 1-1/4" steel vent	3.000	Ea.	8.670	223	395	618
1-1/4" P.V.C. waste, 1-1/4" P.V.C. vent	3.000	Ea.	8.381	124	405	529
1/2" P.V.C. supply, 1-1/2" P.V.C. waste, 1-1/2" P.V.C. vent	3.000	Ea.	9.778	115	475	590
Bathtub, P.E. cast iron 5' long with fittings, white	1.000	Ea.	3.636	830	159	989
Color	1.000	Ea.	3.636	1,025	159	1,184
Steel, enameled 5' long with fittings, white	1.000	Ea.	2.909	345	127	472
Color	1.000	Ea.	2.909	345	127	472
Rough-in, for bathtub						
1/2" copper supply, 4" cast iron waste, 1-1/2" copper vent	1.000	Ea.	2.684	131	125	256
4" P.V.C. waste, 1-1/2" P.V.C. vent	1.000	Ea.	3.210	64	150	214
1/2" steel supply, 4" cast iron waste, 1-1/2" steel vent	1.000	Ea.	3.173	97.50	145	242.50
4" P.V.C. waste, 1-1/2" P.V.C. vent	1.000	Ea.	3.492	53.50	163	216.50
1/2" P.V.C. supply, 4" P.V.C. waste, 1-1/2" P.V.C. vent	1.000	Ea.	3.704	52	174	226
Piping, supply, 1/2" copper	42.000	L.F.	4.148	158	202	360
1/2" steel	42.000	L.F.	5.333	113	258	371
1/2" P.V.C.	42.000	L.F.	6.222	108	300	408
Piping, waste, 4" cast iron, no hub	10.000	L.F.	2.759	107	121	228
4" P.V.C./DWV	10.000	L.F.	3.333	50.50	146	196.50
4" copper/DWV	10.000	L.F.	4.000	515	175	690
Piping, vent, 2" cast iron, no hub	13.000	L.F.	3.105	80.50	136	216.50
2" copper/DWV	13.000	L.F.	2.364	221	115	336
2" P.V.C./DWV	13.000	L.F.	3.525	25	154	179
2" steel, galvanized	13.000	L.F.	3.250	122	142	264
Vanity base cabinet, 2 doors, 24" x 48"	1.000	Ea.	1.400	395	61	456
Vanity top, plastic laminated, square edge	4.170	L.F.	1.112	111	48.50	159.50
Carrier, steel, for studs, no arms	1.000	Ea.	1.143	56	55.50	111.50
Wood, 2" x 8" blocking	1.300	L.F.	.052	1.20	2.26	3.46

8 | MECHANICAL 60 | Gas Heating/Cooling Systems

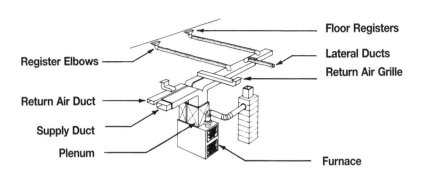

System Description	QUAN.	UNIT	LABOR HOURS	COST PER SYSTEM			
				MAT.	INST.	TOTAL
HEATING ONLY, GAS FIRED HOT AIR, ONE ZONE, 1200 S.F. BUILDING						
Furnace, gas, up flow	1.000	Ea.	5.000	745	218	963
Intermittent pilot	1.000	Ea.		151		151
Supply duct, rigid fiberglass	176.000	S.F.	12.068	119.68	543.84	663.52
Return duct, sheet metal, galvanized	158.000	Lb.	16.137	165.90	728.38	894.28
Lateral ducts, 6" flexible fiberglass	144.000	L.F.	8.862	401.76	385.92	787.68
Register, elbows	12.000	Ea.	3.200	456	139.20	595.20
Floor registers, enameled steel	12.000	Ea.	3.000	264	145.20	409.20
Floor grille, return air	2.000	Ea.	.727	56	35.20	91.20
Thermostat	1.000	Ea.	1.000	29.50	48.50	78
Plenum	1.000	Ea.	1.000	77	43.50	120.50
TOTAL		System	50.994	2,465.84	2,287.74	4,753.58
HEATING/COOLING, GAS FIRED FORCED AIR, ONE ZONE, 1200 S.F. BUILDING						
Furnace, including plenum, compressor, coil	1.000	Ea.	14.720	4,347	639.40	4,986.40
Intermittent pilot	1.000	Ea.		151		151
Supply duct, rigid fiberglass	176.000	S.F.	12.068	119.68	543.84	663.52
Return duct, sheet metal, galvanized	158.000	Lb.	16.137	165.90	728.38	894.28
Lateral duct, 6" flexible fiberglass	144.000	L.F.	8.862	401.76	385.92	787.68
Register elbows	12.000	Ea.	3.200	456	139.20	595.20
Floor registers, enameled steel	12.000	Ea.	3.000	264	145.20	409.20
Floor grille return air	2.000	Ea.	.727	56	35.20	91.20
Thermostat	1.000	Ea.	1.000	29.50	48.50	78
Refrigeration piping, 25 ft. (pre-charged)	1.000	Ea.		224		224
TOTAL		System	59.714	6,214.84	2,665.64	8,880.48

The costs in these systems are based on complete system basis. For larger buildings use the price sheet on the opposite page.

Description	QUAN.	UNIT	LABOR HOURS	COST PER SYSTEM		
				MAT.	INST.	TOTAL

Gas Heating/Cooling Price Sheet	QUAN.	UNIT	LABOR HOURS	COST EACH		
				MAT.	INST.	TOTAL
Furnace, heating only, 100 MBH, area to 1200 S.F.	1.000	Ea.	5.000	745	218	963
120 MBH, area to 1500 S.F.	1.000	Ea.	5.000	745	218	963
160 MBH, area to 2000 S.F.	1.000	Ea.	5.714	1,000	249	1,249
200 MBH, area to 2400 S.F.	1.000	Ea.	6.154	2,400	268	2,668
Heating/cooling, 100 MBH heat, 36 MBH cool, to 1200 S.F.	1.000	Ea.	16.000	4,725	695	5,420
120 MBH heat, 42 MBH cool, to 1500 S.F.	1.000	Ea.	18.462	5,050	835	5,885
144 MBH heat, 47 MBH cool, to 2000 S.F.	1.000	Ea.	20.000	5,825	905	6,730
200 MBH heat, 60 MBH cool, to 2400 S.F.	1.000	Ea.	34.286	6,125	1,550	7,675
Intermittent pilot, 100 MBH furnace	1.000	Ea.		151		151
200 MBH furnace	1.000	Ea.		151		151
Supply duct, rectangular, area to 1200 S.F., rigid fiberglass	176.000	S.F.	12.068	120	545	665
Sheet metal insulated	228.000	Lb.	31.331	360	1,400	1,760
Area to 1500 S.F., rigid fiberglass	176.000	S.F.	12.068	120	545	665
Sheet metal insulated	228.000	Lb.	31.331	360	1,400	1,760
Area to 2400 S.F., rigid fiberglass	205.000	S.F.	14.057	139	635	774
Sheet metal insulated	271.000	Lb.	37.048	425	1,650	2,075
Round flexible, insulated 6" diameter, to 1200 S.F.	156.000	L.F.	9.600	435	420	855
To 1500 S.F.	184.000	L.F.	11.323	515	495	1,010
8" diameter, to 2000 S.F.	269.000	L.F.	23.911	935	1,050	1,985
To 2400 S.F.	248.000	L.F.	22.045	860	960	1,820
Return duct, sheet metal galvanized, to 1500 S.F.	158.000	Lb.	16.137	166	730	896
To 2400 S.F.	191.000	Lb.	19.507	201	880	1,081
Lateral ducts, flexible round 6" insulated, to 1200 S.F.	144.000	L.F.	8.862	400	385	785
To 1500 S.F.	172.000	L.F.	10.585	480	460	940
To 2000 S.F.	261.000	L.F.	16.062	730	700	1,430
To 2400 S.F.	300.000	L.F.	18.462	835	805	1,640
Spiral steel insulated, to 1200 S.F.	144.000	L.F.	20.067	455	860	1,315
To 1500 S.F.	172.000	L.F.	23.952	545	1,025	1,570
To 2000 S.F.	261.000	L.F.	36.352	825	1,550	2,375
To 2400 S.F.	300.000	L.F.	41.825	950	1,800	2,750
Rectangular sheet metal galvanized insulated, to 1200 S.F.	228.000	Lb.	39.056	475	1,725	2,200
To 1500 S.F.	344.000	Lb.	53.966	640	2,375	3,015
To 2000 S.F.	522.000	Lb.	81.926	975	3,625	4,600
To 2400 S.F.	600.000	Lb.	94.189	1,125	4,175	5,300
Register elbows, to 1500 S.F.	12.000	Ea.	3.200	455	139	594
To 2400 S.F.	14.000	Ea.	3.733	530	162	692
Floor registers, enameled steel w/damper, to 1500 S.F.	12.000	Ea.	3.000	264	145	409
To 2400 S.F.	14.000	Ea.	4.308	365	209	574
Return air grille, area to 1500 S.F. 12" x 12"	2.000	Ea.	.727	56	35	91
Area to 2400 S.F. 8" x 16"	2.000	Ea.	.444	49.50	21.50	71
Area to 2400 S.F. 8" x 16"	2.000	Ea.	.727	56	35	91
16" x 16"	1.000	Ea.	.364	40	17.60	57.60
Thermostat, manual, 1 set back	1.000	Ea.	1.000	29.50	48.50	78
Electric, timed, 1 set back	1.000	Ea.	1.000	104	48.50	152.50
2 set back	1.000	Ea.	1.000	229	48.50	277.50
Plenum, heating only, 100 M.B.H.	1.000	Ea.	1.000	77	43.50	120.50
120 MBH	1.000	Ea.	1.000	77	43.50	120.50
160 MBH	1.000	Ea.	1.000	77	43.50	120.50
200 MBH	1.000	Ea.	1.000	77	43.50	120.50
Refrigeration piping, 3/8"	25.000	L.F.		49		49
3/4"	25.000	L.F.		98.50		98.50
7/8"	25.000	L.F.		115		115
Refrigerant piping, 25 ft. (precharged)	1.000	Ea.		224		224
Diffusers, ceiling, 6" diameter, to 1500 S.F.	10.000	Ea.	4.444	205	215	420
To 2400 S.F.	12.000	Ea.	6.000	264	288	552
Floor, aluminum, adjustable, 2-1/4" x 12" to 1500 S.F.	12.000	Ea.	3.000	193	145	338
To 2400 S.F.	14.000	Ea.	3.500	225	169	394
Side wall, aluminum, adjustable, 8" x 4", to 1500 S.F.	12.000	Ea.	3.000	425	145	570
5" x 10" to 2400 S.F.	12.000	Ea.	3.692	560	179	739

8 | MECHANICAL — 64 | Oil Fired Heating/Cooling Systems

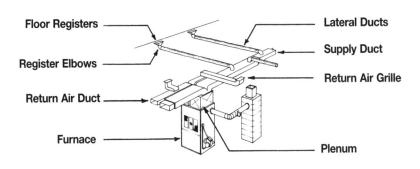

System Description	QUAN.	UNIT	LABOR HOURS	COST PER SYSTEM MAT.	INST.	TOTAL
HEATING ONLY, OIL FIRED HOT AIR, ONE ZONE, 1200 S.F. BUILDING						
Furnace, oil fired, atomizing gun type burner	1.000	Ea.	4.571	1,725	199	1,924
3/8" diameter copper supply pipe	1.000	Ea.	2.759	74.70	134.10	208.80
Shut off valve	1.000	Ea.	.333	10.50	16.20	26.70
Oil tank, 275 gallon, on legs	1.000	Ea.	3.200	365	141	506
Supply duct, rigid fiberglass	176.000	S.F.	12.068	119.68	543.84	663.52
Return duct, sheet metal, galvanized	158.000	Lb.	16.137	165.90	728.38	894.28
Lateral ducts, 6" flexible fiberglass	144.000	L.F.	8.862	401.76	385.92	787.68
Register elbows	12.000	Ea.	3.200	456	139.20	595.20
Floor register, enameled steel	12.000	Ea.	3.000	264	145.20	409.20
Floor grille, return air	2.000	Ea.	.727	56	35.20	91.20
Thermostat	1.000	Ea.	1.000	29.50	48.50	78
TOTAL		System	55.857	3,668.04	2,516.54	6,184.58
HEATING/COOLING, OIL FIRED, FORCED AIR, ONE ZONE, 1200 S.F. BUILDING						
Furnace, including plenum, compressor, coil	1.000	Ea.	16.000	5,050	695	5,745
3/8" diameter copper supply pipe	1.000	Ea.	2.759	74.70	134.10	208.80
Shut off valve	1.000	Ea.	.333	10.50	16.20	26.70
Oil tank, 275 gallon on legs	1.000	Ea.	3.200	365	141	506
Supply duct, rigid fiberglass	176.000	S.F.	12.068	119.68	543.84	663.52
Return duct, sheet metal, galvanized	158.000	Lb.	16.137	165.90	728.38	894.28
Lateral ducts, 6" flexible fiberglass	144.000	L.F.	8.862	401.76	385.92	787.68
Register elbows	12.000	Ea.	3.200	456	139.20	595.20
Floor registers, enameled steel	12.000	Ea.	3.000	264	145.20	409.20
Floor grille, return air	2.000	Ea.	.727	56	35.20	91.20
Refrigeration piping (precharged)	25.000	L.F.		224		224
TOTAL		System	66.286	7,187.54	2,964.04	10,151.58

Description	QUAN.	UNIT	LABOR HOURS	COST EACH MAT.	INST.	TOTAL

Oil Fired Heating/Cooling	QUAN.	UNIT	LABOR HOURS	COST EACH MAT.	COST EACH INST.	COST EACH TOTAL
Furnace, heating, 95.2 MBH, area to 1200 S.F.	1.000	Ea.	4.706	1,750	205	1,955
123.2 MBH, area to 1500 S.F.	1.000	Ea.	5.000	1,800	218	2,018
151.2 MBH, area to 2000 S.F.	1.000	Ea.	5.333	2,100	232	2,332
200 MBH, area to 2400 S.F.	1.000	Ea.	6.154	2,550	268	2,818
Heating/cooling, 95.2 MBH heat, 36 MBH cool, to 1200 S.F.	1.000	Ea.	16.000	5,050	695	5,745
112 MBH heat, 42 MBH cool, to 1500 S.F.	1.000	Ea.	24.000	7,575	1,050	8,625
151 MBH heat, 47 MBH cool, to 2000 S.F.	1.000	Ea.	20.800	6,575	905	7,480
184.8 MBH heat, 60 MBH cool, to 2400 S.F.	1.000	Ea.	24.000	6,925	1,075	8,000
Oil piping to furnace, 3/8" dia., copper	1.000	Ea.	3.412	186	164	350
Oil tank, on legs above ground, 275 gallons	1.000	Ea.	3.200	365	141	506
550 gallons	1.000	Ea.	5.926	1,725	262	1,987
Below ground, 275 gallons	1.000	Ea.	3.200	365	141	506
550 gallons	1.000	Ea.	5.926	1,725	262	1,987
1000 gallons	1.000	Ea.	6.400	2,800	283	3,083
Supply duct, rectangular, area to 1200 S.F., rigid fiberglass	176.000	S.F.	12.068	120	545	665
Sheet metal, insulated	228.000	Lb.	31.331	360	1,400	1,760
Area to 1500 S.F., rigid fiberglass	176.000	S.F.	12.068	120	545	665
Sheet metal, insulated	228.000	Lb.	31.331	360	1,400	1,760
Area to 2400 S.F., rigid fiberglass	205.000	S.F.	14.057	139	635	774
Sheet metal, insulated	271.000	Lb.	37.048	425	1,650	2,075
Round flexible, insulated, 6" diameter to 1200 S.F.	156.000	L.F.	9.600	435	420	855
To 1500 S.F.	184.000	L.F.	11.323	515	495	1,010
8" diameter to 2000 S.F.	269.000	L.F.	23.911	935	1,050	1,985
To 2400 S.F.	269.000	L.F.	22.045	860	960	1,820
Return duct, sheet metal galvanized, to 1500 S.F.	158.000	Lb.	16.137	166	730	896
To 2400 S.F.	191.000	Lb.	19.507	201	880	1,081
Lateral ducts, flexible round, 6", insulated to 1200 S.F.	144.000	L.F.	8.862	400	385	785
To 1500 S.F.	172.000	L.F.	10.585	480	460	940
To 2000 S.F.	261.000	L.F.	16.062	730	700	1,430
To 2400 S.F.	300.000	L.F.	18.462	835	805	1,640
Spiral steel, insulated to 1200 S.F.	144.000	L.F.	20.067	455	860	1,315
To 1500 S.F.	172.000	L.F.	23.952	545	1,025	1,570
To 2000 S.F.	261.000	L.F.	36.352	825	1,550	2,375
To 2400 S.F.	300.000	L.F.	41.825	950	1,800	2,750
Rectangular sheet metal galvanized insulated, to 1200 S.F.	288.000	Lb.	45.183	535	2,000	2,535
To 1500 S.F.	344.000	Lb.	53.966	640	2,375	3,015
To 2000 S.F.	522.000	Lb.	81.926	975	3,625	4,600
To 2400 S.F.	600.000	Lb.	94.189	1,125	4,175	5,300
Register elbows, to 1500 S.F.	12.000	Ea.	3.200	455	139	594
To 2400 S.F.	14.000	Ea.	3.733	530	162	692
Floor registers, enameled steel w/damper, to 1500 S.F.	12.000	Ea.	3.000	264	145	409
To 2400 S.F.	14.000	Ea.	4.308	365	209	574
Return air grille, area to 1500 S.F., 12" x 12"	2.000	Ea.	.727	56	35	91
12" x 24"	1.000	Ea.	.444	49.50	21.50	71
Area to 2400 S.F., 8" x 16"	2.000	Ea.	.727	56	35	91
16" x 16"	1.000	Ea.	.364	40	17.60	57.60
Thermostat, manual, 1 set back	1.000	Ea.	1.000	29.50	48.50	78
Electric, timed, 1 set back	1.000	Ea.	1.000	104	48.50	152.50
2 set back	1.000	Ea.	1.000	229	48.50	277.50
Refrigeration piping, 3/8"	25.000	L.F.		49		49
3/4"	25.000	L.F.		98.50		98.50
Diffusers, ceiling, 6" diameter, to 1500 S.F.	10.000	Ea.	4.444	205	215	420
To 2400 S.F.	12.000	Ea.	6.000	264	288	552
Floor, aluminum, adjustable, 2-1/4" x 12" to 1500 S.F.	12.000	Ea.	3.000	193	145	338
To 2400 S.F.	14.000	Ea.	3.500	225	169	394
Side wall, aluminum, adjustable, 8" x 4", to 1500 S.F.	12.000	Ea.	3.000	425	145	570
5" x 10" to 2400 S.F.	12.000	Ea.	3.692	560	179	739

8 | MECHANICAL — 68 | Hot Water Heating Systems

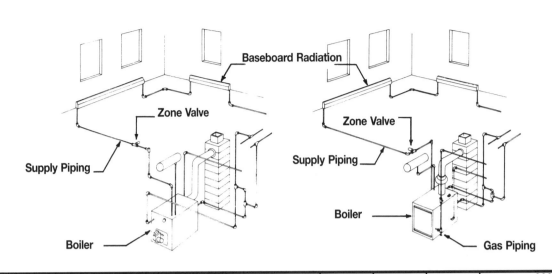

System Description	QUAN.	UNIT	LABOR HOURS	COST EACH MAT.	COST EACH INST.	COST EACH TOTAL
OIL FIRED HOT WATER HEATING SYSTEM, AREA TO 1200 S.F.						
Boiler package, oil fired, 97 MBH, area to 1200 S.F. building	1.000	Ea.	15.000	1,600	640	2,240
3/8" diameter copper supply pipe	1.000	Ea.	2.759	74.70	134.10	208.80
Shut off valve	1.000	Ea.	.333	10.50	16.20	26.70
Oil tank, 275 gallon, with black iron filler pipe	1.000	Ea.	3.200	365	141	506
Supply piping, 3/4" copper tubing	176.000	L.F.	18.526	1,003.20	897.60	1,900.80
Supply fittings, copper 3/4"	36.000	Ea.	15.158	60.12	738	798.12
Supply valves, 3/4"	2.000	Ea.	.800	142	38.90	180.90
Baseboard radiation, 3/4"	106.000	L.F.	35.333	464.28	1,563.50	2,027.78
Zone valve	1.000	Ea.	.400	146	19.60	165.60
TOTAL		Ea.	91.509	3,865.80	4,188.90	8,054.70
OIL FIRED HOT WATER HEATING SYSTEM, AREA TO 2400 S.F.						
Boiler package, oil fired, 225 MBH, area to 2400 S.F. building	1.000	Ea.	25.105	4,700	1,075	5,775
3/8" diameter copper supply pipe	1.000	Ea.	2.759	74.70	134.10	208.80
Shut off valve	1.000	Ea.	.333	10.50	16.20	26.70
Oil tank, 550 gallon, with black iron pipe filler pipe	1.000	Ea.	5.926	1,725	262	1,987
Supply piping, 3/4" copper tubing	228.000	L.F.	23.999	1,299.60	1,162.80	2,462.40
Supply fittings, copper	46.000	Ea.	19.368	76.82	943	1,019.82
Supply valves	2.000	Ea.	.800	142	38.90	180.90
Baseboard radiation	212.000	L.F.	70.666	928.56	3,127	4,055.56
Zone valve	1.000	Ea.	.400	146	19.60	165.60
TOTAL		Ea.	149.356	9,103.18	6,778.60	15,881.78

The costs in this system are on a cost each basis. The costs represent total cost for the system based on a gross square foot of plan area.

Description	QUAN.	UNIT	LABOR HOURS	COST EACH MAT.	COST EACH INST.	COST EACH TOTAL

Hot Water Heating Price Sheet	QUAN.	UNIT	LABOR HOURS	COST EACH		
				MAT.	INST.	TOTAL
Boiler, oil fired, 97 MBH, area to 1200 S.F.	1.000	Ea.	15.000	1,600	640	2,240
118 MBH, area to 1500 S.F.	1.000	Ea.	16.506	3,025	700	3,725
161 MBH, area to 2000 S.F.	1.000	Ea.	18.405	3,800	785	4,585
215 MBH, area to 2400 S.F.	1.000	Ea.	19.704	3,725	795	4,520
Oil piping, (valve & filter), 3/8" copper	1.000	Ea.	3.289	85	150	235
1/4" copper	1.000	Ea.	3.242	101	157	258
Oil tank, filler pipe and cap on legs, 275 gallon	1.000	Ea.	3.200	365	141	506
550 gallon	1.000	Ea.	5.926	1,725	262	1,987
Buried underground, 275 gallon	1.000	Ea.	3.200	365	141	506
550 gallon	1.000	Ea.	5.926	1,725	262	1,987
1000 gallon	1.000	Ea.	6.400	2,800	283	3,083
Supply piping copper, area to 1200 S.F., 1/2" tubing	176.000	L.F.	17.384	660	845	1,505
3/4" tubing	176.000	L.F.	18.526	1,000	900	1,900
Area to 1500 S.F., 1/2" tubing	186.000	L.F.	18.371	700	895	1,595
3/4" tubing	186.000	L.F.	19.578	1,050	950	2,000
Area to 2000 S.F., 1/2" tubing	204.000	L.F.	20.149	765	980	1,745
3/4" tubing	204.000	L.F.	21.473	1,175	1,050	2,225
Area to 2400 S.F., 1/2" tubing	228.000	L.F.	22.520	855	1,100	1,955
3/4" tubing	228.000	L.F.	23.999	1,300	1,175	2,475
Supply pipe fittings copper, area to 1200 S.F., 1/2"	36.000	Ea.	14.400	27	700	727
3/4"	36.000	Ea.	15.158	60	740	800
Area to 1500 S.F., 1/2"	40.000	Ea.	16.000	30	780	810
3/4"	40.000	Ea.	16.842	67	820	887
Area to 2000 S.F., 1/2"	44.000	Ea.	17.600	33	855	888
3/4"	44.000	Ea.	18.526	73.50	900	973.50
Area to 2400, S.F., 1/2"	46.000	Ea.	18.400	34.50	895	929.50
3/4"	46.000	Ea.	19.368	77	945	1,022
Supply valves, 1/2" pipe size	2.000	Ea.	.667	104	32.50	136.50
3/4"	2.000	Ea.	.800	142	39	181
Baseboard radiation, area to 1200 S.F., 1/2" tubing	106.000	L.F.	28.267	780	1,250	2,030
3/4" tubing	106.000	L.F.	35.333	465	1,575	2,040
Area to 1500 S.F., 1/2" tubing	134.000	L.F.	35.734	985	1,575	2,560
3/4" tubing	134.000	L.F.	44.666	585	1,975	2,560
Area to 2000 S.F., 1/2" tubing	178.000	L.F.	47.467	1,300	2,100	3,400
3/4" tubing	178.000	L.F.	59.333	780	2,625	3,405
Area to 2400 S.F., 1/2" tubing	212.000	L.F.	56.534	1,550	2,500	4,050
3/4" tubing	212.000	L.F.	70.666	930	3,125	4,055
Zone valves, 1/2" tubing	1.000	Ea.	.400	146	19.60	165.60
3/4" tubing	1.000	Ea.	.400	147	19.60	166.60

8 | MECHANICAL — 80 | Rooftop Systems

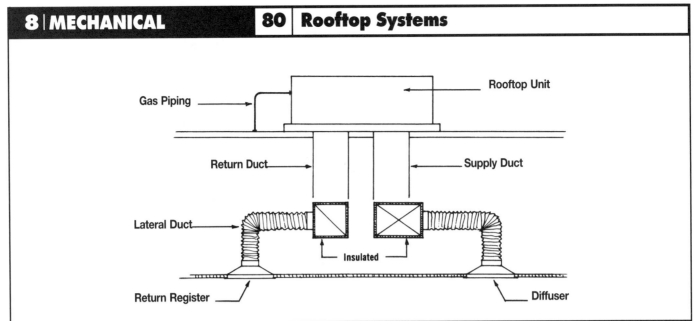

System Description	QUAN.	UNIT	LABOR HOURS	COST EACH MAT.	COST EACH INST.	COST EACH TOTAL
ROOFTOP HEATING/COOLING UNIT, AREA TO 2000 S.F.						
Rooftop unit, single zone, electric cool, gas heat, to 2000 s.f.	1.000	Ea.	28.521	5,400	1,250	6,650
Gas piping	34.500	L.F.	5.207	116.61	253.58	370.19
Duct, supply and return, galvanized steel	38.000	Lb.	3.881	39.90	175.18	215.08
Insulation, ductwork	33.000	S.F.	1.508	22.44	64.02	86.46
Lateral duct, flexible duct 12" diameter, insulated	72.000	L.F.	11.520	367.20	500.40	867.60
Diffusers	4.000	Ea.	4.571	1,320	222	1,542
Return registers	1.000	Ea.	.727	132	35	167
TOTAL		Ea.	55.935	7,398.15	2,500.18	9,898.33
ROOFTOP HEATING/COOLING UNIT, AREA TO 5000 S.F.						
Rooftop unit, single zone, electric cool, gas heat, to 5000 s.f.	1.000	Ea.	42.032	15,800	1,800	17,600
Gas piping	86.250	L.F.	13.019	291.53	633.94	925.47
Duct supply and return, galvanized steel	95.000	Lb.	9.702	99.75	437.95	537.70
Insulation, ductwork	82.000	S.F.	3.748	55.76	159.08	214.84
Lateral duct, flexible duct, 12" diameter, insulated	180.000	L.F.	28.800	918	1,251	2,169
Diffusers	10.000	Ea.	11.429	3,300	555	3,855
Return registers	3.000	Ea.	2.182	396	105	501
TOTAL		Ea.	110.912	20,861.04	4,941.97	25,803.01

Description	QUAN.	UNIT	LABOR HOURS	COST EACH MAT.	COST EACH INST.	COST EACH TOTAL

Rooftop Price Sheet	QUAN.	UNIT	LABOR HOURS	COST EACH		
				MAT.	INST.	TOTAL
Rooftop unit, single zone, electric cool, gas heat to 2000 S.F.	1.000	Ea.	28.521	5,400	1,250	6,650
Area to 3000 S.F.	1.000	Ea.	35.982	10,100	1,525	11,625
Area to 5000 S.F.	1.000	Ea.	42.032	15,800	1,800	17,600
Area to 10000 S.F.	1.000	Ea.	68.376	31,100	3,025	34,125
Gas piping, area 2000 through 4000 S.F.	34.500	L.F.	5.207	117	254	371
Area 5000 to 10000 S.F.	86.250	L.F.	13.019	292	635	927
Duct, supply and return, galvanized steel, to 2000 S.F.	38.000	Lb.	3.881	40	175	215
Area to 3000 S.F.	57.000	Lb.	5.821	60	263	323
Area to 5000 S.F.	95.000	Lb.	9.702	100	440	540
Area to 10000 S.F.	190.000	Lb.	19.405	200	875	1,075
Rigid fiberglass, area to 2000 S.F.	33.000	S.F.	2.263	22.50	102	124.50
Area to 3000 S.F.	49.000	S.F.	3.360	33.50	151	184.50
Area to 5000 S.F.	82.000	S.F.	5.623	56	253	309
Area to 10000 S.F.	164.000	S.F.	11.245	112	505	617
Insulation, supply and return, blanket type, area to 2000 S.F.	33.000	S.F.	1.508	22.50	64	86.50
Area to 3000 S.F.	49.000	S.F.	2.240	33.50	95	128.50
Area to 5000 S.F.	82.000	S.F.	3.748	56	159	215
Area to 10000 S.F.	164.000	S.F.	7.496	112	320	432
Lateral ducts, flexible round, 12" insulated, to 2000 S.F.	72.000	L.F.	11.520	365	500	865
Area to 3000 S.F.	108.000	L.F.	17.280	550	750	1,300
Area to 5000 S.F.	180.000	L.F.	28.800	920	1,250	2,170
Area to 10000 S.F.	360.000	L.F.	57.600	1,825	2,500	4,325
Rectangular, galvanized steel, to 2000 S.F.	239.000	Lb.	24.409	251	1,100	1,351
Area to 3000 S.F.	360.000	Lb.	36.767	380	1,650	2,030
Area to 5000 S.F.	599.000	Lb.	61.176	630	2,750	3,380
Area to 10000 S.F.	998.000	Lb.	101.926	1,050	4,600	5,650
Diffusers, ceiling, 1 to 4 way blow, 24" x 24", to 2000 S.F.	4.000	Ea.	4.571	1,325	222	1,547
Area to 3000 S.F.	6.000	Ea.	6.857	1,975	335	2,310
Area to 5000 S.F.	10.000	Ea.	11.429	3,300	555	3,855
Area to 10000 S.F.	20.000	Ea.	22.857	6,600	1,100	7,700
Return grilles, 24" x 24", to 2000 S.F.	1.000	Ea.	.727	132	35	167
Area to 3000 S.F.	2.000	Ea.	1.455	264	70	334
Area to 5000 S.F.	3.000	Ea.	2.182	395	105	500
Area to 10000 S.F.	5.000	Ea.	3.636	660	175	835

Division 9
Electrical

9 | ELECTRICAL
10 | Electric Service Systems

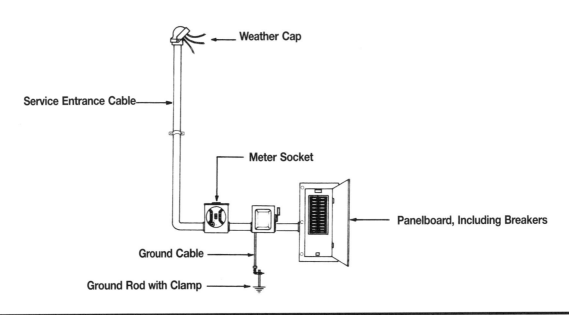

System Description	QUAN.	UNIT	LABOR HOURS	COST EACH MAT.	COST EACH INST.	COST EACH TOTAL
100 AMP SERVICE						
Weather cap	1.000	Ea.	.667	13.40	32	45.40
Service entrance cable	10.000	L.F.	.762	59	36.50	95.50
Meter socket	1.000	Ea.	2.500	41	120	161
Ground rod with clamp	1.000	Ea.	1.455	15.45	69.50	84.95
Ground cable	5.000	L.F.	.250	9.65	12	21.65
Panel board, 12 circuit	1.000	Ea.	6.667	249	320	569
TOTAL		Ea.	12.301	387.50	590	977.50
200 AMP SERVICE						
Weather cap	1.000	Ea.	1.000	37	48	85
Service entrance cable	10.000	L.F.	1.143	69	54.50	123.50
Meter socket	1.000	Ea.	4.211	61.50	202	263.50
Ground rod with clamp	1.000	Ea.	1.818	35.50	87	122.50
Ground cable	10.000	L.F.	.500	19.30	24	43.30
3/4" EMT	5.000	L.F.	.308	5.45	14.75	20.20
Panel board, 24 circuit	1.000	Ea.	12.308	570	490	1,060
TOTAL		Ea.	21.288	797.75	920.25	1,718
400 AMP SERVICE						
Weather cap	1.000	Ea.	2.963	395	142	537
Service entrance cable	180.000	L.F.	5.760	824.40	275.40	1,099.80
Meter socket	1.000	Ea.	4.211	61.50	202	263.50
Ground rod with clamp	1.000	Ea.	2.000	95.50	96	191.50
Ground cable	20.000	L.F.	.485	58.60	23.20	81.80
3/4" greenfield	20.000	L.F.	1.000	11.20	48	59.20
Current transformer cabinet	1.000	Ea.	6.154	193	295	488
Panel board, 42 circuit	1.000	Ea.	33.333	2,925	1,600	4,525
TOTAL		Ea.	55.906	4,564.20	2,681.60	7,245.80

9 | ELECTRICAL — 20 | Electric Perimeter Heating Systems

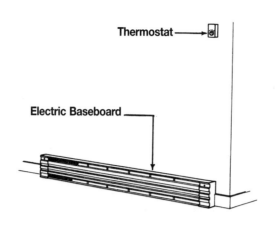

System Description	QUAN.	UNIT	LABOR HOURS	COST EACH MAT.	COST EACH INST.	COST EACH TOTAL
4' BASEBOARD HEATER						
Electric baseboard heater, 4' long	1.000	Ea.	1.194	54	57	111
Thermostat, integral	1.000	Ea.	.500	21.50	24	45.50
Romex, 12-3 with ground	40.000	L.F.	1.600	38.40	76.80	115.20
Panel board breaker, 20 Amp	1.000	Ea.	.300	10.05	14.40	24.45
TOTAL		Ea.	3.594	123.95	172.20	296.15
6' BASEBOARD HEATER						
Electric baseboard heater, 6' long	1.000	Ea.	1.600	71	76.50	147.50
Thermostat, integral	1.000	Ea.	.500	21.50	24	45.50
Romex, 12-3 with ground	40.000	L.F.	1.600	38.40	76.80	115.20
Panel board breaker, 20 Amp	1.000	Ea.	.400	13.40	19.20	32.60
TOTAL		Ea.	4.100	144.30	196.50	340.80
8' BASEBOARD HEATER						
Electric baseboard heater, 8' long	1.000	Ea.	2.000	89.50	96	185.50
Thermostat, integral	1.000	Ea.	.500	21.50	24	45.50
Romex, 12-3 with ground	40.000	L.F.	1.600	38.40	76.80	115.20
Panel board breaker, 20 Amp	1.000	Ea.	.500	16.75	24	40.75
TOTAL		Ea.	4.600	166.15	220.80	386.95
10' BASEBOARD HEATER						
Electric baseboard heater, 10' long	1.000	Ea.	2.424	111	116	227
Thermostat, integral	1.000	Ea.	.500	21.50	24	45.50
Romex, 12-3 with ground	40.000	L.F.	1.600	38.40	76.80	115.20
Panel board breaker, 20 Amp	1.000	Ea.	.750	25.13	36	61.13
TOTAL		Ea.	5.274	196.03	252.80	448.83

The costs in this system are on a cost each basis and include all necessary conduit fittings.

Description	QUAN.	UNIT	LABOR HOURS	COST EACH MAT.	COST EACH INST.	COST EACH TOTAL

9 | ELECTRICAL — 30 | Wiring Device Systems

System Description	QUAN.	UNIT	LABOR HOURS	COST EACH MAT.	COST EACH INST.	COST EACH TOTAL
Air conditioning receptacles						
Using non-metallic sheathed cable	1.000	Ea.	.800	25.50	38.50	64
Using BX cable	1.000	Ea.	.964	36.50	46	82.50
Using EMT conduit	1.000	Ea.	1.194	48	57	105
Disposal wiring						
Using non-metallic sheathed cable	1.000	Ea.	.889	21.50	42.50	64
Using BX cable	1.000	Ea.	1.067	31	51	82
Using EMT conduit	1.000	Ea.	1.333	45	64	109
Dryer circuit						
Using non-metallic sheathed cable	1.000	Ea.	1.455	60	69.50	129.50
Using BX cable	1.000	Ea.	1.739	57	83.50	140.50
Using EMT conduit	1.000	Ea.	2.162	63	104	167
Duplex receptacles						
Using non-metallic sheathed cable	1.000	Ea.	.615	25.50	29.50	55
Using BX cable	1.000	Ea.	.741	36.50	35.50	72
Using EMT conduit	1.000	Ea.	.920	48	44	92
Exhaust fan wiring						
Using non-metallic sheathed cable	1.000	Ea.	.800	23	38.50	61.50
Using BX cable	1.000	Ea.	.964	34.50	46	80.50
Using EMT conduit	1.000	Ea.	1.194	45.50	57	102.50
Furnace circuit & switch						
Using non-metallic sheathed cable	1.000	Ea.	1.333	26.50	64	90.50
Using BX cable	1.000	Ea.	1.600	42	76.50	118.50
Using EMT conduit	1.000	Ea.	2.000	48.50	96	144.50
Ground fault						
Using non-metallic sheathed cable	1.000	Ea.	1.000	56	48	104
Using BX cable	1.000	Ea.	1.212	64.50	58	122.50
Using EMT conduit	1.000	Ea.	1.481	84	71	155
Heater circuits						
Using non-metallic sheathed cable	1.000	Ea.	1.000	29.50	48	77.50
Using BX cable	1.000	Ea.	1.212	34.50	58	92.50
Using EMT conduit	1.000	Ea.	1.481	45	71	116
Lighting wiring						
Using non-metallic sheathed cable	1.000	Ea.	.500	30	24	54
Using BX cable	1.000	Ea.	.602	35.50	29	64.50
Using EMT conduit	1.000	Ea.	.748	43.50	36	79.50
Range circuits						
Using non-metallic sheathed cable	1.000	Ea.	2.000	103	96	199
Using BX cable	1.000	Ea.	2.424	137	116	253
Using EMT conduit	1.000	Ea.	2.963	102	142	244
Switches, single pole						
Using non-metallic sheathed cable	1.000	Ea.	.500	23	24	47
Using BX cable	1.000	Ea.	.602	34.50	29	63.50
Using EMT conduit	1.000	Ea.	.748	45.50	36	81.50
Switches, 3-way						
Using non-metallic sheathed cable	1.000	Ea.	.667	31	32	63
Using BX cable	1.000	Ea.	.800	38	38.50	76.50
Using EMT conduit	1.000	Ea.	1.333	56	64	120
Water heater						
Using non-metallic sheathed cable	1.000	Ea.	1.600	36.50	76.50	113
Using BX cable	1.000	Ea.	1.905	53	91	144
Using EMT conduit	1.000	Ea.	2.353	50.50	113	163.50
Weatherproof receptacle						
Using non-metallic sheathed cable	1.000	Ea.	1.333	141	64	205
Using BX cable	1.000	Ea.	1.600	146	76.50	222.50
Using EMT conduit	1.000	Ea.	2.000	158	96	254

9 | ELECTRICAL — 40 | Light Fixture Systems

DESCRIPTION	QUAN.	UNIT	LABOR HOURS	COST EACH MAT.	COST EACH INST.	COST EACH TOTAL
Fluorescent strip, 4' long, 1 light, average	1.000	Ea.	.941	32.50	45	77.50
Deluxe	1.000	Ea.	1.129	39	54	93
2 lights, average	1.000	Ea.	1.000	35	48	83
Deluxe	1.000	Ea.	1.200	42	57.50	99.50
8' long, 1 light, average	1.000	Ea.	1.194	48.50	57	105.50
Deluxe	1.000	Ea.	1.433	58	68.50	126.50
2 lights, average	1.000	Ea.	1.290	58.50	62	120.50
Deluxe	1.000	Ea.	1.548	70	74.50	144.50
Surface mounted, 4' x 1', economy	1.000	Ea.	.914	65	43.50	108.50
Average	1.000	Ea.	1.143	81	54.50	135.50
Deluxe	1.000	Ea.	1.371	97	65.50	162.50
4' x 2', economy	1.000	Ea.	1.208	82.50	58	140.50
Average	1.000	Ea.	1.509	103	72.50	175.50
Deluxe	1.000	Ea.	1.811	124	87	211
Recessed, 4' x 1', 2 lamps, economy	1.000	Ea.	1.123	44.50	53.50	98
Average	1.000	Ea.	1.404	55.50	67	122.50
Deluxe	1.000	Ea.	1.684	66.50	80.50	147
4' x 2', 4' lamps, economy	1.000	Ea.	1.362	53.50	65	118.50
Average	1.000	Ea.	1.702	67	81.50	148.50
Deluxe	1.000	Ea.	2.043	80.50	98	178.50
Incandescent, exterior, 150W, single spot	1.000	Ea.	.500	21.50	24	45.50
Double spot	1.000	Ea.	1.167	81	56	137
Recessed, 100W, economy	1.000	Ea.	.800	55	38.50	93.50
Average	1.000	Ea.	1.000	69	48	117
Deluxe	1.000	Ea.	1.200	83	57.50	140.50
150W, economy	1.000	Ea.	.800	81.50	38.50	120
Average	1.000	Ea.	1.000	102	48	150
Deluxe	1.000	Ea.	1.200	122	57.50	179.50
Surface mounted, 60W, economy	1.000	Ea.	.800	44	38.50	82.50
Average	1.000	Ea.	1.000	49	48	97
Deluxe	1.000	Ea.	1.194	66	57	123
Metal halide, recessed 2' x 2' 250W	1.000	Ea.	2.500	335	120	455
2' x 2', 400W	1.000	Ea.	2.759	375	132	507
Surface mounted, 2' x 2', 250W	1.000	Ea.	2.963	340	142	482
2' x 2', 400W	1.000	Ea.	3.333	400	160	560
High bay, single, unit, 400W	1.000	Ea.	3.478	405	167	572
Twin unit, 400W	1.000	Ea.	5.000	810	240	1,050
Low bay, 250W	1.000	Ea.	2.500	390	120	510

Location Factors

Costs shown in *RSMeans cost data publications* are based on National Averages for materials and installation. To adjust these costs to a specific location, simply multiply the base cost by the factor for that city. The data is arranged alphabetically by state and postal zip code numbers. For a city not listed, use the factor for a nearby city with similar economic characteristics.

STATE	CITY	Residential
ALABAMA		
350-352	Birmingham	.86
354	Tuscaloosa	.73
355	Jasper	.71
356	Decatur	.76
357-358	Huntsville	.84
359	Gadsden	.73
360-361	Montgomery	.75
362	Anniston	.68
363	Dothan	.74
364	Evergreen	.70
365-366	Mobile	.79
367	Selma	.72
368	Phenix City	.73
369	Butler	.71
ALASKA		
995-996	Anchorage	1.27
997	Fairbanks	1.29
998	Juneau	1.27
999	Ketchikan	1.29
ARIZONA		
850,853	Phoenix	.86
852	Mesa/Tempe	.83
855	Globe	.79
856-857	Tucson	.84
859	Show Low	.81
860	Flagstaff	.86
863	Prescott	.81
864	Kingman	.83
865	Chambers	.80
ARKANSAS		
716	Pine Bluff	.81
717	Camden	.70
718	Texarkana	.75
719	Hot Springs	.70
720-722	Little Rock	.87
723	West Memphis	.81
724	Jonesboro	.79
725	Batesville	.76
726	Harrison	.78
727	Fayetteville	.72
728	Russellville	.77
729	Fort Smith	.79
CALIFORNIA		
900-902	Los Angeles	1.06
903-905	Inglewood	1.05
906-908	Long Beach	1.04
910-912	Pasadena	1.05
913-916	Van Nuys	1.08
917-918	Alhambra	1.09
919-921	San Diego	1.04
922	Palm Springs	1.04
923-924	San Bernardino	1.05
925	Riverside	1.05
926-927	Santa Ana	1.06
928	Anaheim	1.05
930	Oxnard	1.07
931	Santa Barbara	1.06
932-933	Bakersfield	1.03
934	San Luis Obispo	1.08
935	Mojave	1.06
936-938	Fresno	1.09
939	Salinas	1.12
940-941	San Francisco	1.23
942,956-958	Sacramento	1.11
943	Palo Alto	1.18
944	San Mateo	1.22
945	Vallejo	1.15
946	Oakland	1.21
947	Berkeley	1.24
948	Richmond	1.24
949	San Rafael	1.22
950	Santa Cruz	1.15
951	San Jose	1.19
952	Stockton	1.09
953	Modesto	1.08
954	Santa Rosa	1.16

STATE	CITY	Residential
955	Eureka	1.12
959	Marysville	1.10
960	Redding	1.10
961	Susanville	1.09
COLORADO		
800-802	Denver	.94
803	Boulder	.93
804	Golden	.91
805	Fort Collins	.90
806	Greeley	.80
807	Fort Morgan	.93
808-809	Colorado Springs	.90
810	Pueblo	.91
811	Alamosa	.88
812	Salida	.90
813	Durango	.91
814	Montrose	.87
815	Grand Junction	.92
816	Glenwood Springs	.90
CONNECTICUT		
060	New Britain	1.08
061	Hartford	1.08
062	Willimantic	1.08
063	New London	1.08
064	Meriden	1.08
065	New Haven	1.08
066	Bridgeport	1.09
067	Waterbury	1.09
068	Norwalk	1.09
069	Stamford	1.10
D.C.		
200-205	Washington	.95
DELAWARE		
197	Newark	.99
198	Wilmington	1.00
199	Dover	.99
FLORIDA		
320,322	Jacksonville	.77
321	Daytona Beach	.84
323	Tallahassee	.73
324	Panama City	.67
325	Pensacola	.78
326,344	Gainesville	.77
327-328,347	Orlando	.84
329	Melbourne	.86
330-332,340	Miami	.85
333	Fort Lauderdale	.84
334,349	West Palm Beach	.84
335-336,346	Tampa	.86
337	St. Petersburg	.76
338	Lakeland	.83
339,341	Fort Myers	.80
342	Sarasota	.84
GEORGIA		
300-303,399	Atlanta	.90
304	Statesboro	.71
305	Gainesville	.79
306	Athens	.79
307	Dalton	.75
308-309	Augusta	.81
310-312	Macon	.82
313-314	Savannah	.82
315	Waycross	.76
316	Valdosta	.73
317,398	Albany	.79
318-319	Columbus	.83
HAWAII		
967	Hilo	1.21
968	Honolulu	1.23
STATES & POSS.		
969	Guam	.89

Location Factors

STATE	CITY	Residential
IDAHO		
832	Pocatello	.86
833	Twin Falls	.74
834	Idaho Falls	.75
835	Lewiston	.96
836-837	Boise	.87
838	Coeur d'Alene	.94
ILLINOIS		
600-603	North Suburban	1.10
604	Joliet	1.10
605	South Suburban	1.10
606-608	Chicago	1.16
609	Kankakee	1.00
610-611	Rockford	1.04
612	Rock Island	.96
613	La Salle	1.02
614	Galesburg	.99
615-616	Peoria	.98
617	Bloomington	.98
618-619	Champaign	.99
620-622	East St. Louis	1.00
623	Quincy	.98
624	Effingham	.96
625	Decatur	.97
626-627	Springfield	.97
628	Centralia	1.00
629	Carbondale	.95
INDIANA		
460	Anderson	.91
461-462	Indianapolis	.95
463-464	Gary	1.01
465-466	South Bend	.91
467-468	Fort Wayne	.91
469	Kokomo	.92
470	Lawrenceburg	.87
471	New Albany	.86
472	Columbus	.92
473	Muncie	.91
474	Bloomington	.94
475	Washington	.91
476-477	Evansville	.90
478	Terre Haute	.90
479	Lafayette	.91
IOWA		
500-503,509	Des Moines	.91
504	Mason City	.77
505	Fort Dodge	.76
506-507	Waterloo	.79
508	Creston	.81
510-511	Sioux City	.87
512	Sibley	.73
513	Spencer	.74
514	Carroll	.74
515	Council Bluffs	.81
516	Shenandoah	.75
520	Dubuque	.86
521	Decorah	.76
522-524	Cedar Rapids	.94
525	Ottumwa	.84
526	Burlington	.87
527-528	Davenport	.97
KANSAS		
660-662	Kansas City	.99
664-666	Topeka	.80
667	Fort Scott	.85
668	Emporia	.72
669	Belleville	.78
670-672	Wichita	.80
673	Independence	.85
674	Salina	.76
675	Hutchinson	.77
676	Hays	.82
677	Colby	.83
678	Dodge City	.82
679	Liberal	.79
KENTUCKY		
400-402	Louisville	.92
403-405	Lexington	.89
406	Frankfort	.89
407-409	Corbin	.78
410	Covington	1.00
411-412	Ashland	.98

STATE	CITY	Residential
413-414	Campton	.79
415-416	Pikeville	.86
417-418	Hazard	.73
420	Paducah	.90
421-422	Bowling Green	.90
423	Owensboro	.89
424	Henderson	.91
425-426	Somerset	.78
427	Elizabethtown	.87
LOUISIANA		
700-701	New Orleans	.86
703	Thibodaux	.84
704	Hammond	.79
705	Lafayette	.82
706	Lake Charles	.83
707-708	Baton Rouge	.82
710-711	Shreveport	.79
712	Monroe	.74
713-714	Alexandria	.74
MAINE		
039	Kittery	.79
040-041	Portland	.90
042	Lewiston	.89
043	Augusta	.82
044	Bangor	.88
045	Bath	.80
046	Machias	.81
047	Houlton	.85
048	Rockland	.81
049	Waterville	.80
MARYLAND		
206	Waldorf	.85
207-208	College Park	.87
209	Silver Spring	.86
210-212	Baltimore	.90
214	Annapolis	.85
215	Cumberland	.86
216	Easton	.68
217	Hagerstown	.86
218	Salisbury	.75
219	Elkton	.81
MASSACHUSETTS		
010-011	Springfield	1.04
012	Pittsfield	1.01
013	Greenfield	1.00
014	Fitchburg	1.12
015-016	Worcester	1.14
017	Framingham	1.12
018	Lowell	1.13
019	Lawrence	1.13
020-022, 024	Boston	1.19
023	Brockton	1.12
025	Buzzards Bay	1.10
026	Hyannis	1.09
027	New Bedford	1.12
MICHIGAN		
480,483	Royal Oak	1.03
481	Ann Arbor	1.05
482	Detroit	1.07
484-485	Flint	.97
486	Saginaw	.94
487	Bay City	.95
488-489	Lansing	.97
490	Battle Creek	.93
491	Kalamazoo	.92
492	Jackson	.95
493,495	Grand Rapids	.82
494	Muskegon	.89
496	Traverse City	.80
497	Gaylord	.83
498-499	Iron Mountain	.90
MINNESOTA		
550-551	Saint Paul	1.13
553-555	Minneapolis	1.17
556-558	Duluth	1.09
559	Rochester	1.05
560	Mankato	1.03
561	Windom	.83
562	Willmar	.85
563	St. Cloud	1.07
564	Brainerd	.98

Location Factors

STATE	CITY	Residential
565	Detroit Lakes	.96
566	Bemidji	.96
567	Thief River Falls	.95
MISSISSIPPI		
386	Clarksdale	.62
387	Greenville	.69
388	Tupelo	.64
389	Greenwood	.65
390-392	Jackson	.73
393	Meridian	.66
394	Laurel	.63
395	Biloxi	.75
396	McComb	.74
397	Columbus	.65
MISSOURI		
630-631	St. Louis	1.03
633	Bowling Green	.94
634	Hannibal	.87
635	Kirksville	.80
636	Flat River	.94
637	Cape Girardeau	.87
638	Sikeston	.82
639	Poplar Bluff	.82
640-641	Kansas City	1.03
644-645	St. Joseph	.95
646	Chillicothe	.84
647	Harrisonville	.94
648	Joplin	.85
650-651	Jefferson City	.88
652	Columbia	.88
653	Sedalia	.85
654-655	Rolla	.88
656-658	Springfield	.86
MONTANA		
590-591	Billings	.87
592	Wolf Point	.83
593	Miles City	.85
594	Great Falls	.88
595	Havre	.80
596	Helena	.87
597	Butte	.83
598	Missoula	.83
599	Kalispell	.81
NEBRASKA		
680-681	Omaha	.89
683-685	Lincoln	.78
686	Columbus	.70
687	Norfolk	.77
688	Grand Island	.77
689	Hastings	.76
690	Mccook	.70
691	North Platte	.75
692	Valentine	.66
693	Alliance	.66
NEVADA		
889-891	Las Vegas	1.00
893	Ely	.87
894-895	Reno	.94
897	Carson City	.95
898	Elko	.93
NEW HAMPSHIRE		
030	Nashua	.95
031	Manchester	.95
032-033	Concord	.92
034	Keene	.72
035	Littleton	.80
036	Charleston	.70
037	Claremont	.71
038	Portsmouth	.89
NEW JERSEY		
070-071	Newark	1.13
072	Elizabeth	1.15
073	Jersey City	1.11
074-075	Paterson	1.12
076	Hackensack	1.11
077	Long Branch	1.13
078	Dover	1.12
079	Summit	1.12
080,083	Vineland	1.09
081	Camden	1.10

STATE	CITY	Residential
082,084	Atlantic City	1.13
085-086	Trenton	1.11
087	Point Pleasant	1.10
088-089	New Brunswick	1.12
NEW MEXICO		
870-872	Albuquerque	.85
873	Gallup	.85
874	Farmington	.85
875	Santa Fe	.85
877	Las Vegas	.85
878	Socorro	.85
879	Truth/Consequences	.84
880	Las Cruces	.82
881	Clovis	.84
882	Roswell	.85
883	Carrizozo	.85
884	Tucumcari	.85
NEW YORK		
100-102	New York	1.34
103	Staten Island	1.25
104	Bronx	1.27
105	Mount Vernon	1.15
106	White Plains	1.18
107	Yonkers	1.19
108	New Rochelle	1.19
109	Suffern	1.12
110	Queens	1.26
111	Long Island City	1.29
112	Brooklyn	1.31
113	Flushing	1.28
114	Jamaica	1.28
115,117,118	Hicksville	1.19
116	Far Rockaway	1.27
119	Riverhead	1.21
120-122	Albany	.95
123	Schenectady	.96
124	Kingston	1.03
125-126	Poughkeepsie	1.04
127	Monticello	1.05
128	Glens Falls	.89
129	Plattsburgh	.93
130-132	Syracuse	.96
133-135	Utica	.93
136	Watertown	.90
137-139	Binghamton	.95
140-142	Buffalo	1.05
143	Niagara Falls	1.01
144-146	Rochester	.99
147	Jamestown	.89
148-149	Elmira	.86
NORTH CAROLINA		
270,272-274	Greensboro	.85
271	Winston-Salem	.85
275-276	Raleigh	.87
277	Durham	.85
278	Rocky Mount	.75
279	Elizabeth City	.75
280	Gastonia	.86
281-282	Charlotte	.88
283	Fayetteville	.84
284	Wilmington	.83
285	Kinston	.75
286	Hickory	.80
287-288	Asheville	.83
289	Murphy	.74
NORTH DAKOTA		
580-581	Fargo	.79
582	Grand Forks	.76
583	Devils Lake	.79
584	Jamestown	.74
585	Bismarck	.79
586	Dickinson	.77
587	Minot	.80
588	Williston	.77
OHIO		
430-432	Columbus	.94
433	Marion	.91
434-436	Toledo	.99
437-438	Zanesville	.90
439	Steubenville	.95
440	Lorain	.99
441	Cleveland	1.02

Location Factors

STATE		CITY	Residential
	442-443	Akron	.98
	444-445	Youngstown	.96
	446-447	Canton	.94
	448-449	Mansfield	.95
	450	Hamilton	.93
	451-452	Cincinnati	.93
	453-454	Dayton	.91
	455	Springfield	.93
	456	Chillicothe	.96
	457	Athens	.88
	458	Lima	.90
OKLAHOMA			
	730-731	Oklahoma City	.80
	734	Ardmore	.79
	735	Lawton	.82
	736	Clinton	.78
	737	Enid	.78
	738	Woodward	.77
	739	Guymon	.68
	740-741	Tulsa	.79
	743	Miami	.82
	744	Muskogee	.73
	745	Mcalester	.74
	746	Ponca City	.78
	747	Durant	.77
	748	Shawnee	.76
	749	Poteau	.78
OREGON			
	970-972	Portland	1.02
	973	Salem	1.01
	974	Eugene	1.01
	975	Medford	.99
	976	Klamath Falls	1.00
	977	Bend	1.02
	978	Pendleton	.99
	979	Vale	.98
PENNSYLVANIA			
	150-152	Pittsburgh	.99
	153	Washington	.94
	154	Uniontown	.90
	155	Bedford	.88
	156	Greensburg	.94
	157	Indiana	.91
	158	Dubois	.90
	159	Johnstown	.89
	160	Butler	.92
	161	New Castle	.92
	162	Kittanning	.93
	163	Oil City	.90
	164-165	Erie	.95
	166	Altoona	.88
	167	Bradford	.90
	168	State College	.91
	169	Wellsboro	.90
	170-171	Harrisburg	.94
	172	Chambersburg	.89
	173-174	York	.91
	175-176	Lancaster	.91
	177	Williamsport	.84
	178	Sunbury	.91
	179	Pottsville	.91
	180	Lehigh Valley	1.01
	181	Allentown	1.04
	182	Hazleton	.91
	183	Stroudsburg	.92
	184-185	Scranton	.96
	186-187	Wilkes-Barre	.93
	188	Montrose	.90
	189	Doylestown	1.04
	190-191	Philadelphia	1.15
	193	Westchester	1.09
	194	Norristown	1.08
	195-196	Reading	.97
PUERTO RICO			
	009	San Juan	.74
RHODE ISLAND			
	028	Newport	1.06
	029	Providence	1.07
SOUTH CAROLINA			
	290-292	Columbia	.82
	293	Spartanburg	.81

STATE		CITY	Residential
	294	Charleston	.82
	295	Florence	.76
	296	Greenville	.80
	297	Rock Hill	.72
	298	Aiken	.97
	299	Beaufort	.75
SOUTH DAKOTA			
	570-571	Sioux Falls	.76
	572	Watertown	.72
	573	Mitchell	.74
	574	Aberdeen	.75
	575	Pierre	.75
	576	Mobridge	.73
	577	Rapid City	.75
TENNESSEE			
	370-372	Nashville	.85
	373-374	Chattanooga	.77
	375,380-381	Memphis	.83
	376	Johnson City	.72
	377-379	Knoxville	.75
	382	Mckenzie	.73
	383	Jackson	.71
	384	Columbia	.73
	385	Cookeville	.72
TEXAS			
	750	Mckinney	.75
	751	Waxahackie	.75
	752-753	Dallas	.82
	754	Greenville	.69
	755	Texarkana	.73
	756	Longview	.67
	757	Tyler	.74
	758	Palestine	.66
	759	Lufkin	.71
	760-761	Fort Worth	.82
	762	Denton	.77
	763	Wichita Falls	.79
	764	Eastland	.73
	765	Temple	.75
	766-767	Waco	.78
	768	Brownwood	.69
	769	San Angelo	.72
	770-772	Houston	.86
	773	Huntsville	.70
	774	Wharton	.71
	775	Galveston	.84
	776-777	Beaumont	.83
	778	Bryan	.74
	779	Victoria	.75
	780	Laredo	.73
	781-782	San Antonio	.81
	783-784	Corpus Christi	.78
	785	Mc Allen	.76
	786-787	Austin	.80
	788	Del Rio	.67
	789	Giddings	.70
	790-791	Amarillo	.78
	792	Childress	.76
	793-794	Lubbock	.76
	795-796	Abilene	.75
	797	Midland	.76
	798-799,885	El Paso	.75
UTAH			
	840-841	Salt Lake City	.81
	842,844	Ogden	.79
	843	Logan	.80
	845	Price	.72
	846-847	Provo	.81
VERMONT			
	050	White River Jct.	.73
	051	Bellows Falls	.75
	052	Bennington	.82
	053	Brattleboro	.78
	054	Burlington	.80
	056	Montpelier	.81
	057	Rutland	.80
	058	St. Johnsbury	.80
	059	Guildhall	.79
VIRGINIA			
	220-221	Fairfax	1.02
	222	Arlington	1.04

Location Factors

STATE		CITY	Residential
223		Alexandria	1.06
224-225		Fredericksburg	.95
226		Winchester	.93
227		Culpeper	1.00
228		Harrisonburg	.90
229		Charlottesville	.92
230-232		Richmond	1.01
233-235		Norfolk	1.02
236		Newport News	1.01
237		Portsmouth	.92
238		Petersburg	.99
239		Farmville	.91
240-241		Roanoke	.99
242		Bristol	.86
243		Pulaski	.84
244		Staunton	.93
245		Lynchburg	.97
246		Grundy	.85
WASHINGTON			
980-981,987		Seattle	1.02
982		Everett	1.05
983-984		Tacoma	1.01
985		Olympia	1.01
986		Vancouver	.98
988		Wenatchee	.93
989		Yakima	.97
990-992		Spokane	.99
993		Richland	.97
994		Clarkston	.97
WEST VIRGINIA			
247-248		Bluefield	.88
249		Lewisburg	.89
250-253		Charleston	.97
254		Martinsburg	.86
255-257		Huntington	1.01
258-259		Beckley	.90
260		Wheeling	.93
261		Parkersburg	.92
262		Buckhannon	.92
263-264		Clarksburg	.92
265		Morgantown	.93
266		Gassaway	.92
267		Romney	.88
268		Petersburg	.90
WISCONSIN			
530,532		Milwaukee	1.07
531		Kenosha	1.04
534		Racine	1.02
535		Beloit	1.00
537		Madison	.99
538		Lancaster	.97
539		Portage	.96
540		New Richmond	1.00
541-543		Green Bay	1.01
544		Wausau	.95
545		Rhinelander	.95
546		La Crosse	.94
547		Eau Claire	.98
548		Superior	.99
549		Oshkosh	.95
WYOMING			
820		Cheyenne	.84
821		Yellowstone Nat. Pk.	.75
822		Wheatland	.75
823		Rawlins	.76
824		Worland	.75
825		Riverton	.74
826		Casper	.78
827		Newcastle	.74
828		Sheridan	.80
829-831		Rock Springs	.79

STATE		CITY	Residential
CANADIAN FACTORS (reflect Canadian Currency)			
ALBERTA			
		Calgary	1.14
		Edmonton	1.13
		Fort McMurray	1.09
		Lethbridge	1.10
		Lloydminster	1.09
		Medicine Hat	1.10
		Red Deer	1.10
BRITISH COLUMBIA			
		Kamloops	1.08
		Prince George	1.08
		Vancouver	1.09
		Victoria	1.03
MANITOBA			
		Brandon	1.06
		Portage la Prairie	1.06
		Winnipeg	1.05
NEW BRUNSWICK			
		Bathurst	.97
		Dalhousie	.97
		Fredericton	1.05
		Moncton	.98
		Newcastle	.97
		Saint John	1.05
NEWFOUNDLAND			
		Corner Brook	.99
		St. John's	1.01
NORTHWEST TERRITORIES			
		Yellowknife	1.10
NOVA SCOTIA			
		Dartmouth	1.00
		Halifax	1.02
		New Glasgow	1.00
		Sydney	.99
		Yarmouth	1.00
ONTARIO			
		Barrie	1.17
		Brantford	1.19
		Cornwall	1.19
		Hamilton	1.19
		Kingston	1.19
		Kitchener	1.11
		London	1.17
		North Bay	1.15
		Oshawa	1.17
		Ottawa	1.19
		Owen Sound	1.15
		Peterborough	1.16
		Sarnia	1.19
		Sudbury	1.09
		Thunder Bay	1.15
		Toronto	1.20
		Windsor	1.14
PRINCE EDWARD ISLAND			
		Charlottetown	.95
		Summerside	.94
QUEBEC			
		Cap-de-la-Madeleine	1.18
		Charlesbourg	1.18
		Chicoutimi	1.20
		Gatineau	1.16
		Laval	1.17
		Montreal	1.21
		Quebec	1.22
		Sherbrooke	1.17
		Trois Rivieres	1.18
SASKATCHEWAN			
		Moose Jaw	.97
		Prince Albert	.96
		Regina	.99
		Saskatoon	.97
YUKON			
		Whitehorse	.96

Abbreviations

A	Area Square Feet; Ampere	Cab.	Cabinet	Demob.	Demobilization
ABS	Acrylonitrile Butadiene Stryrene; Asbestos Bonded Steel	Cair.	Air Tool Laborer	d.f.u.	Drainage Fixture Units
A.C.	Alternating Current; Air-Conditioning; Asbestos Cement; Plywood Grade A & C	Calc	Calculated	D.H.	Double Hung
		Cap.	Capacity	DHW	Domestic Hot Water
		Carp.	Carpenter	Diag.	Diagonal
		C.B.	Circuit Breaker	Diam.	Diameter
		C.C.A.	Chromate Copper Arsenate	Distrib.	Distribution
A.C.I.	American Concrete Institute	C.C.F.	Hundred Cubic Feet	Dk.	Deck
AD	Plywood, Grade A & D	cd	Candela	D.L.	Dead Load; Diesel
Addit.	Additional	cd/sf	Candela per Square Foot	DLH	Deep Long Span Bar Joist
Adj.	Adjustable	CD	Grade of Plywood Face & Back	Do.	Ditto
af	Audio-frequency	CDX	Plywood, Grade C & D, exterior glue	Dp.	Depth
A.G.A.	American Gas Association			D.P.S.T.	Double Pole, Single Throw
Agg.	Aggregate	Cefi.	Cement Finisher	Dr.	Driver
A.H.	Ampere Hours	Cem.	Cement	Drink.	Drinking
A hr.	Ampere-hour	CF	Hundred Feet	D.S.	Double Strength
A.H.U.	Air Handling Unit	C.F.	Cubic Feet	D.S.A.	Double Strength A Grade
A.I.A.	American Institute of Architects	CFM	Cubic Feet per Minute	D.S.B.	Double Strength B Grade
AIC	Ampere Interrupting Capacity	c.g.	Center of Gravity	Dty.	Duty
Allow.	Allowance	CHW	Chilled Water; Commercial Hot Water	DWV	Drain Waste Vent
alt.	Altitude			DX	Deluxe White, Direct Expansion
Alum.	Aluminum	C.I.	Cast Iron	dyn	Dyne
a.m.	Ante Meridiem	C.I.P.	Cast in Place	e	Eccentricity
Amp.	Ampere	Circ.	Circuit	E	Equipment Only; East
Anod.	Anodized	C.L.	Carload Lot	Ea.	Each
Approx.	Approximate	Clab.	Common Laborer	E.B.	Encased Burial
Apt.	Apartment	Clam	Common maintenance laborer	Econ.	Economy
Asb.	Asbestos	C.L.F.	Hundred Linear Feet	E.C.Y	Embankment Cubic Yards
A.S.B.C.	American Standard Building Code	CLF	Current Limiting Fuse	EDP	Electronic Data Processing
Asbe.	Asbestos Worker	CLP	Cross Linked Polyethylene	EIFS	Exterior Insulation Finish System
A.S.H.R.A.E.	American Society of Heating, Refrig. & AC Engineers	cm	Centimeter	E.D.R.	Equiv. Direct Radiation
		CMP	Corr. Metal Pipe	Eq.	Equation
A.S.M.E.	American Society of Mechanical Engineers	C.M.U.	Concrete Masonry Unit	Elec.	Electrician; Electrical
		CN	Change Notice	Elev.	Elevator; Elevating
A.S.T.M.	American Society for Testing and Materials	Col.	Column	EMT	Electrical Metallic Conduit; Thin Wall Conduit
		CO_2	Carbon Dioxide		
Attchmt.	Attachment	Comb.	Combination	Eng.	Engine, Engineered
Avg.	Average	Compr.	Compressor	EPDM	Ethylene Propylene Diene Monomer
A.W.G.	American Wire Gauge	Conc.	Concrete		
AWWA	American Water Works Assoc.	Cont.	Continuous; Continued	EPS	Expanded Polystyrene
Bbl.	Barrel	Corr.	Corrugated	Eqhv.	Equip. Oper., Heavy
B&B	Grade B and Better; Balled & Burlapped	Cos	Cosine	Eqlt.	Equip. Oper., Light
		Cot	Cotangent	Eqmd.	Equip. Oper., Medium
B.&S.	Bell and Spigot	Cov.	Cover	Eqmm.	Equip. Oper., Master Mechanic
B.&W.	Black and White	C/P	Cedar on Paneling	Eqol.	Equip. Oper., Oilers
b.c.c.	Body-centered Cubic	CPA	Control Point Adjustment	Equip.	Equipment
B.C.Y.	Bank Cubic Yards	Cplg.	Coupling	ERW	Electric Resistance Welded
BE	Bevel End	C.P.M.	Critical Path Method	E.S.	Energy Saver
B.F.	Board Feet	CPVC	Chlorinated Polyvinyl Chloride	Est.	Estimated
Bg. cem.	Bag of Cement	C.Pr.	Hundred Pair	esu	Electrostatic Units
BHP	Boiler Horsepower; Brake Horsepower	CRC	Cold Rolled Channel	E.W.	Each Way
		Creos.	Creosote	EWT	Entering Water Temperature
B.I.	Black Iron	Crpt.	Carpet & Linoleum Layer	Excav.	Excavation
Bit.; Bitum.	Bituminous	CRT	Cathode-ray Tube	Exp.	Expansion, Exposure
Bk.	Backed	CS	Carbon Steel, Constant Shear Bar Joist	Ext.	Exterior
Bkrs.	Breakers			Extru.	Extrusion
Bldg.	Building	Csc	Cosecant	f.	Fiber stress
Blk.	Block	C.S.F.	Hundred Square Feet	F	Fahrenheit; Female; Fill
Bm.	Beam	CSI	Construction Specifications Institute	Fab.	Fabricated
Boil.	Boilermaker			FBGS	Fiberglass
B.P.M.	Blows per Minute	C.T.	Current Transformer	F.C.	Footcandles
BR	Bedroom	CTS	Copper Tube Size	f.c.c.	Face-centered Cubic
Brg.	Bearing	Cu	Copper, Cubic	f'c.	Compressive Stress in Concrete; Extreme Compressive Stress
Brhe.	Bricklayer Helper	Cu. Ft.	Cubic Foot		
Bric.	Bricklayer	cw	Continuous Wave	F.E.	Front End
Brk.	Brick	C.W.	Cool White; Cold Water	FEP	Fluorinated Ethylene Propylene (Teflon)
Brng.	Bearing	Cwt.	100 Pounds		
Brs.	Brass	C.W.X.	Cool White Deluxe	F.G.	Flat Grain
Brz.	Bronze	C.Y.	Cubic Yard (27 cubic feet)	F.H.A.	Federal Housing Administration
Bsn.	Basin	C.Y./Hr.	Cubic Yard per Hour	Fig.	Figure
Btr.	Better	Cyl.	Cylinder	Fin.	Finished
BTU	British Thermal Unit	d	Penny (nail size)	Fixt.	Fixture
BTUH	BTU per Hour	D	Deep; Depth; Discharge	Fl. Oz.	Fluid Ounces
B.U.R.	Built-up Roofing	Dis.;Disch.	Discharge	Flr.	Floor
BX	Interlocked Armored Cable	Db.	Decibel	F.M.	Frequency Modulation; Factory Mutual
c	Conductivity, Copper Sweat	Dbl.	Double		
C	Hundred; Centigrade	DC	Direct Current	Fmg.	Framing
C/C	Center to Center, Cedar on Cedar	DDC	Direct Digital Control	Fndtn.	Foundation

Abbreviations

Fori.	Foreman, Inside	I.W.	Indirect Waste	M.C.F.	Thousand Cubic Feet
Foro.	Foreman, Outside	J	Joule	M.C.F.M.	Thousand Cubic Feet per Minute
Fount.	Fountain	J.I.C.	Joint Industrial Council	M.C.M.	Thousand Circular Mils
FPM	Feet per Minute	K	Thousand; Thousand Pounds; Heavy Wall Copper Tubing, Kelvin	M.C.P.	Motor Circuit Protector
FPT	Female Pipe Thread			MD	Medium Duty
Fr.	Frame			M.D.O.	Medium Density Overlaid
F.R.	Fire Rating	K.A.H.	Thousand Amp. Hours	Med.	Medium
FRK	Foil Reinforced Kraft	KCMIL	Thousand Circular Mils	MF	Thousand Feet
FRP	Fiberglass Reinforced Plastic	KD	Knock Down	M.F.B.M.	Thousand Feet Board Measure
FS	Forged Steel	K.D.A.T.	Kiln Dried After Treatment	Mfg.	Manufacturing
FSC	Cast Body; Cast Switch Box	kg	Kilogram	Mfrs.	Manufacturers
Ft.	Foot; Feet	kG	Kilogauss	mg	Milligram
Ftng.	Fitting	kgf	Kilogram Force	MGD	Million Gallons per Day
Ftg.	Footing	kHz	Kilohertz	MGPH	Thousand Gallons per Hour
Ft. Lb.	Foot Pound	Kip.	1000 Pounds	MH, M.H.	Manhole; Metal Halide; Man-Hour
Furn.	Furniture	KJ	Kiljoule	MHz	Megahertz
FVNR	Full Voltage Non-Reversing	K.L.	Effective Length Factor	Mi.	Mile
FXM	Female by Male	K.L.F.	Kips per Linear Foot	MI	Malleable Iron; Mineral Insulated
Fy.	Minimum Yield Stress of Steel	Km	Kilometer	mm	Millimeter
g	Gram	K.S.F.	Kips per Square Foot	Mill.	Millwright
G	Gauss	K.S.I.	Kips per Square Inch	Min., min.	Minimum, minute
Ga.	Gauge	kV	Kilovolt	Misc.	Miscellaneous
Gal.	Gallon	kVA	Kilovolt Ampere	ml	Milliliter, Mainline
Gal./Min.	Gallon per Minute	K.V.A.R.	Kilovar (Reactance)	M.L.F.	Thousand Linear Feet
Galv.	Galvanized	KW	Kilowatt	Mo.	Month
Gen.	General	KWh	Kilowatt-hour	Mobil.	Mobilization
G.F.I.	Ground Fault Interrupter	L	Labor Only; Length; Long; Medium Wall Copper Tubing	Mog.	Mogul Base
Glaz.	Glazier			MPH	Miles per Hour
GPD	Gallons per Day	Lab.	Labor	MPT	Male Pipe Thread
GPH	Gallons per Hour	lat	Latitude	MRT	Mile Round Trip
GPM	Gallons per Minute	Lath.	Lather	ms	Millisecond
GR	Grade	Lav.	Lavatory	M.S.F.	Thousand Square Feet
Gran.	Granular	lb.; #	Pound	Mstz.	Mosaic & Terrazzo Worker
Grnd.	Ground	L.B.	Load Bearing; L Conduit Body	M.S.Y.	Thousand Square Yards
H	High; High Strength Bar Joist; Henry	L. & E.	Labor & Equipment	Mtd.	Mounted
		lb./hr.	Pounds per Hour	Mthe.	Mosaic & Terrazzo Helper
H.C.	High Capacity	lb./L.F.	Pounds per Linear Foot	Mtng.	Mounting
H.D.	Heavy Duty; High Density	lbf/sq.in.	Pound-force per Square Inch	Mult.	Multi; Multiply
H.D.O.	High Density Overlaid	L.C.L.	Less than Carload Lot	M.V.A.	Million Volt Amperes
Hdr.	Header	L.C.Y.	Loose Cubic Yard	M.V.A.R.	Million Volt Amperes Reactance
Hdwe.	Hardware	Ld.	Load	MV	Megavolt
Help.	Helper Average	LE	Lead Equivalent	MW	Megawatt
HEPA	High Efficiency Particulate Air Filter	LED	Light Emitting Diode	MXM	Male by Male
		L.F.	Linear Foot	MYD	Thousand Yards
Hg	Mercury	Lg.	Long; Length; Large	N	Natural; North
HIC	High Interrupting Capacity	L & H	Light and Heat	nA	Nanoampere
HM	Hollow Metal	LH	Long Span Bar Joist	NA	Not Available; Not Applicable
H.O.	High Output	L.H.	Labor Hours	N.B.C.	National Building Code
Horiz.	Horizontal	L.L.	Live Load	NC	Normally Closed
H.P.	Horsepower; High Pressure	L.L.D.	Lamp Lumen Depreciation	N.E.M.A.	National Electrical Manufacturers Assoc.
H.P.F.	High Power Factor	lm	Lumen		
Hr.	Hour	lm/sf	Lumen per Square Foot	NEHB	Bolted Circuit Breaker to 600V.
Hrs./Day	Hours per Day	lm/W	Lumen per Watt	N.L.B.	Non-Load-Bearing
HSC	High Short Circuit	L.O.A.	Length Over All	NM	Non-Metallic Cable
Ht.	Height	log	Logarithm	nm	Nanometer
Htg.	Heating	L-O-L	Lateralolet	No.	Number
Htrs.	Heaters	L.P.	Liquefied Petroleum; Low Pressure	NO	Normally Open
HVAC	Heating, Ventilation & Air-Conditioning	L.P.F.	Low Power Factor	N.O.C.	Not Otherwise Classified
		LR	Long Radius	Nose.	Nosing
Hvy.	Heavy	L.S.	Lump Sum	N.P.T.	National Pipe Thread
HW	Hot Water	Lt.	Light	NQOD	Combination Plug-on/Bolt on Circuit Breaker to 240V.
Hyd.;Hydr.	Hydraulic	Lt. Ga.	Light Gauge		
Hz.	Hertz (cycles)	L.T.L.	Less than Truckload Lot	N.R.C.	Noise Reduction Coefficient
I.	Moment of Inertia	Lt. Wt.	Lightweight	N.R.S.	Non Rising Stem
I.C.	Interrupting Capacity	L.V.	Low Voltage	ns	Nanosecond
ID	Inside Diameter	M	Thousand; Material; Male; Light Wall Copper Tubing	nW	Nanowatt
I.D.	Inside Dimension; Identification			OB	Opposing Blade
I.F.	Inside Frosted	M^2CA	Meters Squared Contact Area	OC	On Center
I.M.C.	Intermediate Metal Conduit	m/hr; M.H.	Man-hour	OD	Outside Diameter
In.	Inch	mA	Milliampere	O.D.	Outside Dimension
Incan.	Incandescent	Mach.	Machine	ODS	Overhead Distribution System
Incl.	Included; Including	Mag. Str.	Magnetic Starter	O.G.	Ogee
Int.	Interior	Maint.	Maintenance	O.H.	Overhead
Inst.	Installation	Marb.	Marble Setter	O&P	Overhead and Profit
Insul.	Insulation/Insulated	Mat; Mat'l.	Material	Oper.	Operator
I.P.	Iron Pipe	Max.	Maximum	Opng.	Opening
I.P.S.	Iron Pipe Size	MBF	Thousand Board Feet	Orna.	Ornamental
I.P.T.	Iron Pipe Threaded	MBH	Thousand BTU's per hr.	OSB	Oriented Strand Board
		MC	Metal Clad Cable		

Abbreviations

O.S.&Y.	Outside Screw and Yoke	Rsr	Riser	Th.;Thk.	Thick
Ovhd.	Overhead	RT	Round Trip	Thn.	Thin
OWG	Oil, Water or Gas	S.	Suction; Single Entrance; South	Thrded	Threaded
Oz.	Ounce	SC	Screw Cover	Tilf.	Tile Layer, Floor
P.	Pole; Applied Load; Projection	SCFM	Standard Cubic Feet per Minute	Tilh.	Tile Layer, Helper
p.	Page	Scaf.	Scaffold	THHN	Nylon Jacketed Wire
Pape.	Paperhanger	Sch.; Sched.	Schedule	THW.	Insulated Strand Wire
P.A.P.R.	Powered Air Purifying Respirator	S.C.R.	Modular Brick	THWN;	Nylon Jacketed Wire
PAR	Parabolic Reflector	S.D.	Sound Deadening	T.L.	Truckload
Pc., Pcs.	Piece, Pieces	S.D.R.	Standard Dimension Ratio	T.M.	Track Mounted
P.C.	Portland Cement; Power Connector	S.E.	Surfaced Edge	Tot.	Total
P.C.F.	Pounds per Cubic Foot	Sel.	Select	T-O-L	Threadolet
P.C.M.	Phase Contrast Microscopy	S.E.R.; S.E.U.	Service Entrance Cable	T.S.	Trigger Start
P.E.	Professional Engineer; Porcelain Enamel; Polyethylene; Plain End	S.F.	Square Foot	Tr.	Trade
		S.F.C.A.	Square Foot Contact Area	Transf.	Transformer
		S.F. Flr.	Square Foot of Floor	Trhv.	Truck Driver, Heavy
Perf.	Perforated	S.F.G.	Square Foot of Ground	Trlr	Trailer
Ph.	Phase	S.F. Hor.	Square Foot Horizontal	Trlt.	Truck Driver, Light
P.I.	Pressure Injected	S.F.R.	Square Feet of Radiation	TTY	Teletypewriter
Pile.	Pile Driver	S.F. Shlf.	Square Foot of Shelf	TV	Television
Pkg.	Package	S4S	Surface 4 Sides	T.W.	Thermoplastic Water Resistant Wire
Pl.	Plate	Shee.	Sheet Metal Worker		
Plah.	Plasterer Helper	Sin.	Sine	UCI	Uniform Construction Index
Plas.	Plasterer	Skwk.	Skilled Worker	UF	Underground Feeder
Pluh.	Plumbers Helper	SL	Saran Lined	UGND	Underground Feeder
Plum.	Plumber	S.L.	Slimline	U.H.F.	Ultra High Frequency
Ply.	Plywood	Sldr.	Solder	U.L.	Underwriters Laboratory
p.m.	Post Meridiem	SLH	Super Long Span Bar Joist	Unfin.	Unfinished
Pntd.	Painted	S.N.	Solid Neutral	URD	Underground Residential Distribution
Pord.	Painter, Ordinary	S-O-L	Socketolet		
pp	Pages	sp	Standpipe	US	United States
PP; PPL	Polypropylene	S.P.	Static Pressure; Single Pole; Self-Propelled	USP	United States Primed
P.P.M.	Parts per Million			UTP	Unshielded Twisted Pair
Pr.	Pair	Spri.	Sprinkler Installer	V	Volt
P.E.S.B.	Pre-engineered Steel Building	spwg	Static Pressure Water Gauge	V.A.	Volt Amperes
Prefab.	Prefabricated	S.P.D.T.	Single Pole, Double Throw	V.C.T.	Vinyl Composition Tile
Prefin.	Prefinished	SPF	Spruce Pine Fir	VAV	Variable Air Volume
Prop.	Propelled	S.P.S.T.	Single Pole, Single Throw	VC	Veneer Core
PSF; psf	Pounds per Square Foot	SPT	Standard Pipe Thread	Vent.	Ventilation
PSI; psi	Pounds per Square Inch	Sq.	Square; 100 Square Feet	Vert.	Vertical
PSIG	Pounds per Square Inch Gauge	Sq. Hd.	Square Head	V.F.	Vinyl Faced
PSP	Plastic Sewer Pipe	Sq. In.	Square Inch	V.G.	Vertical Grain
Pspr.	Painter, Spray	S.S.	Single Strength; Stainless Steel	V.H.F.	Very High Frequency
Psst.	Painter, Structural Steel	S.S.B.	Single Strength B Grade	VHO	Very High Output
P.T.	Potential Transformer	sst	Stainless Steel	Vib.	Vibrating
P. & T.	Pressure & Temperature	Sswk.	Structural Steel Worker	V.L.F.	Vertical Linear Foot
Ptd.	Painted	Sswl.	Structural Steel Welder	Vol.	Volume
Ptns.	Partitions	St.;Stl.	Steel	VRP	Vinyl Reinforced Polyester
Pu	Ultimate Load	S.T.C.	Sound Transmission Coefficient	W	Wire; Watt; Wide; West
PVC	Polyvinyl Chloride	Std.	Standard	w/	With
Pvmt.	Pavement	STK	Select Tight Knot	W.C.	Water Column; Water Closet
Pwr.	Power	STP	Standard Temperature & Pressure	W.F.	Wide Flange
Q	Quantity Heat Flow	Stpi.	Steamfitter, Pipefitter	W.G.	Water Gauge
Quan.;Qty.	Quantity	Str.	Strength; Starter; Straight	Wldg.	Welding
Q.C.	Quick Coupling	Strd.	Stranded	W. Mile	Wire Mile
r	Radius of Gyration	Struct.	Structural	W-O-L	Weldolet
R	Resistance	Sty.	Story	W.R.	Water Resistant
R.C.P.	Reinforced Concrete Pipe	Subj.	Subject	Wrck.	Wrecker
Rect.	Rectangle	Subs.	Subcontractors	W.S.P.	Water, Steam, Petroleum
Reg.	Regular	Surf.	Surface	WT., Wt.	Weight
Reinf.	Reinforced	Sw.	Switch	WWF	Welded Wire Fabric
Req'd.	Required	Swbd.	Switchboard	XFER	Transfer
Res.	Resistant	S.Y.	Square Yard	XFMR	Transformer
Resi.	Residential	Syn.	Synthetic	XHD	Extra Heavy Duty
Rgh.	Rough	S.Y.P.	Southern Yellow Pine	XHHW; XLPE	Cross-Linked Polyethylene Wire Insulation
RGS	Rigid Galvanized Steel	Sys.	System		
R.H.W.	Rubber, Heat & Water Resistant; Residential Hot Water	t.	Thickness	XLP	Cross-linked Polyethylene
		T	Temperature; Ton	Y	Wye
rms	Root Mean Square	Tan	Tangent	yd	Yard
Rnd.	Round	T.C.	Terra Cotta	yr	Year
Rodm.	Rodman	T & C	Threaded and Coupled	Δ	Delta
Rofc.	Roofer, Composition	T.D.	Temperature Difference	%	Percent
Rofp.	Roofer, Precast	Tdd	Telecommunications Device for the Deaf	~	Approximately
Rohe.	Roofer Helpers (Composition)			Ø	Phase
Rots.	Roofer, Tile & Slate	T.E.M.	Transmission Electron Microscopy	@	At
R.O.W.	Right of Way	TFE	Tetrafluoroethylene (Teflon)	#	Pound; Number
RPM	Revolutions per Minute	T. & G.	Tongue & Groove; Tar & Gravel	<	Less Than
R.S.	Rapid Start			>	Greater Than

Index

A

Acrylic carpet	210
Air conditioning	250, 252
Aluminum flashing	193
siding	148
Asphalt driveway	92
tile	211
Awning window	156

B

Baluster	213
birch	213
pine	213
Baseboard heating	261
Basement stair	213
Bath accessory	219
Bathroom	230, 232, 234, 236, 238, 240, 242, 244, 246, 248
Batt insulation	150
Bay window	160
Bi-fold door	208
Bi-passing door	208
Blind	173
Block wall	140
wall foundation	102
Board siding	144
Bow window	160
Brick driveway	92
veneer	142
wall	142
Building excavation	84, 86
Built-up roofing	192

C

Cabinet	216
Carpet	210
Casement window	154
Cedar shake siding	146
Ceiling	198, 202
Ceramic tile	211
Chain link fence	96
Circuit wiring	262
Closet door	208
Common rafter roof framing	120
Concrete block wall	140
driveway	92
foundation	104
slab	108
Cooling	250, 252

D

Deck wood	226
Disposer garbage	218
Door exterior	164
garage	168
storm	172
wood	206, 208
Dormer framing	132, 134
roofing	186, 188
Double-hung window	152
Driveway	92
Dryer wiring	262
Drywall	196, 198
Duplex receptacle	262
Dutch door	164

E

Electric heating	261
service	260
Entrance door	164
Excavation footing	84
foundation	86
utility	88
Exterior door	164
light	263
wall framing	118

F

Fan wiring	262
Field septic	94
Fireplace masonry	220
prefabricated	222
Fixed window	162
Floor concrete	108
framing	112, 114, 116
Flooring	211
Footing	100
excavation	84
Foundation	100
concrete	104
excavation	86
masonry	102
wood	106
Framing dormer	132, 134
floor	112, 114, 116
partition	136
roof	120, 122, 124, 126, 128, 130
Furnace wiring	262

G

Gable dormer framing	132
dormer roofing	186
roof framing	120
Gambrel roof framing	126
Garage door	168
Gas heating	250, 254
Greenhouse	224
Ground fault	262
Gunite pool	225

H

Handrail oak	213
Hand-split shake	146
Heater wiring	262
Heating	250, 252, 254
electric	261
Hip roof framing	124
Hot-air heating	250, 252
Hot-water heating	254

I

Incandescent light	263
Insulation board	193
Interior door	206, 208
light	263
partition framing	136

K

Kitchen	216

L

Lavatory	230
Leaching field	94
Lighting	262
Louvered door	206, 208

M

Mansard roof framing	128
Masonry fireplace	220
foundation	102
wall	140, 142
Membrane	193
Metal clad window	152, 154, 156, 158, 160, 162
fireplace	222
halide fixture	263
siding	148

N

Newel	213
Nylon carpet	210

O

Oak floor	211
Oil heating	252, 254
Overhead door	168

P

Padding	210
Paneled door	206
Parquet floor	211
Partition	196, 200
framing	136
Passage door	206
Picture window	162
Plaster	200, 202
board	196, 198
Plastic blind	173
clad window	152, 154, 156, 158, 160, 162
siding	148
Poured insulation	150
Prefabricated fireplace	222
Prefinished floor	211

R

Rafter roof framing	120
Range circuit	262
hood	218
Redwood siding	144
Resilient flooring	211
Rigid insulation	150
Riser beech	213
oak	213
Roof framing	120, 122, 124, 126, 128, 130
Roofing	176, 178, 180, 182, 184
gable end	176
gambrel	180
hip roof	178
mansard	182
shed	184

S

Septic system	94
Service electric	260
Shake siding	146
Shed dormer framing	134
dormer roofing	188
roof framing	130
Sheetrock	196, 198
Shingle	176, 178, 180, 182, 184
siding	146
Shutter	173
Siding metal	148
shingle	146
wood	144
Sink	218
Sitework	84
Skim coat	196, 198
Skylight	190
Skywindow	190
Slab concrete	108
Sleeper wood	211
Sliding door	167
window	158
Special stairway	213
Stair tread	213
Stairway	212
Steel fireplace	222
Stone wall	142
Storm door	172
window	172
Stringer	213
Stucco	202
interior	200
wall	140
Stud wall	118
Subfloor plywood	211
Suspended ceiling	204
Swimming pool	225
Swing-up door	168
Switch wiring	262

T

T & G siding	144
Tank septic	94
Teak floor	211
Tennis court fence	96
Thincoat	196, 198
Tongue and groove siding	144
Tread beech	213
oak	213
Trenching	88
Truss roof framing	122

U

Utility	88
excavation	88

V

Veneer brick	142
Vinyl blind	173
flooring	211
shutter	173
siding	148

Index

W

Wall framing 118
   interior 196, 200
   masonry 140, 142
Water heating wiring 262
Weatherproof receptacle 262
Window . 152, 154, 156, 158, 160, 162
   double-hung 152
   storm 172
Wood blind 173
   blocking 193
   deck 226
   door 206, 208
   floor 112, 114, 116, 211
   foundation 106
   shutter 173
   siding 144
   storm door 172
   storm window 172
   wall framing 118

Notes

Contractor's Pricing Guides

For more information visit RSMeans Web site at www.rsmeans.com

Contractor's Pricing Guide: Residential Detailed Costs 2007

Every aspect of residential construction, from overhead costs to residential lighting and wiring, is in here. All the detail you need to accurately estimate the costs of your work with or without markups—labor-hours, typical crews, and equipment are included as well. When you need a detailed estimate, this publication has all the costs to help you come up with a complete, on the money, price you can rely on to win profitable work.

Unit Prices Now Updated to MasterFormat 2004!

$39.95 per copy
Available Nov. 2006
Catalog No. 60337

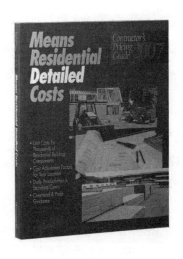

Contractor's Pricing Guide: Residential Repair & Remodeling Costs 2007

This book provides total unit price costs for every aspect of the most common repair and remodeling projects. Organized in the order of construction by component and activity, it includes demolition and installation, cleaning, painting, and more.

With simplified estimating methods, clear, concise descriptions, and technical specifications for each component, the book is a valuable tool for contractors who want to speed up their estimating time, while making sure their costs are on target.

$39.95 per copy
Available Nov. 2006
Catalog No. 60347

Contractor's Pricing Guide: Residential Square Foot Costs 2007

Now available in one concise volume, all you need to know to plan and budget the cost of new homes. If you are looking for a quick reference, the model home section contains costs for over 250 different sizes and types of residences, with hundreds of easily applied modifications. If you need even more detail, the Assemblies Section lets you build your own costs or modify the model costs further. Hundreds of graphics are provided, along with forms and procedures to help you get it right.

$39.95 per copy
Available Nov. 2006
Catalog No. 60327

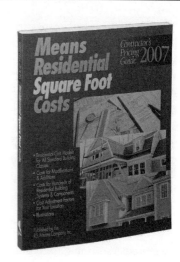

Annual Cost Guides

For more information visit RSMeans Web site at www.rsmeans.com

RSMeans Building Construction Cost Data 2007

Available in Both Softbound and Looseleaf Editions

Many customers enjoy the convenience and flexibility of the looseleaf binder, which increases the usefulness of *RSMeans Building Construction Cost Data 2007* by making it easy to add and remove pages. You can insert your own cost information pages, so everything is in one place. Copying pages for faxing is easier also. Whichever edition you prefer, softbound or the convenient looseleaf edition, you'll be eligible to receive *RSMeans Quarterly Update Service* FREE. Current subscribers can receive *RSMeans Quarterly Update Service* via e-mail.

Unit Prices Now Updated to MasterFormat 2004!

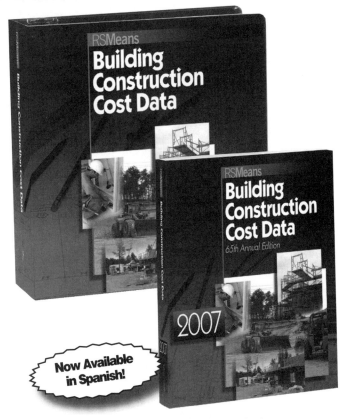

Now Available in Spanish!

$169.95 per copy, Looseleaf
Available Oct. 2006
Catalog No. 61017

RSMeans Building Construction Cost Data 2007

Offers you unchallenged unit price reliability in an easy-to-use arrangement. Whether used for complete, finished estimates or for periodic checks, it supplies more cost facts better and faster than any comparable source. Over 20,000 unit prices for 2007. The City Cost Indexes and Location Factors cover over 930 areas, for indexing to any project location in North America. Order and get *RSMeans Quarterly Update Service* FREE. You'll have year-long access to the RSMeans Estimating **HOTLINE** FREE with your subscription. Expert assistance when using RSMeans data is just a phone call away.

$136.95 per copy (English or Spanish)
Catalog No. 60017 (English) Available Oct. 2006
Catalog No. 60717 (Spanish) Available Jan. 2007

Now Available in Spanish!

Unit Prices Now Updated to MasterFormat 2004!

RSMeans Metric Construction Cost Data 2007

A massive compendium of all the data from both the 2007 *RSMeans Building Construction Cost Data* AND *Heavy Construction Cost Data*, in **metric** format! Access all of this vital information from one complete source. It contains more than 600 pages of unit costs and 40 pages of assemblies costs. The Reference Section contains over 200 pages of tables, charts and other estimating aids. A great way to stay in step with today's construction trends and rapidly changing costs.

$162.95 per copy
Available Dec. 2006
Catalog No. 63017

Annual Cost Guides

For more information visit RSMeans Web site at www.rsmeans.com

RSMeans Mechanical Cost Data 2007

- **HVAC**
- **Controls**

Total unit and systems price guidance for mechanical construction... materials, parts, fittings, and complete labor cost information. Includes prices for piping, heating, air conditioning, ventilation, and all related construction.

Plus new 2007 unit costs for:

- Over 2500 installed HVAC/controls assemblies
- "On Site" Location Factors for over 930 cities and towns in the U.S. and Canada
- Crews, labor, and equipment

$136.95 per copy
Available Oct. 2006
Catalog No. 60027

Unit Prices Now Updated to MasterFormat 2004!

RSMeans Plumbing Cost Data 2007

Comprehensive unit prices and assemblies for plumbing, irrigation systems, commercial and residential fire protection, point-of-use water heaters, and the latest approved materials. This publication and its companion, *RSMeans Mechanical Cost Data*, provide full-range cost estimating coverage for all the mechanical trades.

New for '07: More lines of no-hub CI soil pipe fittings, more flange-type escutcheons, fiberglass pipe insulation in a full range of sizes for 2-1/2" and 3" wall thicknesses, 220 lines of grease duct, and much more.

$136.95 per copy
Available Oct. 2006
Catalog No. 60217

RSMeans Electrical Cost Data 2007

Pricing information for every part of electrical cost planning. More than 13,000 unit and systems costs with design tables; clear specifications and drawings; engineering guides; illustrated estimating procedures; complete labor-hour and materials costs for better scheduling and procurement; and the latest electrical products and construction methods.

- A variety of special electrical systems including cathodic protection
- Costs for maintenance, demolition, HVAC/ mechanical, specialties, equipment, and more

$136.95 per copy
Available Oct. 2006
Catalog No. 60037

Unit Prices Now Updated to MasterFormat 2004!

RSMeans Electrical Change Order Cost Data 2007

RSMeans Electrical Change Order Cost Data 2007 provides you with electrical unit prices exclusively for pricing change orders—based on the recent, direct experience of contractors and suppliers. Analyze and check your own change order estimates against the experience others have had doing the same work. It also covers productivity analysis and change order cost justifications. With useful information for calculating the effects of change orders and dealing with their administration.

$136.95 per copy
Available Nov. 2006
Catalog No. 60237

RSMeans Square Foot Costs 2007

It's Accurate and Easy To Use!

- **Updated 2007 price information**, based on nationwide figures from suppliers, estimators, labor experts, and contractors
- "How-to-Use" sections, with **clear examples** of commercial, residential, industrial, and institutional structures
- Realistic graphics, offering true-to-life illustrations of building projects
- Extensive information on using square foot cost data, including sample estimates and alternate pricing methods

$148.95 per copy
Over 450 pages, illustrated, available Nov. 2006
Catalog No. 60057

RSMeans Repair & Remodeling Cost Data 2007

Commercial/Residential

Use this valuable tool to estimate commercial and residential renovation and remodeling.

Includes: New costs for hundreds of unique methods, materials, and conditions that only come up in repair and remodeling, PLUS:

- Unit costs for over 15,000 construction components
- Installed costs for over 90 assemblies
- Over 930 "On-Site" localization factors for the U.S. and Canada

Unit Prices Now Updated to MasterFormat 2004!

$116.95 per copy
Available Nov. 2006
Catalog No. 60047

Annual Cost Guides

For more information visit RSMeans Web site at www.rsmeans.com

RSMeans Facilities Construction Cost Data 2007

For the maintenance and construction of commercial, industrial, municipal, and institutional properties. Costs are shown for new and remodeling construction and are broken down into materials, labor, equipment, overhead, and profit. Special emphasis is given to sections on mechanical, electrical, furnishings, site work, building maintenance, finish work, and demolition.

More than 43,000 unit costs, plus assemblies costs and a comprehensive Reference Section are included.

$323.95 per copy
Available Dec. 2006
Catalog No. 60207

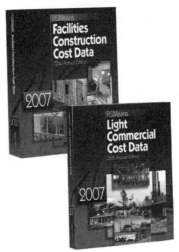

Unit Prices Now Updated to MasterFormat 2004!

RSMeans Light Commercial Cost Data 2007

Specifically addresses the light commercial market, which is a specialized niche in the construction industry. Aids you, the owner/designer/contractor, in preparing all types of estimates—from budgets to detailed bids. Includes new advances in methods and materials.

Assemblies Section allows you to evaluate alternatives in the early stages of design/planning.

Over 11,000 unit costs for 2007 ensure you have the prices you need... when you need them.

$116.95 per copy
Available Dec. 2006
Catalog No. 60187

RSMeans Residential Cost Data 2007

Contains square foot costs for 30 basic home models with the look of today, plus hundreds of custom additions and modifications you can quote right off the page. With costs for the 100 residential systems you're most likely to use in the year ahead. Complete with blank estimating forms, sample estimates, and step-by-step instructions.

Now contains line items for cultured stone and brick, PVC trim lumber, and TPO roofing.

$116.95 per copy
Available Oct. 2006
Catalog No. 60177

Unit Prices Now Updated to MasterFormat 2004!

RSMeans Site Work & Landscape Cost Data 2007

Includes unit and assemblies costs for earthwork, sewerage, piped utilities, site improvements, drainage, paving, trees & shrubs, street openings/repairs, underground tanks, and more. Contains 57 tables of Assemblies Costs for accurate conceptual estimates.

2007 update includes:
- Estimating for infrastructure improvements
- Environmentally-oriented construction
- ADA-mandated handicapped access
- Hazardous waste line items

$136.95 per copy
Available Nov. 2006
Catalog No. 60287

RSMeans Assemblies Cost Data 2007

RSMeans Assemblies Cost Data 2007 takes the guesswork out of preliminary or conceptual estimates. Now you don't have to try to calculate the assembled cost by working up individual component costs. We've done all the work for you.

Presents detailed illustrations, descriptions, specifications, and costs for every conceivable building assembly—240 types in all—arranged in the easy-to-use UNIFORMAT II system. Each illustrated "assembled" cost includes a complete grouping of materials and associated installation costs, including the installing contractor's overhead and profit.

$223.95 per copy
Available Oct. 2006
Catalog No. 60067

RSMeans Open Shop Building Construction Cost Data 2007

The latest costs for accurate budgeting and estimating of new commercial and residential construction... renovation work... change orders... cost engineering.

RSMeans Open Shop "BCCD" will assist you to:
- Develop benchmark prices for change orders
- Plug gaps in preliminary estimates and budgets
- Estimate complex projects
- Substantiate invoices on contracts
- Price ADA-related renovations

Unit Prices Now Updated to MasterFormat 2004!

$136.95 per copy
Available Dec. 2006
Catalog No. 60157

For more information visit RSMeans Web site at www.rsmeans.com

Annual Cost Guides

RSMeans Building Construction Cost Data 2007
Western Edition

This regional edition provides more precise cost information for western North America. Labor rates are based on union rates from 13 western states and western Canada. Included are western practices and materials not found in our national edition: tilt-up concrete walls, glu-lam structural systems, specialized timber construction, seismic restraints, and landscape and irrigation systems.

$136.95 per copy
Available Dec. 2006
Catalog No. 60227

Unit Prices Now Updated to MasterFormat 2004!

RSMeans Heavy Construction Cost Data 2007

A comprehensive guide to heavy construction costs. Includes costs for highly specialized projects such as tunnels, dams, highways, airports, and waterways. Information on labor rates, equipment, and material costs is included. Features unit price costs, systems costs, and numerous reference tables for costs and design.

$136.95 per copy
Available Dec. 2006
Catalog No. 60167

RSMeans Construction Cost Indexes 2007

Who knows what 2007 holds? What materials and labor costs will change unexpectedly? By how much?
- Breakdowns for 316 major cities
- National averages for 30 key cities
- Expanded five major city indexes
- Historical construction cost indexes

$294.00 per year (subscription)
$73.50 individual quarters
Catalog No. 60147 A,B,C,D

RSMeans Interior Cost Data 2007

Provides you with prices and guidance needed to make accurate interior work estimates. Contains costs on materials, equipment, hardware, custom installations, furnishings, and labor costs... for new and remodel commercial and industrial interior construction, including updated information on office furnishings, and reference information.

Unit Prices Now Updated to MasterFormat 2004!

$136.95 per copy
Available Nov. 2006
Catalog No. 60097

RSMeans Concrete & Masonry Cost Data 2007

Provides you with cost facts for virtually all concrete/masonry estimating needs, from complicated formwork to various sizes and face finishes of brick and block—all in great detail. The comprehensive unit cost section contains more than 7,500 selected entries. Also contains an Assemblies Cost section, and a detailed Reference section that supplements the cost data.

$124.95 per copy
Available Dec. 2006
Catalog No. 60117

Unit Prices Now Updated to MasterFormat 2004!

RSMeans Labor Rates for the Construction Industry 2007

Complete information for estimating labor costs, making comparisons, and negotiating wage rates by trade for over 300 U.S. and Canadian cities. With 46 construction trades listed by local union number in each city, and historical wage rates included for comparison. Each city chart lists the county and is alphabetically arranged with handy visual flip tabs for quick reference.

$296.95 per copy
Available Dec. 2006
Catalog No. 60127

RSMeans Facilities Maintenance & Repair Cost Data 2007

RSMeans Facilities Maintenance & Repair Cost Data gives you a complete system to manage and plan your facility repair and maintenance costs and budget efficiently. Guidelines for auditing a facility and developing an annual maintenance plan. Budgeting is included, along with reference tables on cost and management, and information on frequency and productivity of maintenance operations.

The only nationally recognized source of maintenance and repair costs. Developed in cooperation with the Civil Engineering Research Laboratory (CERL) of the Army Corps of Engineers.

$296.95 per copy
Available Dec. 2006
Catalog No. 60307

Reference Books

For more information visit RSMeans Web site at www.rsmeans.com

Home Addition & Renovation Project Costs

This essential home remodeling reference gives you 35 project estimates, and guidance for some of the most popular home renovation and addition projects... from opening up a simple interior wall to adding an entire second story. Each estimate includes a floor plan, color photos, and detailed costs. Use the project estimates as backup for pricing, to check your own estimates, or as a cost reference for preliminary discussion with homeowners.

Includes:
- Case studies—with creative solutions and design ideas.
- Alternate materials costs—so you can match the estimates to the particulars of your projects.
- Location Factors—easy multipliers to adjust the book's costs to your own location.

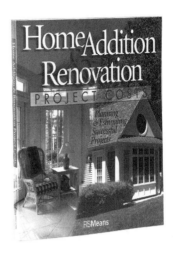

$29.95 per copy
Over 200 pages, illustrated, Softcover
Catalog No. 67349

Kitchen & Bath Project Costs:
Planning & Estimating
Successful Projects

Project estimates for 35 of the most popular kitchen and bath renovations... from replacing a single fixture to whole-room remodels. Each estimate includes:
- All materials needed for the project
- Labor-hours to install (and demolish/remove) each item
- Subcontractor costs for certain trades and services
- An allocation for overhead and profit

PLUS! Takeoff and pricing worksheets—forms you can photocopy or access electronically from the book's Web site; alternate materials—unit costs for different finishes and fixtures; location factors—easy multipliers to adjust the costs to your location; and expert guidance on estimating methods, project design, contracts, marketing, working with homeowners, and tips for each of the estimated projects.

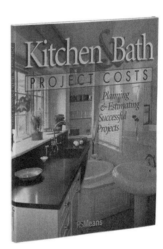

$29.95 per copy
Over 175 pages
Catalog No. 67347

Residential & Light Commercial Construction Standards 2nd Edition
By RSMeans and Contributing Authors

For contractors, subcontractors, owners, developers, architects, engineers, attorneys, and insurance personnel, this book provides authoritative requirements and recommendations compiled from the nation's leading professional associations, industry publications, and building code organizations.

It's an all-in-one reference for establishing a standard for workmanship, quickly resolving disputes, and avoiding defect claims. Includes practical guidance from professionals who are well-known in their respective fields for quality design and construction.

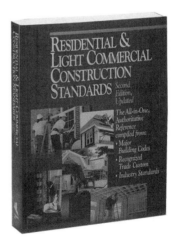

$59.95 per copy
600 pages, illustrated, Softcover
Catalog No. 67322A

For more information
visit RSMeans Web site
at www.rsmeans.com

Reference Books

Value Engineering: Practical Applications
By Alphonse Dell'Isola, PE

For Design, Construction, Maintenance & Operations

A tool for immediate application—for engineers, architects, facility managers, owners, and contractors. Includes: making the case for VE—the management briefing, integrating VE into planning, budgeting, and design, conducting life cycle costing, using VE methodology in design review and consultant selection, case studies, VE workbook, and a life cycle costing program on disk.

$79.95 per copy
Over 450 pages, illustrated, Softcover
Catalog No. 67319

Facilities Operations & Engineering Reference
By the Association for Facilities Engineering and RSMeans

An all-in-one technical reference for planning and managing facility projects and solving day-to-day operations problems. Selected as the official Certified Plant Engineer reference, this handbook covers financial analysis, maintenance, HVAC and energy efficiency, and more.

$54.98 per copy
Over 700 pages, illustrated, Hardcover
Catalog No. 67318

The Building Professional's Guide to Contract Documents
3rd Edition
By Waller S. Poage, AIA, CSI, CVS

A comprehensive reference for owners, design professionals, contractors, and students.

- Structure your documents for maximum efficiency.
- Effectively communicate construction requirements.
- Understand the roles and responsibilities of construction professionals.
- Improve methods of project delivery.

$32.48 per copy, 400 pages
Diagrams and construction forms, Hardcover
Catalog No. 67261A

Building Security: Strategies & Costs
By David Owen

This comprehensive resource will help you evaluate your facility's security needs, and design and budget for the materials and devices needed to fulfill them.

Includes over 130 pages of RSMeans cost data for installation of security systems and materials, plus a review of more than 50 security devices and construction solutions.

$44.98 per copy
350 pages, illustrated, Hardcover
Catalog No. 67339

Cost Planning & Estimating for Facilities Maintenance

In this unique book, a team of facilities management authorities shares their expertise on:

- Evaluating and budgeting maintenance operations
- Maintaining and repairing key building components
- Applying *RSMeans Facilities Maintenance & Repair Cost Data* to your estimating

Covers special maintenance requirements of the ten major building types.

$89.95 per copy
Over 475 pages, Hardcover
Catalog No. 67314

Life Cycle Costing for Facilities
By Alphonse Dell'Isola and Dr. Steven Kirk

Guidance for achieving higher quality design and construction projects at lower costs! Cost-cutting efforts often sacrifice quality to yield the cheapest product. Life cycle costing enables building designers and owners to achieve both. The authors of this book show how LCC can work for a variety of projects — from roads to HVAC upgrades to different types of buildings.

$99.95 per copy
450 pages, Hardcover
Catalog No. 67341

Planning & Managing Interior Projects 2nd Edition
By Carol E. Farren, CFM

Expert guidance on managing renovation & relocation projects.

This book guides you through every step in relocating to a new space or renovating an old one. From initial meeting through design and construction, to post-project administration, it helps you get the most for your company or client. Includes sample forms, spec lists, agreements, drawings, and much more!

$69.95 per copy
200 pages, Softcover
Catalog No. 67245A

Builder's Essentials: Best Business Practices for Builders & Remodelers
An Easy-to-Use Checklist System
By Thomas N. Frisby

A comprehensive guide covering all aspects of running a construction business, with more than 40 user-friendly checklists. Provides expert guidance on: increasing your revenue and keeping more of your profit, planning for long-term growth, keeping good employees, and managing subcontractors.

$29.95 per copy
Over 220 pages, Softcover
Catalog No. 67329

Reference Books

For more information visit RSMeans Web site at www.rsmeans.com

Interior Home Improvement Costs New 9th Edition

Updated estimates for the most popular remodeling and repair projects—from small, do-it-yourself jobs to major renovations and new construction. Includes: Kitchens & Baths; New Living Space from your Attic, Basement, or Garage; New Floors, Paint, and Wallpaper; Tearing Out or Building New Walls; Closets, Stairs, and Fireplaces; New Energy-Saving Improvements, Home Theaters, and More!

$24.95 per copy
250 pages, illustrated, Softcover
Catalog No. 67308E

Exterior Home Improvement Costs New 9th Edition

Updated estimates for the most popular remodeling and repair projects—from small, do-it-yourself jobs, to major renovations and new construction. Includes: Curb Appeal Projects—Landscaping, Patios, Porches, Driveways, and Walkways; New Windows and Doors; Decks, Greenhouses, and Sunrooms; Room Additions and Garages; Roofing, Siding, and Painting; "Green" Improvements to Save Energy & Water.

$24.95 per copy
Over 275 pages, illustrated, Softcover
Catalog No. 67309E

Builder's Essentials: Plan Reading & Material Takeoff

By Wayne J. DelPico

For Residential and Light Commercial Construction

A valuable tool for understanding plans and specs, and accurately calculating material quantities. Step-by-step instructions and takeoff procedures based on a full set of working drawings.

$35.95 per copy
Over 420 pages, Softcover
Catalog No. 67307

Builder's Essentials: Framing & Rough Carpentry 2nd Edition

By Scot Simpson

Develop and improve your skills with easy-to-follow instructions and illustrations. Learn proven techniques for framing walls, floors, roofs, stairs, doors, and windows. Updated guidance on standards, building codes, safety requirements, and more. Also available in Spanish!

$24.95 per copy
Over 150 pages, Softcover
Catalog No. 67298A
Spanish Catalog No. 67298AS

Concrete Repair and Maintenance Illustrated

By Peter Emmons

Hundreds of illustrations show users how to analyze, repair, clean, and maintain concrete structures for optimal performance and cost effectiveness. From parking garages to roads and bridges to structural concrete, this comprehensive book describes the causes, effects, and remedies for concrete wear and failure. Invaluable for planning jobs, selecting materials, and training employees, this book is a must-have for concrete specialists, general contractors, facility managers, civil and structural engineers, and architects.

$34.98 per copy
300 pages, illustrated, Softcover
Catalog No. 67146

Means Unit Price Estimating Methods

New 3rd Edition

This new edition includes up-to-date cost data and estimating examples, updated to reflect changes to the CSI numbering system and new features of RSMeans cost data. It describes the most productive, universally accepted ways to estimate, and uses checklists and forms to illustrate shortcuts and timesavers. A model estimate demonstrates procedures. A new chapter explores computer estimating alternatives.

$29.98 per copy
Over 350 pages, illustrated, Hardcover
Catalog No. 67303A

Total Productive Facilities Management

By Richard W. Sievert, Jr.

Today, facilities are viewed as strategic resources... elevating the facility manager to the role of asset manager supporting the organization's overall business goals. Now, Richard Sievert Jr., in this well-articulated guidebook, sets forth a new operational standard for the facility manager's emerging role... a comprehensive program for managing facilities as a true profit center.

$29.98 per copy
275 pages, Softcover
Catalog No. 67321

Means Environmental Remediation Estimating Methods 2nd Edition

By Richard R. Rast

Guidelines for estimating 50 standard remediation technologies. Use it to prepare preliminary budgets, develop estimates, compare costs and solutions, estimate liability, review quotes, negotiate settlements.

$49.98 per copy
Over 750 pages, illustrated, Hardcover
Catalog No. 64777A

For more information visit RSMeans Web site at www.rsmeans.com

Reference Books

Means Illustrated Construction Dictionary Condensed, 2nd Edition
Recognized in the industry as the best resource of its kind.

This essential tool has been further enhanced with updates to existing terms and the addition of hundreds of new terms and illustrations—in keeping with recent developments. For contractors, architects, insurance and real estate personnel, homeowners, and anyone who needs quick, clear definitions for construction terms.

$59.95 per copy
Over 500 pages, Softcover
Catalog No. 67282A

Means Repair & Remodeling Estimating New 4th Edition
By Edward B. Wetherill & RSMeans

This important reference focuses on the unique problems of estimating renovations of existing structures, and helps you determine the true costs of remodeling through careful evaluation of architectural details and a site visit.

New section on disaster restoration costs.

$69.95 per copy
Over 450 pages, illustrated, Hardcover
Catalog No. 67265B

Facilities Planning & Relocation
New, lower price and user-friendly format.
By David D. Owen

A complete system for planning space needs and managing relocations. Includes step-by-step manual, over 50 forms, and extensive reference section on materials and furnishings.

$89.95 per copy
Over 450 pages, Softcover
Catalog No. 67301

Means Square Foot & Assemblies Estimating Methods 3rd Edition

Develop realistic square foot and assemblies costs for budgeting and construction funding. The new edition features updated guidance on square foot and assemblies estimating using UNIFORMAT II. An essential reference for anyone who performs conceptual estimates.

$34.98 per copy
Over 300 pages, illustrated, Hardcover
Catalog No. 67145B

Means Electrical Estimating Methods 3rd Edition

Expanded edition includes sample estimates and cost information in keeping with the latest version of the CSI MasterFormat and UNIFORMAT II. Complete coverage of fiber optic and uninterruptible power supply electrical systems, broken down by components, and explained in detail. Includes a new chapter on computerized estimating methods. A practical companion to *RSMeans Electrical Cost Data.*

$64.95 per copy
Over 325 pages, Hardcover
Catalog No. 67230B

Means Mechanical Estimating Methods 3rd Edition

This guide assists you in making a review of plans, specs, and bid packages, with suggestions for takeoff procedures, listings, substitutions, and pre-bid scheduling. Includes suggestions for budgeting labor and equipment usage. Compares materials and construction methods to allow you to select the best option.

$64.95 per copy
Over 350 pages, illustrated, Hardcover
Catalog No. 67294A

Means ADA Compliance Pricing Guide New Second Edition
By Adaptive Environments and RSMeans

Completely updated and revised to the new 2004 *Americans with Disabilities Act Accessibility Guidelines,* this book features more than 70 of the most commonly needed modifications for ADA compliance. Projects range from installing ramps and walkways, widening doorways and entryways, and installing and refitting elevators, to relocating light switches and signage.

$79.95 per copy
Over 350 pages, illustrated, Softcover
Catalog No. 67310A

Project Scheduling & Management for Construction
New 3rd Edition
By David R. Pierce, Jr.

A comprehensive yet easy-to-follow guide to construction project scheduling and control—from vital project management principles through the latest scheduling, tracking, and controlling techniques. The author is a leading authority on scheduling, with years of field and teaching experience at leading academic institutions. Spend a few hours with this book and come away with a solid understanding of this essential management topic.

$64.95 per copy
Over 300 pages, illustrated, Hardcover
Catalog No. 67247B

Reference Books

For more information visit RSMeans Web site at www.rsmeans.com

The Practice of Cost Segregation Analysis
by Bruce A. Desrosiers and Wayne J. DelPico

This expert guide walks you through the practice of cost segregation analysis, which enables property owners to defer taxes and benefit from "accelerated cost recovery" through depreciation deductions on assets that are properly identified and classified.

With a glossary of terms, sample cost segregation estimates for various building types, key information resources, and updates via a dedicated Web site, this book is a critical resource for anyone involved in cost segregation analysis.

$99.95 per copy
Over 225 pages
Catalog No. 67345

Preventive Maintenance for Multi-Family Housing
by John C. Maciha

Prepared by one of the nation's leading experts on multi-family housing.

This complete PM system for apartment and condominium communities features expert guidance, checklists for buildings and grounds maintenance tasks and their frequencies, a reusable wall chart to track maintenance, and a dedicated Web site featuring customizable electronic forms. A must-have for anyone involved with multi-family housing maintenance and upkeep.

$89.95 per copy
225 pages
Catalog No. 67346

Means Landscape Estimating Methods
4th Edition
By Sylvia H. Fee

This revised edition offers expert guidance for preparing accurate estimates for new landscape construction and grounds maintenance. Includes a complete project estimate featuring the latest equipment and methods, and chapters on Life Cycle Costing and Landscape Maintenance Estimating.

$62.95 per copy
Over 300 pages, illustrated, Hardcover
Catalog No. 67295B

Job Order Contracting
Expediting Construction Project Delivery
by Allen Henderson

Expert guidance to help you implement JOC—fast becoming the preferred project delivery method for repair and renovation, minor new construction, and maintenance projects in the public sector and in many states and municipalities. The author, a leading JOC expert and practitioner, shows how to:

- Establish a JOC program
- Evaluate proposals and award contracts
- Handle general requirements and estimating
- Partner for maximum benefits

$89.95 per copy
192 pages, illustrated, Hardcover
Catalog No. 67348

Builder's Essentials: Estimating Building Costs
For the Residential & Light Commercial Contractor
By Wayne J. DelPico

Step-by-step estimating methods for residential and light commercial contractors. Includes a detailed look at every construction specialty—explaining all the components, takeoff units, and labor needed for well-organized, complete estimates. Covers correctly interpreting plans and specifications, and developing accurate and complete labor and material costs.

$29.95 per copy
Over 400 pages, illustrated, Softcover
Catalog No. 67343

Building & Renovating Schools

This all-inclusive guide covers every step of the school construction process—from initial planning, needs assessment, and design, right through moving into the new facility. A must-have resource for anyone concerned with new school construction or renovation. With square foot cost models for elementary, middle, and high school facilities, and real-life case studies of recently completed school projects.

The contributors to this book—architects, construction project managers, contractors, and estimators who specialize in school construction—provide start-to-finish, expert guidance on the process.

$99.95 per copy
Over 425 pages, Hardcover
Catalog No. 67342

For more information visit RSMeans Web site at www.rsmeans.com

Reference Books

Historic Preservation: Project Planning & Estimating
By Swanke Hayden Connell Architects

Expert guidance on managing historic restoration, rehabilitation, and preservation building projects and determining and controlling their costs. Includes:
- How to determine whether a structure qualifies as historic
- Where to obtain funding and other assistance
- How to evaluate and repair more than 75 historic building materials

$49.98 per copy
Over 675 pages, Hardcover
Catalog No. 67323

Means Illustrated Construction Dictionary
Unabridged 3rd Edition, with CD-ROM

Long regarded as the industry's finest, *Means Illustrated Construction Dictionary* is now even better. With the addition of over 1,000 new terms and hundreds of new illustrations, it is the clear choice for the most comprehensive and current information. The companion CD-ROM that comes with this new edition adds many extra features: larger graphics, expanded definitions, and links to both CSI MasterFormat numbers and product information.

$99.95 per copy
Over 790 pages, illustrated, Hardcover
Catalog No. 67292A

Designing & Building with the IBC 2nd Edition
By Rolf Jensen & Associates, Inc.

This updated, comprehensive guide helps building professionals make the transition to the 2003 International Building Code®. Includes a side-by-side code comparison of the IBC 2003 to the IBC 2000 and the three primary model codes, a quick-find index, and professional code commentary. With illustrations, abbreviations key, and an extensive Resource section.

$99.95 per copy
Over 875 pages, Softcover
Catalog No. 67328A

Means Plumbing Estimating Methods 3rd Edition
By Joseph Galeno and Sheldon Greene

Updated and revised! This practical guide walks you through a plumbing estimate, from basic materials and installation methods through change order analysis. *Plumbing Estimating Methods* covers residential, commercial, industrial, and medical systems, and features sample takeoff and estimate forms and detailed illustrations of systems and components.

$29.98 per copy
330+ pages, Softcover
Catalog No. 67283B

Builder's Essentials: Advanced Framing Methods
By Scot Simpson

A highly illustrated, "framer-friendly" approach to advanced framing elements. Provides expert, but easy to interpret, instruction for laying out and framing complex walls, roofs, and stairs, and special requirements for earthquake and hurricane protection. Also helps bring framers up to date on the latest building code changes, and provides tips on the lead framer's role and responsibilities, how to prepare for a job, and how to get the crew started.

$24.95 per copy
250 pages, illustrated, Softcover
Catalog No. 67330

Means Estimating Handbook
2nd Edition

Updated Second Edition answers virtually any estimating technical question—all organized by CSI MasterFormat. This comprehensive reference covers the full spectrum of technical data required to estimate construction costs. The book includes information on sizing, productivity, equipment requirements, code-mandated specifications, design standards, and engineering factors.

$99.95 per copy
Over 900 pages, Hardcover
Catalog No. 67276A

Preventive Maintenance Guidelines for School Facilities
By John C. Maciha

A complete PM program for K-12 schools that ensures sustained security, safety, property integrity, user satisfaction, and reasonable ongoing expenditures.

Includes schedules for weekly, monthly, semiannual, and annual maintenance in hard copy and electronic format.

$149.95 per copy
Over 225 pages, Hardcover
Catalog No. 67326

Preventive Maintenance for Higher Education Facilities
By Applied Management Engineering, Inc.

An easy-to-use system to help facilities professionals establish the value of PM, and to develop and budget for an appropriate PM program for their college or university. Features interactive campus building models typical of those found in different-sized higher education facilities, and PM checklists linked to each piece of equipment or system in hard copy and electronic format.

$149.95 per copy
150 pages, Hardcover
Catalog No. 67337

For more information visit RSMeans Web site at www.rsmeans.com

Seminars

Means CostWorks® Training

This one-day seminar has been designed with the intention of assisting both new and existing users to become more familiar with the *Means CostWorks* program. The class is broken into two unique sections: (1) A one-half day presentation on the function of each icon; and each student will be shown how to use the software to develop a cost estimate. (2) Hands-on estimating exercises that will ensure that each student thoroughly understands how to use *CostWorks*. You must bring your own laptop computer to this course.

Means CostWorks Benefits/Features:
- Estimate in your own spreadsheet format
- Power of RSMeans National Database
- Database automatically regionalized
- Save time with keyword searches
- Save time by establishing common estimate items in "Bookmark" files
- Customize your spreadsheet template
- Hot Key to Product Manufacturers' listings and specs
- Merge capability for networking environments
- View crews and assembly components
- AutoSave capability
- Enhanced sorting capability

Unit Price Estimating

This interactive two-day seminar teaches attendees how to interpret project information and process it into final, detailed estimates with the greatest accuracy level.

The single most important credential an estimator can take to the job is the ability to visualize construction in the mind's eye, and thereby estimate accurately.

Some Of What You'll Learn:
- Interpreting the design in terms of cost
- The most detailed, time-tested methodology for accurate "pricing"
- Key cost drivers—material, labor, equipment, staging, and subcontracts
- Understanding direct and indirect costs for accurate job cost accounting and change order management

Who Should Attend: Corporate and government estimators and purchasers, architects, engineers… and others needing to produce accurate project estimates.

Square Foot and Assemblies Estimating

This two-day course teaches attendees how to quickly deliver accurate square foot estimates using limited budget and design information.

Some Of What You'll Learn:
- How square foot costing gets the estimate done faster
- Taking advantage of a "systems" or "assemblies" format
- The RSMeans "building assemblies/square foot cost approach"
- How to create a very reliable preliminary and systems estimate using bare-bones design information

Who Should Attend: Facilities managers, facilities engineers, estimators, planners, developers, construction finance professionals… and others needing to make quick, accurate construction cost estimates at commercial, government, educational, and medical facilities.

Repair and Remodeling Estimating

This two-day seminar emphasizes all the underlying considerations unique to repair/remodeling estimating and presents the correct methods for generating accurate, reliable R&R project costs using the unit price and assemblies methods.

Some Of What You'll Learn:
- Estimating considerations—like labor-hours, building code compliance, working within existing structures, purchasing materials in smaller quantities, unforeseen deficiencies
- Identifying problems and providing solutions to estimating building alterations
- Rules for factoring in minimum labor costs, accurate productivity estimates, and allowances for project contingencies
- R&R estimating examples calculated using unit price and assemblies data

Who Should Attend: Facilities managers, plant engineers, architects, contractors, estimators, builders… and others who are concerned with the proper preparation and/or evaluation of repair and remodeling estimates.

Mechanical and Electrical Estimating

This two-day course teaches attendees how to prepare more accurate and complete mechanical/electrical estimates, avoiding the pitfalls of omission and double-counting, while understanding the composition and rationale within the RSMeans Mechanical/Electrical database.

Some Of What You'll Learn:
- The unique way mechanical and electrical systems are interrelated
- M&E estimates–conceptual, planning, budgeting, and bidding stages
- Order of magnitude, square foot, assemblies, and unit price estimating
- Comparative cost analysis of equipment and design alternatives

Who Should Attend: Architects, engineers, facilities managers, mechanical and electrical contractors… and others needing a highly reliable method for developing, understanding, and evaluating mechanical and electrical contracts.

Plan Reading and Material Takeoff

This two-day program teaches attendees to read and understand construction documents and to use them in the preparation of material takeoffs.

Some of What You'll Learn:
- Skills necessary to read and understand typical contract documents—blueprints and specifications
- Details and symbols used by architects and engineers
- Construction specifications' importance in conjunction with blueprints
- Accurate takeoff of construction materials and industry-accepted takeoff methods

Who Should Attend: Facilities managers, construction supervisors, office managers… and others responsible for the execution and administration of a construction project, including government, medical, commercial, educational, or retail facilities.

Facilities Maintenance and Repair Estimating

This two-day course teaches attendees how to plan, budget, and estimate the cost of ongoing and preventive maintenance and repair for existing buildings and grounds.

Some Of What You'll Learn:
- The most financially favorable maintenance, repair, and replacement scheduling and estimating
- Auditing and value engineering facilities
- Preventive planning and facilities upgrading
- Determining both in-house and contract-out service costs
- Annual, asset-protecting M&R plan

Who Should Attend: Facility managers, maintenance supervisors, buildings and grounds superintendents, plant managers, planners, estimators… and others involved in facilities planning and budgeting.

Scheduling and Project Management

This two-day course teaches attendees the most current and proven scheduling and management techniques needed to bring projects in on time and on budget.

Some Of What You'll Learn:
- Crucial phases of planning and scheduling
- How to establish project priorities and develop realistic schedules and management techniques
- Critical Path and Precedence Methods
- Special emphasis on cost control

Who Should Attend: Construction project managers, supervisors, engineers, estimators, contractors… and others who want to improve their project planning, scheduling, and management skills.

Assessing Scope of Work for Facility Construction Estimating

This two-day course is a practical training program that addresses the vital importance of understanding the SCOPE of projects in order to produce accurate cost estimates in a facility repair and remodeling environment.

Some Of What You'll Learn:
- Discussions on site visits, plans/specs, record drawings of facilities, and site-specific lists
- Review of CSI divisions, including means, methods, materials, and the challenges of scoping each topic
- Exercises in SCOPE identification and SCOPE writing for accurate estimating of projects
- Hands-on exercises that require SCOPE, take-off, and pricing

Who Should Attend: Corporate and government estimators, planners, facility managers… and others needing to produce accurate project estimates.

Seminars

2007 RSMeans Seminar Schedule

For more information visit RSMeans Web site at www.rsmeans.com

Location	Dates
Las Vegas, NV	March
Washington, DC	April
Phoenix, AZ	April
Denver, CO	May
San Francisco, CA	June
Philadelphia, PA	June
Washington, DC	September
Dallas, TX	September
Las Vegas, NV	October
Orlando, FL	November
Atlantic City, NJ	November
San Diego, CA	December

Note: Call for exact dates and details.

Registration Information

Register Early... Save up to $100! Register 30 days before the start date of a seminar and save $100 off your total fee. *Note: This discount can be applied only once per order. It cannot be applied to team discount registrations or any other special offer.*

How to Register Register by phone today! RSMeans' toll-free number for making reservations is: **1-800-334-3509.**

Individual Seminar Registration Fee $935. *Means CostWorks®* **Training Registration Fee $375.** To register by mail, complete the registration form and return with your full fee to: Seminar Division, Reed Construction Data, RSMeans Seminars, 63 Smiths Lane, Kingston, MA 02364.

Federal Government Pricing All federal government employees save 25% off regular seminar price. Other promotional discounts cannot be combined with Federal Government discount.

Team Discount Program Two to four seminar registrations, call for pricing: 1-800-334-3509, Ext. 5115

Multiple Course Discounts When signing up for two or more courses, call for pricing.

Refund Policy Cancellations will be accepted up to ten days prior to the seminar start. There are no refunds for cancellations received later than ten working days prior to the first day of the seminar. A $150 processing fee will be applied for all cancellations. Written notice of cancellation is required. Substitutions can be made at any time before the session starts. **No-shows are subject to the full seminar fee.**

AACE Approved Courses Many seminars described and offered here have been approved for 14 hours (1.4 recertification credits) of credit by the AACE International Certification Board toward meeting the continuing education requirements for recertification as a Certified Cost Engineer/Certified Cost Consultant.

AIA Continuing Education We are registered with the AIA Continuing Education System (AIA/CES) and are committed to developing quality learning activities in accordance with the CES criteria. Many seminars meet the AIA/CES criteria for Quality Level 2. AIA members may receive (14) learning units (LUs) for each two-day RSMeans course.

NASBA CPE Sponsor Credits We are part of the National Registry of CPE Sponsors. Attendees may be eligible for (16) CPE credits.

Daily Course Schedule The first day of each seminar session begins at 8:30 A.M. and ends at 4:30 P.M. The second day is 8:00 A.M.–4:00 P.M. Participants are urged to bring a hand-held calculator since many actual problems will be worked out in each session.

Continental Breakfast Your registration includes the cost of a continental breakfast, a morning coffee break, and an afternoon break. These informal segments will allow you to discuss topics of mutual interest with other members of the seminar. (You are free to make your own lunch and dinner arrangements.)

Hotel/Transportation Arrangements RSMeans has arranged to hold a block of rooms at most host hotels. To take advantage of special group rates when making your reservation, be sure to mention that you are attending the RSMeans Seminar. You are, of course, free to stay at the lodging place of your choice. **(Hotel reservations and transportation arrangements should be made directly by seminar attendees.)**

Important Class sizes are limited, so please register as soon as possible.

Note: Pricing subject to change.

Registration Form

Call 1-800-334-3509 to register or FAX 1-800-632-6732. Visit our Web site: www.rsmeans.com

Please register the following people for the RSMeans Construction Seminars as shown here. We understand that we must make our own hotel reservations if overnight stays are necessary.

☐ Full payment of $_____ enclosed.

☐ Bill me

Name of Registrant(s)
(To appear on certificate of completion)

P.O. #: _____
GOVERNMENT AGENCIES MUST SUPPLY PURCHASE ORDER NUMBER OR TRAINING FORM.

Firm Name _____
Address _____
City/State/Zip _____
Telephone No. _____ Fax No. _____
E-mail Address _____
Charge our registration(s) to: ☐ MasterCard ☐ VISA ☐ American Express ☐ Discover
Account No. _____ Exp. Date _____
Cardholder's Signature _____
Seminar Name _____ City _____ Dates _____

Please mail check to: Seminar Division, Reed Construction Data, RSMeans Seminars, 63 Smiths Lane, P.O. Box 800, Kingston, MA 02364 USA

MeansData™
CONSTRUCTION COSTS FOR SOFTWARE APPLICATIONS
Your construction estimating software is only as good as your cost data.

A proven construction cost database is a mandatory part of any estimating package. The following list of software providers can offer you MeansData™ as an added feature for their estimating systems. See the table below for what types of products and services they offer (match their numbers). Visit online at **www.rsmeans.com/demosource/** for more information and free demos. Or call their numbers listed below.

1. **3D International**
   713-871-7000
   venegas@3di.com

2. **4Clicks-Solutions, LLC**
   719-574-7721
   mbrown@4clicks-solutions.com

3. **Aepco, Inc.**
   301-670-4642
   blueworks@aepco.com

4. **Applied Flow Technology**
   800-589-4943
   info@aft.com

5. **ArenaSoft Estimating**
   888-370-8806
   info@arenasoft.com

6. **Ares Corporation**
   925-299-6700
   sales@arescorporation.com

7. **Beck Technology**
   214-303-6293
   stewartcarroll@beckgroup.com

8. **BSD - Building Systems Design, Inc.**
   888-273-7638
   bsd@bsdsoftlink.com

9. **CMS - Computerized Micro Solutions**
   800-255-7407
   cms@proest.com

10. **Corecon Technologies, Inc.**
    714-895-7222
    sales@corecon.com

11. **CorVet Systems**
    301-622-9069
    sales@corvetsys.com

12. **Earth Tech**
    303-771-3103
    kyle.knudson@earthtech.com

13. **Estimating Systems, Inc.**
    800-967-8572
    esipulsar@adelphia.net

14. **HCSS**
    800-683-3196
    info@hcss.com

15. **MC^2 - Management Computer**
    800-225-5622
    vkeys@mc2-ice.com

16. **Maximus Asset Solutions**
    800-659-9001
    assetsolutions@maximus.com

17. **Sage Timberline Office**
    800-628-6583
    productinfo.timberline@sage.com

18. **Shaw Beneco Enterprises, Inc.**
    877-719-4748
    inquire@beneco.com

19. **US Cost, Inc.**
    800-372-4003
    sales@uscost.com

20. **Vanderweil Facility Advisors**
    617-451-5100
    info@VFA.com

21. **WinEstimator, Inc.**
    800-950-2374
    sales@winest.com

TYPE	1	2	3	4	5	6	7	8	9	10	11	12	13	14	15	16	17	18	19	20	21
BID					•			•	•		•			•	•		•				•
Estimating		•			•	•	•	•	•	•	•	•	•	•	•		•	•	•		•
DOC/JOC/SABER		•			•			•			•		•			•	•	•			•
ID/IQ		•									•		•			•		•			•
Asset Mgmt.																•				•	•
Facility Mgmt.	•		•													•				•	
Project Mgmt.	•	•					•			•	•			•			•	•			
TAKE-OFF					•					•	•	•	•	•	•		•		•		
EARTHWORK												•		•	•						
Pipe Flow				•										•							
HVAC/Plumbing					•				•			•									
Roofing					•				•			•									
Design	•				•		•										•	•			•
Other Offers/Links:																					
Accounting/HR		•			•											•	•	•			
Scheduling						•	•										•		•		•
CAD							•										•				
PDA																	•		•		
Lt. Versions		•							•								•				
Consulting	•	•			•		•		•	•	•		•			•	•	•	•	•	•
Training		•			•		•		•	•	•	•	•	•	•	•	•	•	•	•	•

Qualified re-seller applications now being accepted. Call Carol Polio, Ext. 5107.

FOR MORE INFORMATION
CALL 1-800-448-8182, EXT. 5107 OR FAX 1-800-632-6732

For more information
visit RSMeans Web site
at www.rsmeans.com

New Titles

Understanding & Negotiating Construction Contracts

By Kit Werremeyer

Take advantage of the author's 30 years' experience in small-to-large (including international) construction projects. Learn how to identify, understand, and evaluate high risk terms and conditions typically found in all construction contracts—then negotiate to lower or eliminate the risk, improve terms of payment, and reduce exposure to claims and disputes. The author avoids "legalese" and gives real-life examples from actual projects.

$69.95 per copy
300 pages, Softcover
Catalog No. 67350

Green Building: Project Planning & Cost Estimating, 2nd Edition

This new edition has been completely updated with the latest in green building technologies, design concepts, standards, and costs. Now includes a 2007 Green Building *CostWorks* CD with more than 300 green building assemblies and over 5,000 unit price line items for sustainable building. The new edition is also full-color with all new case studies—plus a new chapter on deconstruction, a key aspect of green building.

$129.95 per copy
350 pages, Softcover
Catalog No. 67338A

How to Estimate with Means Data & CostWorks
New 3rd Edition

By RSMeans and Saleh A. Mubarak, Ph.D.

New 3rd Edition—fully updated with new chapters, plus new CD with updated *CostWorks* cost data and MasterFormat organization. Includes all major construction items—with more than 300 exercises and two sets of plans that show how to estimate for a broad range of construction items and systems—including general conditions and equipment costs.

$59.95 per copy
272 pages, Softcover
Includes CostWorks CD
Catalog No. 67324B

Means Spanish/English Construction Dictionary
2nd Edition

By RSMeans and the International Code Council

This expanded edition features thousands of the most common words and useful phrases in the construction industry with easy-to-follow pronunciations in both Spanish and English. Over 800 new terms, phrases, and illustrations have been added. It also features a new stand-alone "Safety & Emergencies" section, with colored pages for quick access. Unique to this dictionary are the systems illustrations showing the relationship of components in the most common building systems for all major trades.

$23.95 per copy
Over 400 pages
Catalog No. 67327A

Construction Business Management

By Nick Ganaway

Only 43% of construction firms stay in business after four years. Make sure your company thrives with valuable guidance from a pro with 25 years of success as a commercial contractor. Find out what it takes to build all aspects of a business that is profitable, enjoyable, and enduring. With a bonus chapter on retail construction.

$49.95 per copy
200 pages, Softcover
Catalog No. 67352

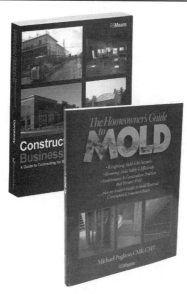

The Homeowner's Guide to Mold

Expert guidance to protect your health and your home.

Mold, whether caused by leaks, humidity or flooding, is a real health and financial issue—for homeowners and contractors. This full-color book explains:

- Construction and maintenance practices to prevent mold
- How to inspect for and remove mold
- Mold remediation procedures and costs
- What to do after a flood
- How to deal with insurance companies if you're thinking of submitting a mold damages claim

$21.95 per copy
144 pages, Softcover
Catalog No. 67344

2007 Order Form

ORDER TOLL FREE 1-800-334-3509
OR FAX 1-800-632-6732

Qty.	Book No.	COST ESTIMATING BOOKS	Unit Price	Total
	60067	Assemblies Cost Data 2007	$223.95	
	60017	Building Construction Cost Data 2007	136.95	
	61017	Building Const. Cost Data–Looseleaf Ed. 2007	169.95	
	60717	Building Const. Cost Data–Spanish 2007	136.95	
	60227	Building Const. Cost Data–Western Ed. 2007	136.95	
	60117	Concrete & Masonry Cost Data 2007	124.95	
	50147	Construction Cost Indexes 2007 (subscription)	294.00	
	60147A	Construction Cost Index–January 2007	73.50	
	60147B	Construction Cost Index–April 2007	73.50	
	60147C	Construction Cost Index–July 2007	73.50	
	60147D	Construction Cost Index–October 2007	73.50	
	60347	Contr. Pricing Guide: Resid. R & R Costs 2007	39.95	
	60337	Contr. Pricing Guide: Resid. Detailed 2007	39.95	
	60327	Contr. Pricing Guide: Resid. Sq. Ft. 2007	39.95	
	60237	Electrical Change Order Cost Data 2007	136.95	
	60037	Electrical Cost Data 2007	136.95	
	60207	Facilities Construction Cost Data 2007	323.95	
	60307	Facilities Maintenance & Repair Cost Data 2007	296.95	
	60167	Heavy Construction Cost Data 2007	136.95	
	60097	Interior Cost Data 2007	136.95	
	60127	Labor Rates for the Const. Industry 2007	296.95	
	60187	Light Commercial Cost Data 2007	116.95	
	60027	Mechanical Cost Data 2007	136.95	
	63017	Metric Construction Cost Data 2007	162.95	
	60157	Open Shop Building Const. Cost Data 2007	136.95	
	60217	Plumbing Cost Data 2007	136.95	
	60047	Repair and Remodeling Cost Data 2007	116.95	
	60177	Residential Cost Data 2007	116.95	
	60287	Site Work & Landscape Cost Data 2007	136.95	
	60057	Square Foot Costs 2007	148.95	
	62017	Yardsticks for Costing (2007)	136.95	
	62016	Yardsticks for Costing (2006)	126.95	
		REFERENCE BOOKS		
	67310A	ADA Compliance Pricing Guide, 2nd Ed.	79.95	
	67330	Bldrs Essentials: Adv. Framing Methods	24.95	
	67329	Bldrs Essentials: Best Bus. Practices for Bldrs	29.95	
	67298A	Bldrs Essentials: Framing/Carpentry 2nd Ed.	24.95	
	67298AS	Bldrs Essentials: Framing/Carpentry Spanish	24.95	
	67307	Bldrs Essentials: Plan Reading & Takeoff	35.95	
	67342	Building & Renovating Schools	99.95	
	67261A	Bldg. Prof. Guide to Contract Documents, 3rd Ed.	32.48	
	67339	Building Security: Strategies & Costs	44.98	
	67146	Concrete Repair & Maintenance Illustrated	34.98	
	67352	Construction Business Management	49.95	
	67314	Cost Planning & Est. for Facil. Maint.	89.95	
	67328A	Designing & Building with the IBC, 2nd Ed.	99.95	
	67230B	Electrical Estimating Methods, 3rd Ed.	64.95	
	64777A	Environmental Remediation Est. Methods, 2nd Ed.	49.98	
	67343	Estimating Bldg. Costs for Resi. & Lt. Comm.	29.95	

Qty.	Book No.	REFERENCE BOOKS (Cont.)	Unit Price	Total
	67276A	Estimating Handbook, 2nd Ed.	$99.95	
	67318	Facilities Operations & Engineering Reference	54.98	
	67301	Facilities Planning & Relocation	89.95	
	67338A	Green Building: Proj. Planning & Cost Est., 2nd Ed.	129.95	
	67323	Historic Preservation: Proj. Planning & Est.	49.98	
	67349	Home Addition & Renovation Project Costs	29.95	
	67308E	Home Improvement Costs–Int. Projects, 9th Ed.	24.95	
	67309F	Home Improvement Costs–Ext. Projects, 9th Ed.	24.95	
	67344	Homeowner's Guide to Mold	21.95	
	67324B	How to Est. w/Means Data & CostWorks, 3rd Ed.	59.95	
	67282A	Illustrated Const. Dictionary, Condensed, 2nd Ed.	59.95	
	67292A	Illustrated Const. Dictionary, w/CD-ROM, 3rd Ed.	99.95	
	67348	Job Order Contracting	89.95	
	67347	Kitchen & Bath Project Costs	29.95	
	67295B	Landscape Estimating Methods, 4th Ed.	62.95	
	67341	Life Cycle Costing for Facilities	99.95	
	67294A	Mechanical Estimating Methods, 3rd Ed.	64.95	
	67245A	Planning & Managing Interior Projects, 2nd Ed.	69.95	
	67283B	Plumbing Estimating Methods, 3rd Ed.	29.98	
	67345	Practice of Cost Segregation Analysis	99.95	
	67337	Preventive Maint. for Higher Education Facilities	149.95	
	67346	Preventive Maint. for Multi-Family Housing	89.95	
	67326	Preventive Maint. Guidelines for School Facil.	149.95	
	67247B	Project Scheduling & Management for Constr. 3rd Ed.	64.95	
	67265B	Repair & Remodeling Estimating Methods, 4th Ed.	69.95	
	67322A	Resi. & Light Commercial Const. Stds., 2nd Ed.	59.95	
	67327A	Spanish/English Construction Dictionary, 2nd Ed.	23.95	
	67145B	Sq. Ft. & Assem. Estimating Methods, 3rd Ed.	34.98	
	67321	Total Productive Facilities Management	29.98	
	67350	Understanding and Negotiating Const. Contracts	69.95	
	67303A	Unit Price Estimating Methods, 3rd Ed.	29.98	
	67319	Value Engineering: Practical Applications	79.95	

MA residents add 5% state sales tax

Shipping & Handling**

Total (U.S. Funds)*

Prices are subject to change and are for U.S. delivery only. *Canadian customers may call for current prices. **Shipping & handling charges: Add 7% of total order for check and credit card payments. Add 9% of total order for invoiced orders.

Send Order To: ADDV-1000

Name (Please Print) _____

Company _____

☐ Company
☐ Home Address _____
_____

City/State/Zip _____

Phone # _____ P.O. # _____

(Must accompany all orders being billed)

Mail To: RSMeans, P.O. Box 800, Kingston, MA 02364-0800